THE MULLAN ROAD

Carving a Passage through the Frontier Northwest, 1859–62

EDITED BY

Paul D. McDermott, Ronald E. Grim, and Philip Mobley

2015
Mountain Press Publishing Company
Missoula, Montana

Cover image "Field sketch showing the side-cut by which the 1st and 2nd crossings of the Hell Gate River are [avoided]" by Gustavus Sohon, 1862. Courtesy National Archives and Records Administration, Cartographic Section, College Park, MD

Back cover map, "Overview of the Mullan Road" by Philip Mobley

Library of Congress Cataloging-in-Publication Data

McDermott, Paul D.
The Mullan Road : carving a passage through the frontier Northwest, 1859 to 1862 / Paul D. McDermott, Ronald Grim, and Philip Mobley. – First edition.
pages cm
Includes bibliographical references and index.
ISBN 978-0-87842-632-4 (pbk. : alk. paper)
1. Mullan Road–History. 2. Mullan, John, 1830-1909. 3. Military roads–Montana. 4. Military roads–Idaho. 5. Military roads–Washington Territory. 6. Military engineers–United States–Biography. 7. United States. Army. Corps of Engineers–Officers–Biography. 8. Explorers–Northwest, Pacific–Biography. 9. Surveyors–Northwest, Pacific–Biography. 10. Montana–History–19th century. I. Grim, Ronald E. II. Mobley, Philip, 1955- III. Title.
UA963.M43 2015
623'.6209795–dc23

2015004995

PRINTED IN HONG KONG

P.O. Box 2399 • Missoula, MT 59806 • 406-728-1900
800-234-5308 • info@mtnpress.com
www.mountain-press.com

To the memory of
Gayl Brown McDermott (1941–2014)

TABLE OF CONTENTS

LIST OF ILLLUSTRATIONS

ACKNOWLEDGMENTS

During the years that Ron Grim, Phil Mobley, and I have been collaborating, we have been fortunate to have met and worked with numerous fine scholars, many of whom have contributed to making this book possible. All have shared a common interest in the Mullan Road and in the men who played significant roles in its construction. It has indeed been a pleasure to work with all of the individuals involved in writing and publishing this book, including Bill Weikel, Ken Robison, Bill Youngs, Don Popejoy, Kim Briggeman, Alex McGregor, Richard Scheuerman, Keith Petersen, and the newest member of our group, Major Ryan Shaw, with whom I became acquainted at the 2010 Mullan Road conference in Fort Benton. Most particularly, Ron and I would like to acknowledge the fine cartography and graphics of our colleague Philip Mobley. Phil drew the original maps for our previous collaboration, *Eye of the Explorer,* as well as for this book, continuing as the major cartographer for our historical and geographical works. In addition, I thank our fine editor at Mountain Press, Gwen McKenna, and book designer Jeannie Painter.

I cannot leave out mentioning the descendants of John Mullan and Gustavus Sohon who helped us enormously in our research. Particularly supportive were Nancy and Julian Sohon, whom we met over twenty years ago. The Sohons provided access to privately held materials relating not only to Gustavus Sohon but also to his friend John Mullan. Ron and I have also been fortunate to work with members of the Mullan family, including Anthony Mullan, at the Library of Congress, and Joan Mullan, an English instructor at Montgomery College, now retired.

On a personal level, I would like to thank my mentor at the University of Washington, Dr. John Sherman, who supported me in becoming a graduate student in the geography department. It was poetic that at my last meeting with Dr. Sherman in the 1970s, I presented him with a copy of a map depicting the Mullan Road. Little did I know then how important the road would become to my career as a historical geographer.

Finally and most importantly, I wish to acknowledge the support of my beloved wife of fifty-four years, Gayl, whom I lost just weeks before this book's publication.

PAUL D. MCDERMOTT

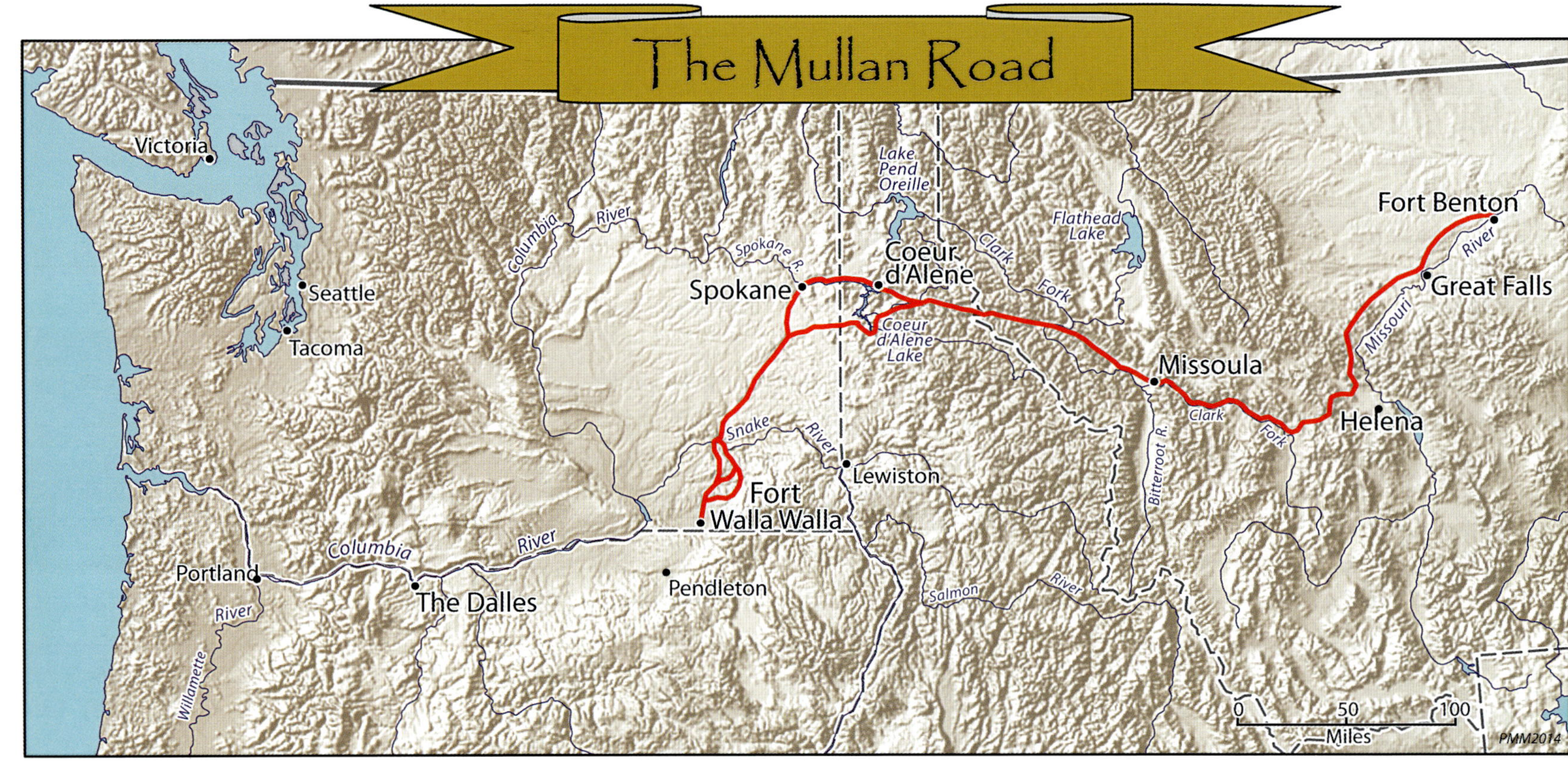

Overview of the Mullan Road –Map by Philip Mobley

Introduction

BY PAUL D. MCDERMOTT AND RONALD E. GRIM

During the first half of the ninteenth century, the geography of the United States changed dramatically, transformed by the acquisition of all the territory west of the Mississippi River, by westward migration, and by a transportation revolution. The century began with extensive migration from the Atlantic seaboard across the Appalachian Mountains, initially to Kentucky and Tennessee, then to the Old Northwest (Ohio, Indiana, Illinois, Michigan, and Wisconsin) and the Old Southwest (Alabama and Mississippi). As trade opportunities developed in these regions, the nation focused on developing transportation in areas that had recently been wilderness–first roads, then canals, and finally railroads.

While navigable rivers and roads were the primary components of the nation's transportation network during the late eighteenth and early nineteenth centuries, canals soon became a cause célèbre, especially after the completion of the 364-mile-long Erie Canal across New York state in 1824. This canal's immediate success motivated other urban centers–including Philadelphia, Baltimore, Washington, D.C., and Richmond–to create their own canal systems, though none was as successful as the Erie. The canal-building era lasted from 1824 to about 1840, after which time the appeal of canals began to decline, especially with the advent and rapid success of the railroad.

Meanwhile, in the 1840s, migration to the Far West began in earnest. Travel and commerce were initially promoted through the development of the Santa Fe and Oregon Trails. A further burst of energy resulted from the discovery of gold in California in 1849. As population grew along the Pacific Coast, the nation was motivated to connect the East with the West by creating a transcontinental railroad. Most politicians and businessmen supported this goal, but they could not agree on a particular route. Congress sought to resolve this question by sponsoring extensive explorations of the nation's interior. Five major expeditions, known as the Pacific Railroad Surveys, were authorized. From this effort, the northern survey, under the leadership of Washington territorial governor Isaac Stevens, was the most successful in terms of data acquired. This survey investigated a broad region from St. Paul to the Puget Sound.

Significantly, one of Stevens's younger assistants was Lieutenant John Mullan, a recent (1852) graduate of West Point. Mullan's work with Stevens began in 1853 and continued through 1854. During this period, Mullan became enamored, if not obsessed, with the idea of building a railroad through the region. To further this goal, Mullan came to the conclusion that a road should be built to connect the Missouri River with the Columbia River across the northern Rockies. At the time, only two significant settlements existed along the proposed route, Fort Walla Walla and Fort Benton. A few years later, in 1859, John Mullan would begin construction of a 624-mile military wagon road between these two forts. Upon its completion in 1862, Mullan's road became, effectively, a new Northwest Passage.

Building the Mullan Road

Mullan initiated construction of the military wagon road in 1858, but the effort was temporarily curtailed by a Native American uprising in May 1858. Construction was resumed at Fort Walla Walla in 1859, and by 1860 the construction crews had reached Fort Benton, the road's terminus. However, it was apparent to Mullan that the road still needed considerable improvement. Although he had thoroughly explored the region, even as he was building the road, he was not always convinced of the precise route to be used. Again and again, he changed his mind about the road's location. From 1860 to 1862, he spent considerable effort in rerouting several sections of the road. Early on, he had determined that the route he had selected around the southern end of Lake Coeur d'Alene was subject to massive spring flooding; subsequently, he rerouted the road around the northern shore of the lake. Another significant change he made was moving the road's path from the floodplain of the Hell Gate (Clark Fork) River to the adjacent mountain slopes. As late as 1862, as the road was nearing completion, Mullan sent out four different parties to determine the best route from the Sun River crossing to Fort Benton.

As he began the final phase of road construction in 1861, Mullan also saw the need to both repair previously built bridges and build new ones in order to avoid excessive river crossings. In 1862 he organized three crews to construct bridges over the Hell Gate River. So focused was he on bridge building that he was sometimes called the "Duke of Bridgewater."[1]

Immediately upon its completion, the military wagon road provided a connection between the navigable portions of the Missouri and Columbia Rivers. But how successful was it in the long run? A great deal of money and labor was expended to create the road in approximately four years. Unfortunately, the timing was bad. In 1862, the Civil War was well under way in the East, and the nation's energy turned from settlement of the West to the outcome of the war. Thus the newly constructed road was not maintained properly, and it rapidly fell into disrepair. Spring washouts, bridge deterioration, and the reemergence of troublesome tree stumps along the route disabled the road's functionality for wagons. At best, it could be used by pack trains. All of that could have been avoided if Congress had committed more funds to maintaining the road.

Even if the usefulness of the road itself was limited, Mullan had selected a good route for future transportation systems. Significant portions of his route still function as segments of major pathways for railroads, highways, and today's interstates. Certainly, it must have been a memorable experience for Mullan to travel on the first Northern Pacific Railway train in 1883. What started out as an exploration conducted by Isaac Stevens through wilderness was ultimately transformed into an operational transcontinental connection. In the years that followed, two other transcontinental railroads were constructed through the same general region–the Great Northern Railway and the Milwaukee Railroad.

Origins of this Book

At the 2010 Mullan Road Conference in Fort Benton, Montana, coordinated by historian Ken Robison, a number of interested scholars presented papers focusing on the Mullan Road. From this conference, the idea originated to create a comprehensive publication discussing the road in its many facets–including material relating to Lieutenant John Mullan, the project's leader. Up to this time, most of our knowledge about the construction of the Mullan Road had been garnered from the official report Mullan wrote for the U.S. Congress, published in 1863. In 1994 the official report was republished by Ye Galleon Press with an introduction and notes by Kimberly Rice Brown. That book is now out of print.

Other than the official report, a number of journals written by Mullan and other major players in the construction project are available, but many still need to be transcribed. All of this material, which is handwritten, is important because it documents in detail the daily activities of the construction crews. It was from these journals that Mullan created the narrative for his final report. Unfortunately, Mullan's handwriting is not easily transcribed. The clarity of his penmanship changes from day to day–sometimes it is legible, but frequently the script is literally a scrawl. Until someone takes time to convert this material into typescript, the full story of the Mullan Road will not be known.

While Mullan's official report to Congress certainly provides a good overview of the expedition, details that Mullan thought irrelevant or unimportant for the report were excluded. For example, the journals reveal that as the outbreak of the Civil

War approached, Mullan was forced to accept resignations from many of his officers and soldiers so they could return to the East to fight in the war. Another interesting but officially unreported incident was the party's confrontation with rattlesnakes at a campsite north of Walla Walla. The notes in Mullan's journals testify to his meticulous organization of construction crews to build bridges over and dig side cuts along fast-running western rivers. These supplementary materials provide new insights into the official report.

In recent years, very few published works have focused directly on John Mullan or on his military wagon road except Keith C. Petersen's *John Mullan: The Tumultuous Life of a Western Road Builder*, published in 2014. An expert on Mullan's life, Petersen contributed this book's opening chapter, a brief biography of the explorer. Other than Petersen's biography, however, the last major book dealing with the Mullan Road was privately published in 1959. Titled *The Mullan Road: Captain John Mullan, His Life, Building the Mullan Road, As It Is Today and Interesting Tales of Occurrences along the Road*, this volume exploited material from Mullan's official report of 1863, with additional information extracted from a diary kept by Mullan's wife, Rebecca. For a number of years, the diary was in the possession of Mullan's daughter May, who eventually allowed Helen Addison Howard to access it for her historical writings. After May's death, her parents' handwritten papers were donated to Georgetown University, where they are preserved today.

Organization and Content of This Book

It is the purpose of this collection to fill the information void about the Mullan Road with essays derived not only from Mullan's published reports but also from materials that have long been neglected or have recently come to light. Including the three editors, twelve diverse authors—from amateur Mullan buffs to credentialed academics—have here contributed their individual takes on the people, places, and events related to John Mullan's historic road-building project, allowing readers to view its multiple dimensions from various perspectives.

We have divided our volume into three main sections. The three essays in Part One focus on John Mullan himself. The first provides a condensed biography of Mullan, the second is a close examination of his experience as a cadet at West Point, and the third is an account of his 1854 explorations in the Flathead Lake region of Montana. Part Two looks at the actual construction of the Mullan Road. Each of the five essays in this section focuses on a geographical region traversed by the road and describes the ups and downs the expedition experienced building the road in each region. Part Three comprises essays devoted to specific aspects of the expedition, including the maps and illustrations the expedition produced, the economic impact the road had in the Northwest, the expedition's relationship with Northwestern Indians, and finally, a fresh look at the style and substance of Mullan's 1863 congressional report.

The opening piece, "John Mullan: Into the Unknown," was written by Idaho historian Keith C. Petersen, author of the acclaimed recent biography *John Mullan: The Tumultuous Life of a Western Road Builder*. In researching the book, Petersen traveled around the United States gathering materials that pertain to Mullan's life and activities, making extensive use of the Mullan family papers at Georgetown University.

Major Ryan L. Shaw, an actively serving U.S. Army officer and new history scholar, contributed the second essay, "The Making of an Explorer: John Mullan at West Point," as well as two others, which appear in Part Two. In "The Making of an Explorer," Shaw paints a vivid portrait of West Point in the mid-nineteenth century and the effects of that institution on an eager young cadet named John Mullan. Originally from Idaho, Major Shaw came to know West Point as a military cadet and later as an instructor. We are indeed fortunate to have his work incorporated here.

In the third essay, "Journey to the Flathead Country: John Mullan's April 1854 Expedition," we, as historical geographers, explore an aspect of John Mullan's seminal experience working on the Northern Pacific Railroad Survey with Washington territorial governor Isaac Stevens in 1853–54, an assignment that served as a prelude to his road-building project. During this expedition, Mullan, accompanied by artist, interpreter, and guide Gustavus Sohon and five other men, explored the Flathead Lake region in what is now northwestern Montana. The essay documents the party's observations and the difficulties they encountered. In our previous collaboration with Philip Mobley, *Eye of the Explorer,* we examined the Stevens survey in depth. In addition, we have written a book and several articles about Sohon and his artwork, which, incidentally, is the focus of our essay "Illustrating the Mullan Road" in Part Three.[2]

Part Two begins with "Across the Plateau and into the Mountains: The Western Section of the Mullan Road," by historian and educator Don Popejoy. This absorbing essay covers Mullan's route from Fort Walla Walla, Washington, to the Coeur d'Alene Mission in present-day Idaho. A native of Spokane, Washington, Popejoy is very familiar with the eastern Columbia River Plateau and its history. Particularly interesting is his discussion of the Snake and Spokane River crossings.

The next essay is part 1 of "'A Creditable Piece of Mountain Work': Building the Mullan Road in the Rockies," written by Major Ryan L. Shaw. Both parts 1 and 2 of this study reflect on Mullan's harrowing experiences building through the Bitterroot Range of the northern Rocky Mountains, the most troublesome region of the entire expedition. Here, Mullan dealt with severe winter conditions, starving animals, springtime floods, and wary local Indians.

Between parts 1 and 2 of Major Shaw's essay is "John Mullan on the Hell Gate River" by Missoula journalist and historian Kim Briggeman. Here, Briggeman homes in on the explorers' challenges along the Hell Gate River (today identified as part of the Clark Fork River). Much of Mullan's work in this region began at Cantonment Wright, located near the junction of the Big Blackfoot and Hell Gate Rivers, near today's Missoula, Montana. The expedition's primary challenge here was rerouting the road from the Hell Gate floodplain to higher ground along the adjacent mountain slopes.

The last piece in Part Two is Ken Robison's "Completing the Mullan Road from Mullan Pass to Fort Benton: A Harbinger of Change," which looks at the eastern section of the road and reflects on the last days of the project. Robison, resident historian at Fort Benton's Heritage Complex, describes the issues the expedition struggled with as they worked from Mullan Pass to Fort Benton, Montana. He notes that late in the project, Mullan remained indecisive about which route to use from the Sun River crossing to the fort, investigating four different possibilities before making his choice. Robison also analyzes the completed road's impact on subsequent development in the West, especially its impact on its eastern terminus, Fort Benton.

The final group of essays, Part Three, opens with William ("Montana Bill") Weikel's look at surveying in the mid-1800s. For a number of years, Weikel, a professional surveyor, has been giving living-history presentations on the instruments and techniques used by nineteenth-century surveyors. "Surveying the Mullan Road," based on a presentation Weikel gave in Walla Walla in April 2012 and coauthored by us, offers an unusually detailed look at the nuts and bolts of the surveyor's work, particularly the work of the surveyors on the Mullan Road expedition.

Our next essay, "Maps of the Mullan Road," discusses one of the most significant contributions of the Mullan Road project–its detailed maps. Mullan and his cartographers charted a broad region from Fort Dalles to Fort Benton, creating a wide variety of maps, drawings, and topographical notes. When we began our research on the Mullan Road in the mid-1970s, Dr. Grim was working at the Cartographic Division of the National Archives, where he discovered the Mullan Road maps filed among the Records of the Office of the Chief of Engineers (Record Group 77). This collection of maps created by Mullan and his cartographers as they built the road was an invaluable source for preparing this volume. Some of these maps were prints–reproductions of finished hand-drawn or etched maps–while others were sketched on tissue paper and tracing cloth. Line work on the maps was often drawn in ink, but pencil was also sometimes used. Most of these maps, some of which appear as illustrations in this book, can be found in the Cartographic Division of the National Archives in College Park, Maryland.

In the next essay, "Illustrating the Mullan Road," we take an in-depth look at the images Mullan used to illustrate his report to Congress; we also provide the historical context that reveals their significance. In contrast to Isaac Stevens's report on the Northern Pacific Railroad Survey, Mullan's final published report incorporated relatively few illustrations. This is not surprising considering that the document was published in the midst of the Civil War, meaning that resources for printing were limited. Furthermore, Mullan was in a hurry to complete the report because he wanted to return to the West, marry his sweetheart, and settle into life as a civilian. The images he did include, however, are telling. Many of the illustrations in Mullan's final report originally appeared in the Stevens railroad survey report, but the report also incorporated new scenes and different perspectives of previously published ones; examples include the Great Falls of the Missouri, the arrival of steamboats at Fort Benton, and Mullan's winter camp, Cantonment Wright. Besides the report illustrations, we also discuss some additional manuscript artwork we located in the aforementioned Record Group 77.

The next essay is Alexander C. McGregor's "Merchants, Mule Packers, and a Rugged Path to Success: Walla Walla and the Mullan Road, 1860–83." Here McGregor notes that the majority of the traffic on the road was composed not of wagons but of pack trains, as much of the road was not sufficiently maintained to allow smooth passage for wagons. He considers the reasons for this circumstance and the effects it had on both suppliers and consumers in the region. The essay then tackles the bigger topic of the road's impact on the economic development of the Northwest, particularly of the new settlement of Walla Walla, Washington.

The history and geography of the American Northwest were forever transformed when the Mullan Road became a reality. The region's Native tribes, in particular, were well aware that would the road would irreversibly change their lives, their culture, and their place in the world. Most significantly, Indian society would be nudged from a hunting and gathering mode of subsistance to one increasingly focused on farming and raising livestock. This subject is masterfully investigated in Richard D. Scheuerman's article, titled "'Through the Indian Country'": John Mullan and Native American Encounters on the Northern Overland Road."

The final piece in this collection was written by Eastern Washington University history professor J. William T. Youngs. This provocative essay, titled "Precision and Poetry: The Utility and Elegance of the Mullan Report," explores Mullan's writings as literature. In this paper, Youngs asserts that the 1863 report "is the work of an exacting reporter and a literary craftsman." Among the numerous examples of Mullan's distinctive writing style that Youngs presents is this lyrical passage: "A military occupation of the Walla-Walla valley became necessary, and the rude mud walls of old Walla-Walla, that year after year had listened to nought but the jargon of Indians and traders, caught a new sound in the tramp of the march of civilization." Later, in his book *Miners' and Travelers' Guide to Oregon, Washington, Idaho, Montana, Wyoming and Colorado via the Missouri and Columbia Rivers*, Mullan engagingly advises readers on handling their mules: "Never maltreat them, but govern them as you would a woman, with kindness, affection, and caresses, and you will be repaid by their docility and easy management."

A poem I (McDermott) wrote, titled "Pathways," expresses my feelings about the writers and historians who contributed to this volume: "Each has a path that begins and ends in eternity. Sometimes long and sometimes short, sometimes rough and sometimes smooth, at various points the path of one crosses another and from these crossings we are forever changed."

---

1. Louis C. Coleman and Leo Rieman, *Captain John Mullan, His Life, Building the Mullan Road, as It Is Today and Interesting Tales of Occurrences along the Road* (Montreal: B. C. Payette, 1968).

2. Paul D. McDermott and Ronald E. Grim, *Gustavus Sohon's Cartographic and Artistic Works: An Annotated Bibliography.* Philip Lee Phillips Society Occasional Paper No. 4 (Washington, DC: Geography and Map Division of the Library of Congress, 2002).

Part One

John Mullan as a young man, c. 1857 –Courtesy Overholser Historical Research Center, Fort Benton, MT

John Mullan

Into the Unknown

BY KEITH C. PETERSEN

John Mullan arrived in Walla Walla on a hot August day in 1862 after four years of road building. He was escorting eleven wagons of emigrants, the vanguard of what he envisioned would soon be a flood of settlers traveling his 624-mile military highway from Fort Benton "towards the setting sun."[1]

Until recently a dozy town known mostly for the military fort of the same name, from which Mullan had commenced the road-building expedition in 1859, Walla Walla now sprawled along dirt lanes dotted with a jumble of single-story, false-fronted buildings. Freight wagons, pack animals, and a "considerable army of rough men" crowded the streets as the town boomed, having become the supply point for newly discovered gold mines to the east, in a place soon to be called Idaho. Hundreds of bonanza seekers arrived weekly, tarrying only long enough to pay exorbitant prices for supplies.[2]

But where others saw only ramshackle buildings and rampant hucksterism, John Mullan envisioned his future. "With its mild . . . climate, with its rich soil, abundance of timber of the finest quality on the mountain tops and with its numberless streams affording mill sites at any point of the valley," the place would soon "fully develop . . . to our nation's notice!" He had already determined he would settle there, banking his future on the community that his road had helped to swell.[3]

Mullan's arrival in Walla Walla marked the completion of his greatest accomplishment and the work for which he would be remembered–the construction of the first engineered highway in the Northwest. Then age thirty-two, Mullan would live nearly another half century, much of it spent in an effort to capitalize on his fame. Mullan's experience in constructing the highway is one of the most dramatic stories in Northwest history, and that is the subject of this book. But published biographical accounts describing Mullan's life before and after his great achievement are scarce. To know nothing of him before the expedition, and likewise to leave him as a thirty-two-year-old who built a road and then disappeared into obscurity, is to understand little about the complexities of his life.

John Mullan was born on July 31, 1830, in Norfolk, Virginia, the eldest of John and Mary Bright Mullan's eleven children. When John Jr. was three years old, the family moved to Annapolis, Maryland's capital. The boy grew up on East Street, an eclectic neighborhood of modest cottages owned by whites and free blacks that sat alongside spacious homes and gardens. Each day he would have looked out upon the grand Brice Mansion. The two-and-a-half-story Georgian-style home would have been an imposing structure to a young boy who shared a modest dwelling with a large family. Around the corner, on Prince George Street, he would have walked past the even more imposing Paca House with its elegant garden. William Paca, a signer of the Declaration of Independence, constructed the home in the 1760s. Mullan would have been equally familiar with other mansions, and in this small community of a few thousand, he likely made the acquaintance of many of the owners. This setting had a lasting effect on Mullan. For his entire life, he aspired to wealth and public stature; his youthful familiarity with the lifestyles of the Annapolis elite helped drive this lifelong obsession.[4]

The Mullan family valued education, and young John attended St. John's College, a combination preparatory school

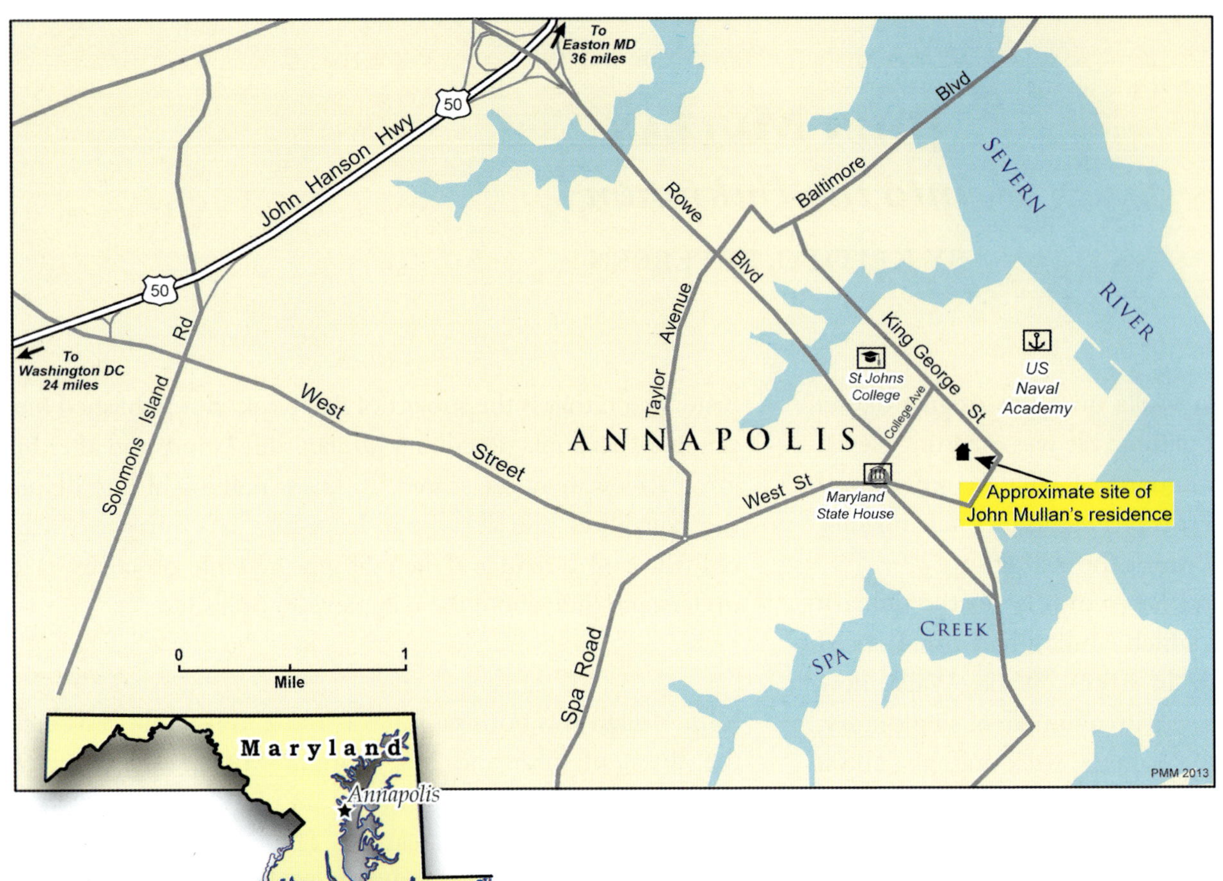

John Mullan's Annapolis. Mullan lived his early life in Annapolis, the state capital of Maryland. Upon his death in 1909, he was buried in the city.
–Map by Philip Mobley

and college in Annapolis, from 1839 to 1847. Six brothers would follow him to the institution. St. John's, established in 1789, consisted of two buildings on the edge of town. During Mullan's time, the school's curriculum emphasized science, including mineralogy and geology, disciplines Mullan later employed in his engineering and construction endeavors.[5]

Annapolis became home to the United States Naval Academy in 1845. Several of Mullan's brothers enrolled there. But when John was ready to further his education in 1848, the academy, consisting of only a few old buildings, lacked esteem. Because the Mullan family's financial circumstances prohibited anything other than free schooling beyond St. John's, John decided to enroll at one of the best free institutions in the nation–the U.S. Military Academy at West Point, New York.

In the scant biographical literature about Mullan, one story recurs. Helen Addison Howard, Mullan's first biographer, told it this way:

> Armed with letters of recommendation from influential friends [Mullan] went to Washington to apply in person to President James K. Polk for admittance to the Point.
>
> He found the dignitary alone and recumbent in an armchair with his feet elevated on the mantel, smoking his homely clay pipe. Shyly the youth approached him and accepted the President's extended hand.
>
> "Well, my little man, what can I do for you?" he asked in a genial tone.

Young John, losing some of his timidity, replied that he desired "to enter the army by graduating at West Point."

"Don't you think you are rather small to want to be a soldier?" inquired the President with the flicker of a smile at the corners of his mouth.

"I may be somewhat small, sir," John answered stoutly, "but can't a small man be a soldier as well as a large one?"

There being no evidence to the contrary, President Polk said, "Well, my young friend, leave your address, and I will see what I can do for you."[6]

A few weeks later, Mullan received papers containing his appointment to West Point.

Despite having all the trappings of an apocryphal tale, the story turns out to be essentially true. Mullan did travel to Washington, D.C., in early 1848 to seek an appointment to West Point. While Polk's papers do not verify that Mullan gained an audience with the president, they do not disprove it either: getting access to presidents was relatively easy in the nineteenth century, and Polk saw many visitors who remain anonymous today, being unlisted in official records. Mullan's West Point application packet contains supporting letters from influential Maryland residents and a petition from all the Democrats in the state assembly, supporting the son of this strongly Democratic family. The petition and letters no doubt influenced Polk, who took a more active interest in West Point than did any other pre–Civil War president. He fully used his authority to appoint at-large cadets, making John Mullan one of ten at-large selections to the class of 1852. Mullan later indicated the extent of his gratitude to Polk. "I was taken by the hand by President Polk," Mullan wrote, "and by him educated at West Point."[7]

Mullan traveled to West Point by boat, arriving on June 3, 1848. Upon disembarking, he put his bags on a horse-drawn cart and followed on foot, walking up a steep hill to the campus, which accommodated about two hundred cadets. He signed the campus registry and handed over all of his money—thirty-five dollars—to the treasurer; cadets could not carry cash on campus, but their "treasurer's accounts" allowed them to purchase sweets, soft drinks, and pickles from the barracks soda shop.[8]

Due to the sensationalism surrounding California's gold rush, the American West would have been much on Mullan's mind as a cadet. From the academy library he borrowed Washington Irving's works on Benjamin Bonneville and Astoria; accounts of the newly ended Mexican War; and tales of the Santa Fe expedition. He graduated from West Point as a brevet second lieutenant in the summer of 1852 and was assigned to Fort Columbus, in New York Harbor. As the closest major army facility to West Point, it was a familiar first posting for recently graduated cadets.[9]

Being fascinated by the West, Mullan must have enthusiastically received the news of his assignment to the Northern Pacific Railroad Survey in 1853. The expedition would be headed by 1839 West Point graduate Isaac I. Stevens, recently appointed by President Franklin Pierce as the first governor of the newly formed Washington Territory. Now the twenty-three-year-old John Mullan would experience the West firsthand. Moreover, he would come under the tutelage of Stevens, who was to have a more profound influence on his development than anyone, save perhaps his parents.

Stevens did not know Mullan prior to the expedition, but he soon gained confidence in the young lieutenant and assigned him one of the most significant operations of the survey: exploring western mountain passes for possible railroad routes in the winter of 1853–54. In isolated areas, military wagon roads were built in advance of railroad construction as a means of transporting supplies to the workers. After a year undertaking some of the most significant explorations in the American West, Mullan reported to Stevens on his preferred route over the mountains. Stevens, ever the advocate of a northern transcontinental rail line, began promoting the construction of a wagon road shortly after completing the railroad survey, and he lobbied for Mullan to lead the effort.

Congress authorized $30,000 for the project in 1855, but this was hardly enough to undertake the work, as Stevens understood. Once Mullan began road construction, he, too, recognized the inadequacy of the appropriation and encouraged additional congressional funding. After Stevens won election as the Washington territorial delegate to Congress in

1857, he had a more prominent platform from which to promote the wagon road, and in 1859 Congress complied with his wishes, providing an additional $100,000 appropriation.[10]

Mullan hired a crew and had already begun construction in the spring of 1858 when his work was interrupted. In May, word arrived of Colonel Edward Steptoe's ignominious defeat by combined Inland Northwest tribes, and Mullan volunteered to serve with Colonel George Wright in his expedition to avenge the Steptoe disaster. After Wright's scorched-earth campaign forced permanent peace among inland tribes in September, Mullan returned to the road project. He began construction in earnest in 1859 and completed the job in 1862.[11]

Mullan's march into Walla Walla in August 1862 must have seemed to him like the triumphal beginning of profitable ventures to come. The road project had garnered him a promotion to captain, and Northwest newspapers gushed over his road-building accomplishment. Since at least 1859, he had viewed his work on the road as a stepping stone to greater wealth and public acclaim. That year, he revealed to his future wife, Rebecca Williamson of Baltimore, his intention to leave the army, a notion that greatly concerned her. She wrote in her diary, "Ah, Mr. Mullan, take care, remember, 'Vaulting ambition often overleaps itself'—do not attempt more than you can achieve." She also secretly stated her hope that he would remain in the army, where "you can always maintain your position as a gentleman." In hindsight, Mullan might have wished he had taken Rebecca's advice, for the next decade would be a roller coaster.[12]

As he prepared to leave Walla Walla for Washington, D.C., in September 1862 to complete his report on the road, Mullan would not yet have heard the news that his mentor, Isaac Stevens, then a Union general, had been killed in action while leading a charge at the Battle of Chantilly. Stevens had taught Mullan much about the ways of Washington, but as Mullan found himself pulled ever more deeply into the capital's politics, he could have used more help from the savvy Stevens.[13]

Mullan left Walla Walla on September 7 with Gustavus Sohon, the fine artist who had accompanied him on the Stevens survey before embarking on the Mullan Road expedition. The two journeyed overland to San Francisco before boarding a ship to the East Coast via the Isthmus of Panama. Along the way, Mullan noted "the extensive and rich fields" of John Bidwell's Rancho Chico, one of California's largest ranches. The scene greatly affected Mullan, who in a couple of years would partner with Bidwell to create a stagecoach line from California to new goldfields in southern Idaho.[14]

Moving to Washington, D.C., in 1862, Mullan began writing his report under the direction of the Army Corps of Topographical Engineers. As he came under the strict supervision of the Topographical Bureau, Mullan chafed, having grown used to relative autonomy in the West. The bureau chastised him for employing assistants without authorization; for working on a map outside the scope of his report; and for making unsanctioned plans to build yet another military road. "I am led to believe you mistake the scope of your duties," scolded a commanding officer. When Mullan sought permission to complete his report from his residence rather than at the army offices—probably to avoid daily contact with irritating superiors—the bureau promptly denied his request.[15]

Incensed by what he saw as the army's intractability, Mullan increasingly turned his energies to politics. He spent a good portion of the winter of 1862–63 assisting Representative James Ashley of Ohio in drawing boundaries for a proposed new western territory called Idaho. It is possible that Ashley sought out Mullan's assistance, since Mullan knew more about western geography than anyone else in Washington. Yet it is just as likely that Mullan, grating under military discipline and eager to use his reputation to gain public stature, courted Ashley.

When a party headed by Elias D. Pierce—who would later become a business partner of Mullan and Bidwell—discovered gold in future Idaho in 1860, it set off a population shift to what was then the eastern part of Washington Territory. Later gold discoveries in today's Montana further increased demand for new territorial boundaries, as travel from these gold centers to Washington's territorial capital at Olympia proved difficult in the best of times, and virtually impossible at others. The region's new residents clamored for a territorial capital close enough to their settlements to allow reasonable representation. Ashley, who chaired the House Committee on Territories, powerfully influenced the shape of new territories.

Because Mullan assisted Ashley in developing Idaho's boundaries, he had the opportunity to potentially shape the future of Walla Walla, where he intended to eventually make his fortune. The town was booming at this time, largely because Mullan's road provided access to the new mines. Walla Walla residents hoped to wrest the Washington territorial capital from Olympia, but to do so would require a significant population base east of Walla Walla to legitimize the town's claim to being centrally located. Not surprisingly, Mullan catered to the Walla Walla faction. The boundaries he proposed kept the Idaho panhandle, along with its mines, and western Montana in Washington Territory. The plan greatly concerned Olympia residents, for Walla Walla would then sit at the virtual center of Washington Territory's population base, giving it a solid claim to become the capital. Despite Olympia's opposition, the Ashley-Mullan bill passed the House of Representatives in February 1863. In the Senate, however, the proposal was progressing differently.[16]

William Wallace, Washington's territorial delegate to Congress and a resident of Puget Sound, heeded the concerns of those in Olympia that new boundaries along the lines Mullan proposed would all but guarantee the capital's move to Walla Walla. Quietly, Wallace worked with senators on alternative boundaries. Wallace's bill—which passed the Senate late on March 3, 1863, the last night of the 37th Congress—gave Washington its current boundaries and created Idaho as a huge new territory that included not only all of present-day Idaho but also Montana and virtually all of Wyoming. In this scenario, since there were very few voters in eastern Washington to contest the issue, Olympia was practically assured of remaining the capital of Washington. Over the vigorous protests of Ashley, the House caved to the Senate and approved Wallace's version of the boundary. President Abraham Lincoln signed the bill the next morning.

Wallace had completely outmaneuvered Mullan in the latter's first attempt to reap political advantage from his reputation as the Northwest's road builder. Now, with his boundary proposal rejected, Mullan's chances of fulfilling his next goal—being appointed governor of the new territory—were severely diminished. Because Wallace had no political connections east of the panhandle, Mullan's plan—which placed the panhandle in Washington Territory—would have put Idaho's boundaries in a region that Mullan knew much better than Wallace, and Mullan would have staked his claim to become its governor based on that knowledge. But, having lost the boundary fight, Mullan's position was weakened. Still, on the very day that Lincoln signed the bill creating Idaho Territory, Mullan wrote to him seeking the position. Naturally, he reminded the president of his road:

> For the last ten years I have resided within the region now organized into a Territory, and . . . by my labors and Explorations have added to the development and progress of that region by opening a communication from the Head Waters of the Missouri to the Head Waters of the Columbia River by which Immigrants from the North West have found a new route to the Pacific.

Mullan understood that Wallace had the inside track for the position, so he ended his appeal by requesting that "at the least," the president undertake "a special investigation in regard to my case before any appointment is made." An impressive number of politicians supported Mullan's appointment. Even so, Mullan, a lifelong Democrat, perhaps never had a realistic chance of appointment by the Republican Lincoln.[17]

Wallace's own request for appointment arrived with a list of supporters equally as impressive as Mullan's, and Wallace had the strongest advantage of all—he was a Republican and he knew the president. On March 10, 1863, Lincoln appointed William Wallace governor of the new Territory of Idaho. Regarding Mullan's fruitless quest, the *Sacramento Union* noted, "It would appear that the redoubtable Captain Mullan has beat the bush and somebody else has caught the bird in the creation of this new Territory."[18]

A frustrated Mullan reacted to Wallace's appointment by resigning from the army in May 1863, indicating a clear break from his past. His resignation came at a time when other West Point graduates were seeking promotion and recognition by serving in the Civil War. But Mullan had long planned to leave the army upon completing his road, as he had indicated to Rebecca as early as 1859. Lincoln's snub merely speeded things up.

Now a civilian, Mullan prepared for the next phase of his life. In April 1863, a month after losing his bid for governor, he married Rebecca Williamson. The couple honeymooned at West Point. Afterward, the newlyweds continued on to New York City, where the American Geographical and Statistical Society had invited Mullan to give a brief talk on the resources of the Northwest Territories as warm-up to precede the main attraction of the evening, an address on Arctic explorations. In typical Mullan fashion, his short summary turned into a full-fledged lecture. The *New York Times* charitably called his hour-and-a-half talk "lengthy but still interesting." In it, Mullan described the Northwest as a region of "pleasant homes and well-tilled fields, grand mountains and useful rivers, forests of orchards and oceans of grain, miles of sluice-boxes and tons of gold, and the beauty of a region reclaimed from the wilderness, and now enlivened by the merry songs of industry, contentment, and thrift."[19]

But the talk was not Mullan's primary reason for visiting New York; he had come mostly to court investors for the Walla Walla Railroad Company. In 1862 Mullan had remarked on the need for a short-line railroad to connect Wallula, W.T. [Washington Territory], on the Columbia River, with his road in Walla Walla. "One needs but see the long line of wagons and pack trains" between the two points, he wrote, "to become convinced that . . . some more rapid and economical means is positively demanded, in order to connect the heart of the valley with the Columbia River." The Walla Walla Railroad Company was incorporated later that year. In January 1863, the company elected Mullan commissioner and tasked him with finding capital for the project.[20]

In the summer of 1863, John and Rebecca Mullan moved to Walla Walla, ostensibly to farm. In partnership with two of his brothers, John owned a farm in the area, and when Rebecca wrote her reminiscences in 1892, she stated that they had moved there so that John could begin "cultivating this tract." But farming was not Mullan's true calling, and he had greater aspirations–he intended to get rich as an entrepreneur. In addition to the farm, he purchased a lumberyard and a livery stable. But Mullan pinned his highest hopes on the short-line railroad. He predicted that, with completion of the line, Walla Walla would "become the commercial centre which will supply the large and rich mining region recently discovered."[21]

The railroad's incorporators believed that Mullan, with his reputation as the Northwest road builder and his connections in the East, would successfully lead the financing effort. Certainly Mullan tried, appealing to investors to purchase $600,000 worth of stock. "Capitalists will do well to look into this project, which appears to possess elements of profit and public usefulness in no ordinary degree," enthused the *Railway Times*. "Capt. Mullan has been employed in surveying the route for the Pacific Railway and a wagon road to the Pacific for the past five or six years, and is in every way well informed upon the subject." Still, financing for the railroad never arrived and the company failed.[22]

Meanwhile, John Mullan brought three of his brothers into his other Walla Walla business ventures. The Mullans proved to be poor managers, however, and the businesses, poorly capitalized, suffered from unrelenting lawsuits successfully brought by those to whom the brothers owed money. Finally the family businesses totally collapsed in a series of ugly lawsuits pitting brother against brother, the upshot being that John lost his stake in the farm and the businesses. His great hope of growing wealthy in Walla Walla spiraled into the reality of bankruptcy in a remarkably short period of time–by 1864 John and Rebecca had returned to Maryland.[23]

In Maryland, Mullan attempted to capitalize on his fame by writing a guide for people immigrating to the West. Unfortunately, his book missed most of the market; by the time it was published in 1865, the great migration to the Northwest mines had subsided. In the meantime he also contemplated creating a supply route from California to the gold-mining camps in southern Idaho. This idea soon brought him into contact with a man he had admired since he rode past his Chico ranch in 1862–John Bidwell.[24]

As one of California's wealthiest and most innovative agriculturalists and the founder of the city of Chico, Bidwell hoped to enhance the markets for his Rancho Chico crops, as well as growth in his newly established town, by constructing a stage and freight line to the booming mines of southern Idaho. Soon Bidwell would find the perfect person to lay out a route into that territory. In January 1865, Elias D. Pierce, reputed discoverer of gold in Idaho, visited Rancho Chico to discuss with Bidwell his own plans for a freight line out of Chico. The two formed a partnership, and with Bidwell's backing, Pierce scouted the course through heavy snows. In the spring of 1865, he led the first pack train over the new wagon road into the Idaho goldfields.[25]

In Maryland, John Mullan was keeping abreast of these developments. He, too, thought the Chico freight line offered opportunity, and, upon meeting Bidwell, he became a partner in the line. But Bidwell soon had second thoughts about Mullan, believing he lacked the organizational skills necessary to make a success of the venture. In 1865 he wrote to his Rancho Chico manager, D. D. Harris, asking him if he would like to take charge of the enterprise. "Capt. Mullan . . . may be in the new concern, perhaps, but I prefer, if possible, to get the whole thing out of his hands." Getting rid of Mullan, however, would prove to be impossible. As Bidwell would soon learn, Mullan approached projects with dogged determination, and he would not be denied a leadership position in the new operation.[26]

Mullan accompanied the first group of stage passengers from Chico to Ruby City, Idaho, in September 1865. He saw the possibilities of hauling not only passengers and freight but also mail. Government mail contracts came with a federal subsidy, cash the fledgling company desperately needed. Mullan traveled to Washington bent on gaining the contract, which, with help from Bidwell, he won. Mullan became so much the public face of the company that by the spring of 1866, much to Bidwell's consternation, he had emerged as president of the newly reorganized California and Idaho Stage and Fast Freight Company. "I had to make a fight for a mail contract for someone—and Mullan edged himself in continually," Bidwell wrote. "Mullan kept waging his cause and was so omnipresent, that I was obliged to let him in—and then, once in—nobody else would come in. There were plenty of Idaho people here—they all said that Mullan was not the right kind of man—too visionary, etc."[27]

Despite Bidwell's reservations, Mullan worked hard. The Chico newspaper described the energetic entrepreneur as a "small steam engine." He made every effort to ensure the venture's success—buying stock, purchasing equipment, hiring two hundred Chinese laborers to construct stage stations and improve the road, and seeking military assistance to counter Indian attacks on the stagecoaches.[28]

In addition, Mullan diligently sought investors, though without much success. Even Bidwell refused to put up money. "Everything must be done to aid him, without incurring pecuniary liability," Bidwell wrote. Bidwell recognized the consequences of failure: "It is life or death with us," he wrote. "If this does not succeed, then everything is to be drawn away from our section of the country." Yet he refused to provide the money that might have brought the operation success. Bidwell's lack of confidence in Mullan's management capability no doubt influenced his decision. Though Bidwell offered Mullan advice and encouragement, without an infusion of cash, the line suffered.[29]

Besides lack of capital, other factors contributed to the demise of the stage company. Indian attacks, though exaggerated in the press, did present serious difficulties. At Mullan's urging, the army stationed soldiers "at convenient distances along all that part of the line that stands in any danger of the savages." Nevertheless, stock stealing at the stations and raids on stages continued, discouraging potential customers.[30]

The stage line's main challenge, however, proved to be the railroad. By 1867, the Central Pacific Railroad, moving east toward its eventual junction with the Union Pacific at Promontory Point in Utah, had reached the Humboldt River in Nevada. There it connected with a rival stage line operated by Idahoan Hill Beachey. The combined service of the railroad and Beachey's company could transport mail less expensively than could Mullan's stage line, so the government did not renew his contract. Without the federal subsidy, Mullan's venture quickly collapsed. The company sold its stock and equipment at auction in Chico. Afterward, Idaho's *Owyhee Avalanche* bid Mullan's company a scathing adieu: "Thus concludes the farce commenced nearly two years ago—a continually distracting subject."[31]

Rebecca Mullan estimated that her husband lost $12,000 on the stagecoach venture. Once again the Mullans, out of money, sought another opportunity. In 1866 they moved to San Francisco. Both took jobs in a local bank while John studied law. After Mullan passed the bar exam, the city hired him, primarily because of his engineering experience, to provide an opinion on a water-rights issue. The case helped establish Mullan's legal reputation. He soon opened an office on Montgomery Street, specializing in real estate and land law. He also became managing agent of the California Immigration Association, established to encourage settlement in California's vast lands by small-time farmers.[32]

By the early 1870s, individual law practices had begun to give way to firms providing specialized legal services, so Mullan sought a partner. He joined with Frederick A. Hyde, and the Mullans, with their two young daughters, moved into "a very handsome house" at 717 Post Street. Mullan's choice of partners proved financially wise but ethically dubious. Frederick Hyde would eventually spend time in prison for defrauding the federal government of vast tracts of western land. Although Hyde's legal problems came after he broke with Mullan, the Mullan & Hyde firm often skirted the boundaries of legality, taking advantage of lax laws in California.[33]

When California became a state in 1850, the federal government turned over more than nine million acres to support internal improvements, common schools, higher education, and public buildings. The government intended for these lands to be sold to small farmers, with the proceeds benefitting the new state, as had occurred elsewhere in the country. In most states, legislatures had established a land-administering agency to oversee the sales of these properties, but California did not create such an agency until 1860, and even then, the chronically understaffed bureau was given little authority. The California system proved ready-made for swindlers, who used dummy purchasers to game the system and amass huge tracts of the land intended for small farmers.

No firm proved more adept at this scheme than Mullan & Hyde, flooding the state land office with dubious claims. The "notorious land-grabbing firm" corralled thousands of acres that the partners then sold for personal gain. The Mullan & Hyde law office was next door to the land office, facilitating these raids on the books. A legislative committee in 1876 investigated the favoritism shown toward Mullan & Hyde. "The state land office . . . enabled them to file applications on huge acreages of lieu and school land without the $5.00 required payment for each application," noted historian Paul Gates. "The firm was also permitted to have access to the land books out of hours, a privilege denied all others. Through these means Mullan & Hyde could tie up great acreages at little or no cost to themselves until they determined which were certain to increase in demand, at which point they might complete the purchase or sell their reservation to others. Their applications were so large—71,557 [acres] in one district alone—that the office had special forms printed at state expense for them.[34]

On the one hand, Mullan worked for the California Immigration Association, which lured landless settlers to California. The association's sole purpose, it proudly advertised, was the "filling [of] our valleys and foot-hills with a class of population most needed by the State, and not for the accomplishment of any mercenary or selfish end." Mullan perhaps did not read that last clause, for, on the other hand, his law firm worked at cross-purposes. Once in California, new immigrants found it hard to get land because firms like Mullan & Hyde monopolized the property intended for them. And as a sideline of their practice, Mullan & Hyde also represented large landowners in legal suits against squatters, many of whom moved onto private property after being shut out of public lands. The firm worked virtually every nefarious angle of California land dealing, but in the process the partners created a public-relations nightmare that eventually came to haunt them. Still, despite its questionable ethics, the prominence of the firm established Mullan as a lawyer to be reckoned with, and for a time it brought him the prosperity he had yearned for.[35]

Meanwhile, Mullan became actively involved in Democratic Party politics. He supported the election of William

Irwin as governor in California in 1875 and campaigned for Samuel Tilden for president in 1876. During the Tilton campaign, Mullan spoke out on the Chinese issue, so hot in California in the 1870s. At a mass rally in San Francisco and again in a well-publicized debate, he asserted his belief, according to newspaper coverage, that "this was a white man's government, that its laws were to be administered by white men, for the benefit of white men, for all time to come." Mullan's views reflected the platform of the California Democratic Party.[36]

In 1878 Governor Irwin appointed Mullan counsel for the state of California and tasked him with seeking federal compensation of 5 percent of the proceeds from the sale of federal lands within its borders. This arrangement had been standard in other states, but when California entered the Union, no allowance had been made for such compensation. The state had been ineffectively battling with the federal government over the issue since the 1850s. Now Mullan took up the fight. He gradually closed up his San Francisco practice with Hyde and moved to Washington, D.C., where he established an office on Connecticut Avenue. From there he would practice law for the rest of his life.[37]

Mullan proved very successful in this position. Within a few years, the states of Oregon and Nevada, as well as Washington Territory, had also appointed Mullan their official counsel "for the collection of certain claims alleged to be due from the U.S. government." In addition to the "5 percent" claims, Mullan sought refunds for state money expended during the Civil War and the various Indian wars and for expenses incurred establishing federal land grants that were later cancelled by the U.S. government, but not before states had incurred costs. For his efforts, Mullan received a percentage of all the money he gained for the states—generally a hefty 20 percent.[38]

For a time the Mullans lived well. The 1880 census showed they had two domestic servants living in their D.C. home. They were prominent in the capital's social world as well. After Democrat Grover Cleveland was elected president for the first time, the Mullans received an invitation to the inaugural ball, where Rebecca wore "a blue and white striped silk with front of yellow satin covered with lace." A few months after the March 4, 1885, inauguration, Rebecca took the three children on a two-year grand tour of Europe, where John joined them the following summer. Upon their return, their eldest daughter, Emma, noted for her singing, was presented to Washington society as a debutante. But by then, John had worries about the family's finances, as states had begun reneging on their commitments to pay his fees.[39]

When Mullan's efforts brought small sums of money into the states, no one complained about his fees. But once he began winning large settlements against the federal government, critics blasted his 20 percent share as outrageous. According to some, Mullan was asking as much as $700,000 from California alone. Most of the critics were Republicans, attacking Mullan because of party politics. But Mullan's earlier land shenanigans in California had left many, Republican and Democrat alike, hoping for his comeuppance. In 1886, to justify his fees, which had been previously approved by two legislatures and two governors, Mullan wrote to California's Democratic governor George Stoneman with a meticulously documented description of his efforts on behalf of the state. Detractors called the document "a clear waste of [the] time, money, and the raw material" required to print it.[40]

When Republicans gained power in California in 1887, the new governor, Robert Waterman, refused to pay Mullan's fees. "I do not recognize your authority to act on behalf of California," he wrote in a tersely worded letter. Mullan, he said, had not "been appointed by any competent authority" and was "acting entirely in a private capacity, without any sanction." Waterman's refusal to recognize Mullan's legitimate appointment as state counsel by earlier Democratic governors and legislatures set off a legal and public-relations battle that remained unresolved at the time of Mullan's death more than twenty years later. Worse, California's bold refusal established a precedent; soon Oregon and Nevada similarly reneged on their commitments to Mullan. These connivances would not only embitter Mullan but also ensure that he would die destitute.[41]

At first, Mullan attempted conciliation, going to Sacramento to meet with Waterman in his capitol office. He presented the governor with a letter in which he painstakingly

Washington, D.C., in 1897. Mullan conducted his law practice at his residence on Connecticut Avenue, a few blocks northwest of the White House. –Map by Philip Mobley

explained that he had legitimately been appointed to oversee state interests not only by Governor Irwin in 1878 but also by Governor George Perkins in 1882, and that both appointments had been duly approved by the legislatures. Consequently, he had proceeded "in good faith . . . with due diligence under said contract, and at my own expense, and never at any time or place at any expense of the State of California." He concluded with a reasonable request for compensation. But Waterman refused to accept the letter, and Mullan's only recourse became legal action.[42]

The details of Mullan's various lawsuits are complicated, but his 1888 struggle to get his 20 percent commission from California epitomized his overall problem. The state's proceeds from federal land sales went to benefit public schools. When Mullan successfully brought in a million dollars, his cut stood to be $200,000. In trying to collect, however, Mullan was painted, particularly in the Republican press, as a villain trying to rob schoolchildren of their rightful due. The "Public School Fund of the state is the subject of insidious attack," railed one critic. After a Democratic paper defended Mullan, its Republican counterpart harangued, "Governor Waterman has discharged this man Mullan from the service of the State; and, lo and behold, on the authority of our Democratic contemporary, the world is to come to an end, and chaos to reign. Well, the State can afford to take some chances in order to keep $200,000 out of Mullan's pockets."[43]

Mullan's suit against the state came before the Supreme Court of California. The state's case essentially hinged on a legal technicality. Though two governors and two legislatures had approved Mullan's appointment as state counsel, California argued that the appointments had been confirmed by legislative resolutions, not laws, so the agreements were not legally binding. The court ruled against Mullan.

Despite the legal setback, Mullan continued his efforts to win federal funds for the states, and he often prevailed, but the states continued to renege on payments. Finally, in the 1890s, Mullan suffered a series of debilitating illnesses that limited his ability to further pursue legal recourse. "I am very weak and have fallen off to a shadow of my former self," he wrote to his three children in May 1898.[44]

As Mullan's health deteriorated, so too did his financial situation. By 1905 he was reduced to asking an associate for the small sum of $250. "I would not have written had I not needed the amount asked for," he pleaded. Mullan had fallen a long way since the 1880s, when his family had mingled with the social elite of the nation's capital.[45]

After Rebecca's death in 1898, Mullan lived most of the rest of his life with daughters Emma and Mary (better known as May). To help shore up family finances, the two sisters

John Mullan on his 78th birthday, July 31, 1908, a year and a half before his death on December 28, 1909 –Courtesy John Mullan Papers, Georgetown University Library, Special Collections Research Center, Washington, DC

opened the DeSales Hand Laundry next to the Mullan family home on Connecticut Avenue. The opening of a laundry by former socialites caused a stir: "The laundry . . . is . . . in the sacred precincts of Dupont circle, about which gather the crème de la crème of Washington's smart set. The Mullans have themselves entertained 'Washington's best people.' But poverty came, as it will to the most aristocratic of army people, and it was up to the girls to marry for money or to do something. They chose to do something."[46]

The laundry proved short-lived, however. Mullan found a better use for Emma's talents, sending her to the West Coast to further pursue efforts to obtain the payments the states owed him. Mullan's friends in California acquainted Emma with the ways of lobbying. "I will endeavor to . . . introduce you personally to such members of the Legislature as I am acquainted with," wrote attorney Isaac Frohman. "I feel that your father has not received the consideration at the hands of this state which he is entitled to, and I will gladly render whatever assistance I can in this matter."[47]

Emma took to her new task with talent and skill. "Your daughter . . . is the boss lobbyist of the world," wrote one California legislator to Mullan. She made such a splash in Oregon that one admirer wrote to her, "I really think posterity will be the loser if you do not carry out your oft repeated threat of writing a book on your legislative experiences."[48]

One of those entranced by Emma was California senator George Lukens. Lukens took up Mullan's cause in the legislature, worked with the governor, and even handled some of Mullan's legal cases as Mullan grew increasingly feeble. In the spring of 1907 Lukens proposed to Emma, and the two married in June at the Mullan home in D.C. The bride was given away by her invalid father, and May served as maid of honor. Together, Emma and George continued to pursue Mullan's payment claims, winning some contested fees in Oregon and California, though never in the amount rightfully owed to him. In California, the legal battle continued for years after Mullan's death.[49]

As Mullan's health continued to decline, May and Emma took charge of virtually all his personal and business affairs. In his 1907 will, Mullan granted to them all future proceeds they might win in claims against the states. "I take this opportunity to express to my two daughters the great love and affection which I bear toward them as well as my appreciation of the tender care which they have bestowed upon me during all the latter years of my life, and the filial manner in which they have been ever mindful of my welfare in providing a comfortable home for me after I met with severe financial losses."[50]

Missing from that paean was a mention of son Frank. In August 1903 Frank had married Mary McDonald Knapp in South Dakota, a state notorious for its easy divorces. The bride had obtained her divorce only a few days before her marriage to Frank. Her former husband later questioned the legality of the South Dakota divorce, and the entire affair scandalized the Mullan family. "I married whom I pleased & how I pleased," Frank defiantly wrote his father, who had expressed grave misgivings upon hearing the news. "I have neither disgraced you [nor] insulted you." Although Frank continued to see his father and sisters occasionally, for all practical purposes he had become estranged from his family.[51]

Although Mullan's will bequeathed one-third of his estate to Frank, he well knew it had no value–he died "possessed of no real or personal property." Frank's main inheritance was a gold watch that had belonged to Rebecca's grandfather. Mullan's only assets of potential value were his claims against the western states, and these he left to Emma and May, who continued to recoup some funds. The women did not share the proceeds with their brother, and when Emma died in 1915, she left only $500 to Frank; virtually all the rest of her estate went to May.[52]

John Mullan died on December 28, 1909, at age seventy-nine, of "primary senility," according to his official death record. Although Rebecca had been interred in her hometown of Baltimore, Mullan had chosen to be buried in his own hometown of Annapolis, in St. Mary's Catholic Cemetery.[53]

Mullan had lived a long life during a time of dramatic change. On the same page as his obituary, the *San Francisco Call* carried a story of a flight attempt by an "aeroplane." In another paper, his obituary appeared below an advertisement noting telephone numbers of various businesses. A man who had constructed a wagon road through Indian country had lived into the era of airplanes and telephones. Along his old

John Mullan's gravestone in Annapolis. He was buried next to his father. Mullan's wife, Rebecca, was buried in Baltimore. –Photograph by Paul D. McDermott

road, many of the posts his crew had meticulously set in the ground to mark each mile could still be seen in 1909, though "the storms of nearly fifty winters" had washed away most of the black-painted numerals. Later, as residents of Washington, Idaho, and Montana clamored for improvements to Mullan's road, construction crews destroyed nearly all of the markers. In 1911, the first automobile crossed Idaho's Fourth of July Pass, so named for the Mullan road crew's boisterous holiday celebration there in 1861. The famous "Mullan Tree" on that pass retained the last visible Mullan Road blaze to be found along the 624-mile route, until the tree fell to disease and lightning strikes in 1988.[54]

May Mullan, later May Flather, in an afterword to her mother's reminiscences, remembered the human side of her father: "My father, as I recall him, had a masterful air, a keen sense of humor, loved a joke and a good story, his blue eyes twinkling with merriment." In the 1920s, May worked closely with Mullan's first biographer, Helen Addison Howard, in an effort to preserve her father's legacy. When May died in 1962, she left her house and all its contents–including her father's papers from the last decades of his life–to Georgetown University. Although the papers remained unprocessed until 2003, they were carefully preserved and, thanks to May's foresight, are now permanently accessible to researchers.[55]

For only a brief period did Mullan achieve the fortune he so ardently sought. He realized a measure of fame–and in some places, notoriety. Few histories about the Northwest fail to mention the military road he constructed, which played a significant role in the development of western Montana, northern Idaho, and eastern Washington. And while some criticized Mullan for his choice of route, which passed through the often snow-clogged Bitterroot Mountains, when later engineers sought the best course for a modern highway, they chose the path pioneered by Mullan. Today, thousands of cars a day pass over Interstate 90, which in Idaho is officially named the "Captain John Mullan Highway."

In much western literature, John Mullan has been a caricature: he came to the West, explored possible railroad routes, and built a military highway, then disappeared from the written record. His reputation has grown shinier since his death, primarily because historians have focused almost exclusively on his early military adventures, omitting or glossing over the often unsavory and disappointing career paths he chose afterward. But Mullan lived a full, if oftentimes contentious, life. His story is in many ways the story of western development, of settlers in search of a better livelihood who sometimes, like Mullan, skirted the edge of legality and who frequently failed. To leave John Mullan as a young lieutenant entering Walla Walla in triumph in 1862, unaccompanied by controversy, is an injustice not only to his personal story, but also to the story of the emerging West.

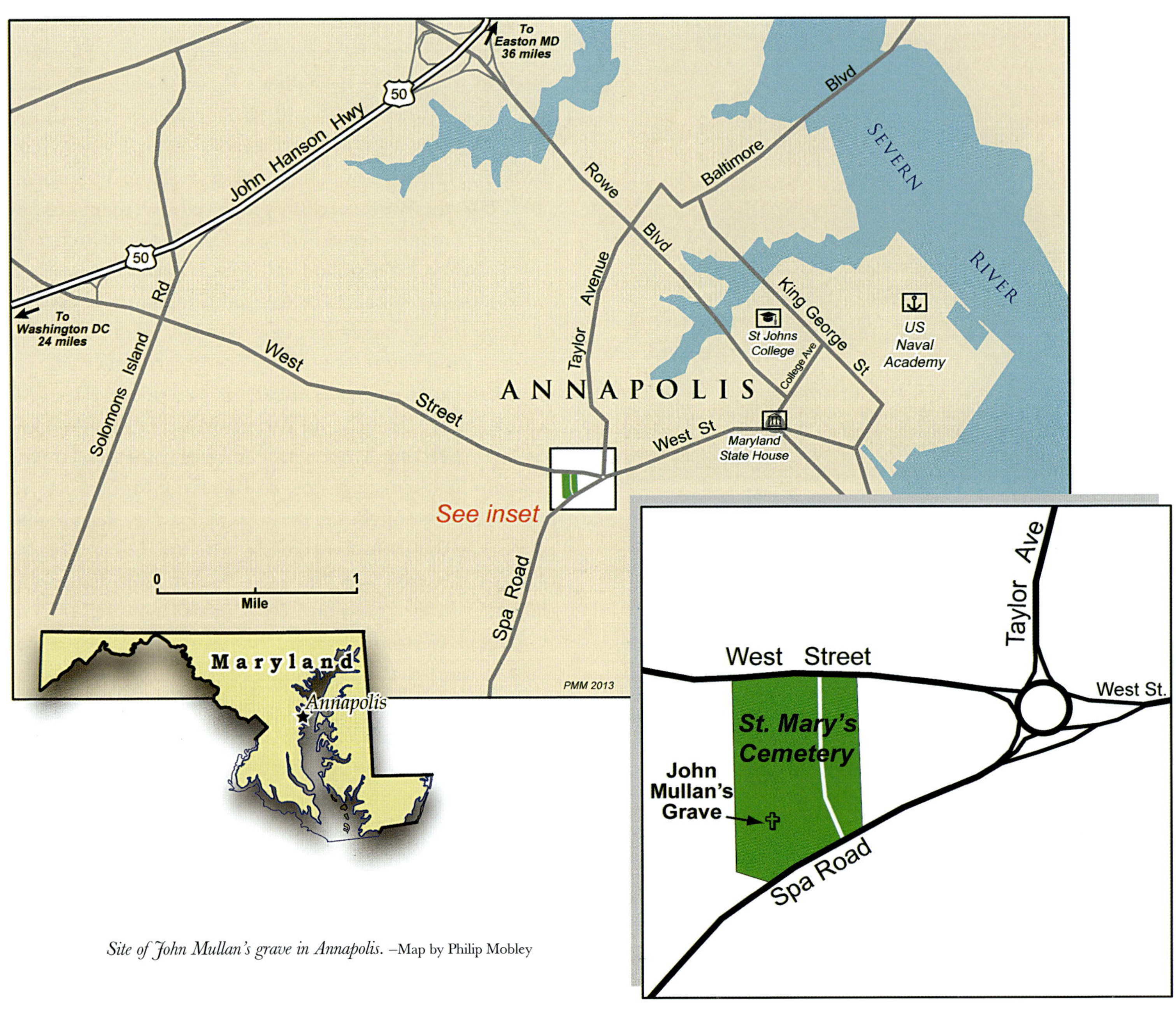

Site of John Mullan's grave in Annapolis. –Map by Philip Mobley

1 Information on Mullan's arrival in Walla Walla from *Washington Statesman*, 26 July 1862 and 16 Aug. 1862. Quotation from John Mullan, *Report on the Construction of a Military Road from Fort Walla Walla to Fort Benton* (1863; reprint, Fairfield, WA.: Ye Galleon Press, 1994), p. 36.

2 Quotation from Randall Hewett, in Oscar Osburn Winther, *The Old Oregon Country: A History of Frontier Trade, Transportation, and Travel* (Palo Alto: Stanford University Press, 1950), p. 282.

3 Mullan quotation from John Mullan, *Journal from Fort Dalles O. T. to Fort Walla Walla W.T. July 1858; Lieut. John Mullan, U.S. Army*, edited by Pal Clark, Sources of Northwest History No. 18 (Missoula: State University of Montana, c. 1933), p. 10. Although this report came from an earlier period, Mullan's enthusiasm had not dampened by the 1860s. For another glowing report of the area's potential see "Remarks of John Mullan, 7 May 1863, on the Geography, Topography and Resources of the Northwestern Territories" in *Proceedings of the American Geographical & Statistical Society* 2:1 (1863–64), p. 12, Yale Collection of Western Americana, ZC10/863mu, Beinecke Rare Book and Manuscript Library, Yale University, New Haven, CT (hereafter cited as Yale Collection).

4 I am indebted to historian Jean Russo for the time she took to take me through historic Annapolis, describing how it would have appeared in Mullan's time; additional helpful resources on the city's history include Jane Wilson McWilliams, *Annapolis: City on the Severn–A History* (Baltimore: Johns Hopkins University Press, 2011); Marion E. and Mary Elizabeth Warren, *"The Train's Done Been and Gone": An Annapolis Portrait, 1859–1910* (Boston: David R. Godine, 1976); Clarence Marbury White, *The Years Between: A Chronicle of Annapolis, Maryland, 1800–1900 and Memoirs of Clarence Marbury White, Sr., and Evangeline Kaiser White* (New York: Exposition Press, 1957); and Elihu S. Riley, *The Ancient City: A History of Annapolis, in Maryland, 1649–1887* (reprint, Baltimore: Clearfield Company, 1995).

5 St. John's College, Matriculation Book, 1789–1860, Matriculation Records for St. John's College, SC5698-2-1, Maryland State Archives, Annapolis; and Tench Francis Tilghman, *The Early History of St. John's College in Annapolis* (Annapolis: St. John's College Press, 1984).

6 Quotation from Helen Addison Howard, *Northwest Trail Blazers* (Caldwell, ID: Caxton Printers, 1963), p. 149; Howard utilized the 1892 reminiscences of Mullan's wife, Rebecca, as well as information supplied by the Mullans' daughter May. Rebecca's reminiscences are in Mullan Papers, Box 8, Fldr. 9, Special Collections Division, Georgetown University Library, Washington, DC (hereafter cited as Rebecca reminiscences; the Mullan Papers at Georgetown University are hereafter cited as Mullan Papers).

7 Regarding Polk's visitors, I am indebted to Bruce Kirby, Manuscript Reference Librarian at the Library of Congress, which houses the Polk papers (email to the author, 8 Jan. 2010). Regarding applicants gaining an audience with the president, see Benson J. Lossing (George Crockett Strong), *Cadet Life at West Point, by an Officer of the United States Army with a Descriptive Sketch of West Point* (Boston: T.O.H. Burnham, 1862), pp. 37–43; Strong graduated from the school in 1857. Mullan's West Point application papers are in the Special Collections and Archives Division, U.S. Military Academy Library, West Point, NY (hereafter cited as West Point Archives). Regarding Polk's personal interest in West Point, see James L. Morrison, Jr., *The Best School: West Point, 1833–1866* (Kent, OH: Kent State University Press, 1986), pp. 26–28. Mullan's quotation about Polk from a letter, Mullan to Lincoln, 4 March 1863, State Department Applications and Recommendations for Public Office Series, RG 59, National Archives, Washington, DC.

8 For descriptions of West Point in Mullan's time see "Descriptive List of New Cadets for the Year 1848," West Point Archives; see also Morrison, *The Best School*; Lossing, *Cadet Life at West Point*; Theodore J. Crackel, *West Point: A Bicentennial History* (Lawrence: University Press of Kansas, 2002) and *An Illustrated History of West Point, 1848–1952* (New York: Henry N. Adams, 1991).

9 U.S. Military Academy Library Circulation Records, 1844–1850, West Point Archives. The school's library at this time was used for recreational reading; standard course work did not require library materials. Thus circulation records give some evidence of a cadet's areas of interest, though in 1852 the entire collection consisted of only 15,500 books, leaning heavily toward "devotional material, military history, heroic novels, and biographies of the 'Great Captains'" (Morrison, *The Best School*, p. 76; see also pp. 90–91).

10 Mullan's efforts to increase the appropriation are summarized in his letters of 21 June 1858 and 2 Sept. 1858 to Captain A. A. Humphreys of the Topographical Engineers, Records of the Office of Chief of Engineers, Office of Explorations and Surveys, RG 77, Box 3, Entry 359, National Archives.

11 Between his time on the railroad survey and his return to the Northwest in 1858, Mullan was posted in Louisiana, Maryland, and Kansas.

12 Quotations from 1859 diary of Mullan's future wife, Rebecca Williamson, in Mullan Papers, Box 8, Fldr. 14. It is clear from the diary that Mullan had made up his mind to leave the Army upon completion of his road and return west as an entrepreneur at least since 1859.

13 For description of Stevens's death and elaborate funeral see Kent D. Richards, *Isaac I. Stevens: Young Man in a Hurry* (reprint, Pullman: Washington State University Press, 1993), pp. 387–90.

14 Regarding Sohon and Mullan leaving Walla Walla together, see Walla Walla's *Washington Statesman*, 6 Sept. 1862. Quotation regarding Bidwell ranch from John Mullan, "From Walla Walla to San Francisco," reprinted in *Oregon Historical Quarterly* 4:3 (Sept. 1903), pp. 202–26.

15 For reprimands of Mullan see Col. S. W. Long, Corps of Topographical Engineers, to Mullan, 11 July 1862, 7 Nov. 1862, 10 Nov. 1862, and 17 Nov. 1862, all in RG 66, vols. 22, 23, 24, National Archives.

16 The best analysis of the territorial boundary issues of 1862–63 can be found in Merle Wells, "The Creation of the Territory of Idaho," *Pacific Northwest Quarterly* 40:2 (Apr. 1949), pp. 106–23, and Wells (unattributed), "Idaho's Centennial: How Idaho Was Created," *Idaho Yesterdays* 7:1 (Spring 1963), pp. 44–58.

17 Mullan's letter to Lincoln, 4 March 1863, National Archives.

18 *Sacramento Union* quotation in Wells, "Idaho's Centennial," p. 56. For the best analysis of Wallace in this period see Anne Laurie Bird, "Portrait of a Frontier Politician," Part 4, *Idaho Yesterdays* 2:3 (Fall 1958), pp. 12–23, and Parts 5 and 6, *Idaho Yesterdays* 3:1 (Spring 1959), pp. 8–20. See also David H. Leroy, "Lincoln and Idaho: A Rocky Mountain Legacy," *Idaho Yesterdays* 42:2 (Summer 1998), pp. 8–25, particularly pp. 12–13.

19 For description of the Mullans' honeymoon, see Rebecca reminiscences. *New York Times* quotation from 10 May 1863. Speech quotation from "Remarks of John Mullan, 7 May 1863," Yale Collection. I estimated the length of time it took Mullan to deliver the talk based upon a reading of the speech.

20 Mullan quotation from "From Walla Walla to San Francisco," pp. 202–03.

21 Rebecca's quotation from Rebecca reminiscences. Mullan quotation from *Railway Times*, 28 March 1863. Details about the Mullan brothers' Walla Walla business ventures can be found in a series of legal documents from the U.S. District Court of Walla Walla, years 1861–66, Washington State Archives, Eastern Region Branch, Cheney (hereafter cited as Washington Archives East).

22 Quotation from *Railway Times*, 28 May 1863. In 1872 Walla Walla capitalist Dorsey Baker—one of the original incorporators in the Walla Walla Railroad Company—did construct a short-line railroad between Wallula and Walla Walla, but by that time Mullan had long since left the area; see W. W. Baker, "The Building of the Walla Walla & Columbia River Railroad," *Washington Historical Quarterly* 14:1 (Jan. 1923).

23 Details of the Mullan brothers' legal actions against each other can be found in U.S. District Court of Walla Walla records, Washington Archives East.

24 John Mullan, *Miners' and Travelers' Guide to Oregon, Washington, Idaho, Montana, Wyoming and Colorado via the Missouri and Columbia Rivers* (New York: Wm. M. Franklin, 1865).

25 For a good overview of the Chico to Idaho route, see *Reproduction of Fariss and Smith's History of Plumas, Lassen & Sierra Counties, California, 1882* (Berkeley: Howell-North Books, 1971), pp. 386–87; Clarence F. McIntosh, "The Chico and Red Bluff Route," *Idaho Yesterdays* 6:3 (Fall 1962), pp. 12–19; and Larry Francis Bourdeau, "The Historic Archaeology of Cabo's Tavern" (master's thesis, California State University-Chico, 1982), particularly pp. 123–37. For background on Bidwell, Rancho Chico, and Chico see Marcus Benjamin, *John Bidwell, Pioneer: A Sketch of His Career* (Washington, D.C., 1907). Pierce's involvement is best detailed in his memoir, J. Gary Williams and Ronald W. Stark, eds., *The Pierce Chronicles: Personal Reminiscences of E. D. Pierce* (Moscow: Idaho Research Foundation, 1976), pp. 103–9.

26 Quotation from letter, Bidwell to Harris, 27 Nov. 1865, John Bidwell Papers, Box 3, Fldr. 1, MS 2, California State University, Chico (hereafter cited as Bidwell Papers, Chico).

27 Quotation from letter, Bidwell to Harris, 13 Apr. 1866 and 28 Apr. 1866, Bidwell Papers, Box 3, Fldr. 4, Bidwell Papers, Chico. Regarding Mullan on the first stage into Ruby City see Howard, *Northwest Trail Blazers*, p. 167.

28 Newspaper quotation from McIntosh, "The Chico and Red Bluff Route," p. 19, which also has a good summary of Mullan's yeoman efforts.

29 Quotations from letters, Bidwell to Harris, 4 July 1866, and Bidwell to Harris, 13 Apr. 1866, Bidwell Papers, Box 3, Fldr. 4, Bidwell Papers, Chico.

30 Quotation from *Idaho Tri-Weekly Statesman*, 5 Sept. 1865.

31 Newspaper quotation from McIntosh, "The Chico and Red Bluff Route," p. 19; the article also has a synopsis of Beachey's competitive line.

32 Mullan actually began the study of law earlier, as noted in a letter to President Lincoln in 1863: "I have been a student of law for the past five years." (Mullan to Lincoln, 4 March 1863, National Archives). For Rebecca's comments, including details on Mullan's losses, the Mullans' working in a bank, and Mullan's work on the water-rights case, see Rebecca reminiscences. For background on the California Immigration Association see *New York Times*, 10 Aug. 1868. Mullan's career can be traced through San Francisco city directories: the 1870 census lists him as a "public land dealer" with real estate valued at $5,000, employing both a servant and a nurse—no doubt for daughter Emma, then one year old.

33 For more on California's legal climate at this time see Gordon Morris Bakken, *Practicing Law in Frontier California* (Lincoln: University of Nebraska Press, 1991). For a description of the Mullans' "handsome house," see Rebecca reminiscences. The press carried many stories about Hyde's land dealings; for a summary see the *Washington Times,* 19 Nov. 1908; President Woodrow Wilson later partially commuted Hyde's sentence (see *Klamath Falls Evening Herald*, 24 Oct. 1913).

34 First quotation from Gerald Nash, "The California Land Office, 1858–1898," *Huntington Library Quarterly*, 27 (Aug. 1964), p. 353; second quotation from Paul W. Gates, "Public Lands Disposal in California," *Agricultural History* 49:1 (Jan. 1975), pp. 167–68. See also Gates, "California's Agricultural College Lands," *Pacific Historical Review*, 30:2 (1961), pp. 103–22.

35 Quotation from *New York Times*, 10 Aug. 1868. Regarding Mullan & Hyde working for large landowners against squatters, see Bakken, *Practicing Law in Frontier California*, p. 79.

36 Rebecca's reminiscences discuss Mullan's support of Irwin. The 1876 campaign issues in California, including the Chinese issue and Mullan's role as an alternate elector for Tilden, are covered in Winfield J. Davis, *History of Political Conventions in California, 1849–1892*, Publications of the California State Library, No. 1 (Sacramento: 1893), pp. 356–62. For the debate quotation, see *San Francisco Alta California*, 18 Oct. 1876.

37 For background on the complicated 5 percent issue, see William C. Fankhauser, *A Financial History of California: Public Revenues, Debts, and Expenditures* (Berkeley: University of California Press, 1913), pp. 313–14, and for Mullan's role, John Mullan, *Reports to Hon. George Stoneman, Governor of California, on Certain Claims of the State of California Against the United States* (Sacramento: State Office, 1886).

38 Mullan's *Reports to Stoneman* goes into great detail about the various claims, which are briefly summarized on pp. 5–6. See also Fankhauser, *A Financial History of California*, pp. 313–17. Mullan's work was basically the same for Oregon, Nevada, and Washington Territory. The quotation is from Watson Squire, Governor of Washington Territory, to Mullan, 15 Apr. 1886, Mullan Papers, Box 1, Fldr. 42.

39 Quotation about Rebecca's dress from *Washington National Republican*, 5 March 1885. For description of European trip, see *Washington Sunday Herald*, 14 June 1885, 24 Jan. 1886. For description of Emma as debutante, see *Sacramento Daily Record-Union*, 3 Nov. 1887.

40 Quotation from "The Direct Tax Bill," an 1888 clipping in Mullan Papers, Box 5, Fldr. 8. For the $700,000 figure see "Finishing Captain Mullan," another 1888 clipping in the same location.

41 Waterman quotation from *Sacramento Daily Record-Union*, 6 Feb. 1888.

42 Mullan quotation from *Sacramento Daily Record-Union*, 19 Jan. 1889.

43 First quotation from *San Francisco Call*, 31 Jan. 1888; second quotation from *San Francisco Chronicle*, 11 Feb. 1888. The Mullan Papers at Georgetown University contain a large number of clippings dealing with Mullan's various legal difficulties.

44 Quotation from letter, Mullan to "Dear Children," 30 May 1898, Mullan Papers, Box 2, Fldr. 2.

45 Quotation from letter, Mullan to William King, 2 Oct. 1905, Mullan Papers, Box 2, Fldr. 4.

46 Quotation from *Minneapolis Journal*, 6 Jan. 1905. For more on the laundry, see also *St. Paul Globe,* 10 Jan. 1905 and 19 March 1905, and *San Francisco Chronicle*, 8 Dec. 1904. The laundry continued until at least 1906; various letters from Mullan to Emma in 1906 make passing references to it (see Mullan Papers, Box 8, Fldr. 11).

47 Emma was in California in early 1905; later that year, Mullan requested that she go to Oregon to pursue his claim against that state (see Mullan to Emma and

May, 14 Nov. 1905, Mullan Papers, Box 2, Fldr. 3). Frohman quotation from Isaac Frohman to Emma, 19 Jan. 1905, Mullan Papers, Box 6, Fldr. 3.

48 Both quotations are from Mullan Papers: first from Grove Johnson to Mullan, 21 Dec. 1906, Box 1, Fldr. 25; second from W. W. Douglas to Emma, 7 Feb. 1907, Box 6, Fldr. 6.

49 Emma's engagement announcement is in *San Francisco Call,* 19 Apr. 1907; *Washington Herald*, 2 May 1907; and *Daily Capital Journal* (Salem, Oregon), 4 June 1907. Wedding description in *New York Tribune,* 19 June 1907. For Lukens's growing involvement in the Mullan family's legal issues, see Mullan to Lukens, 7 Sept. 1906, Mullan Papers, Box 2, Fldr. 7; W. W. Douglas, Deputy Controller, State of California, to Emma, 22 March 1905, Mullan Papers, Box 6, Fldr. 3; and various letters from Lukens to Mullan's clients in 1908, Mullan Papers, Box 6, Fldr. 6.

50 "Last Will and Testament of John Mullan," 12 June 1907, Mullan Papers, Box 7, Fldr. 3; see also Mullan's deeds and assignments to his daughters, Mullan Papers, Box 5, Fldr. 7.

51 Quotation from Frank to "My Dear Father," 10 Aug. 1903, Mullan Papers, Box 1, Fldr. 37; see also letters from Frank to Mullan, 6 Aug. and 12 Aug. 1903, in Mullan Papers, same folder; also Frank to Emma, 18 Aug. 1903, Mullan Papers, Box 8, Fldr. 6; and Frank to May, undated, 1903, Mullan Papers, Box 8, Fldr. 11. For details on Frank's marriage see *Sioux Falls American*, 4 Aug. 1903. For Alexander Knapp's testing of the legality of the divorce, see *New York Times*, 25 Feb. 1905. For controversy over South Dakota divorces in this period, see Glenda Riley, *Divorce: An American Tradition* (Lincoln: University of Nebraska Press, 1991), p. 100.

52 Quotation about Mullan possessing no property from May Mullan, statement about funeral expense, Mullan Papers, Box 7, Fldr. 23. "Last Will and Testament of Emma Mullan Lukens," 1 March 1910, Mullan Papers, Box 6, Fldr. 7.

53 St. Mary's Parish, Annapolis, Record of Internment, Microfilm 2345, "John Mullan," Maryland State Archives, Annapolis.

54 Several newspapers around the country carried the story of Mullan's death; for examples see Mullan obituaries in *San Francisco Call*, 29 Dec. 1909, and *Richmond Times Dispatch*, 29 Dec. 1909. Quotation about the mile markers in Day Allen Willey, "Building the Mullan Road," *Sunset* 24:6 (June 1910), p. 636. For road improvement requests, see, for example, the *Missoulian* (Missoula, MT), 31 Aug. 1910. Regarding the first automobile over Fourth of July Pass, see Thomas Flanagan, "The Fourth of July Tunnel," *Museum of North Idaho Quarterly Newsletter* 25:3 (Summer 2004), p. 1. Regarding the removal of the Mullan Tree, when the U.S. Forest Service removed it in 1988, the agency donated the stump with the mile marker blaze to the Museum of North Idaho in Coeur d'Alene, where it remains on exhibit.

55 May's quotation is from the afterword in Rebecca reminiscences. Correspondence between Helen Addison Howard and May can be found in Howard Papers, MC 188, Archives and Special Collections, Mansfield Library, University of Montana, Missoula.

Historical map of West Point, 1867 –Courtesy West Point Museum Collection, USMA

The Making of an Explorer

John Mullan at West Point

BY MAJ. RYAN L. SHAW

If the 1850s West Point curriculum seems like it was tailor-made to equip a graduate for a project like the Mullan Road, that is because it was. Mullan graduated in 1852, the fiftieth anniversary of the founding of the United States Military Academy. Arguably, it was just such projects as Mullan's that Thomas Jefferson had in mind when he founded the academy in 1802. It is certainly the case that by the time of Mullan's matriculation, West Point was purposefully and effectively producing officers who would not only defend the nation in wartime, but would, like Mullan, undertake projects to develop and expand it during peacetime.

Legend has it that Mullan applied in person to President James K. Polk for his appointment to the military academy. Polk allegedly looked the sixteen-year-old up and down and asked if he wasn't rather small for a career in the army. "I may be small, sir," young Mullan replied, "but can't a small man be a soldier as well as a large one?"[1] Though the story matches well the reputation for pugnacity that Mullan acquired on his road project, no record of the conversation exists in the archives of the admissions department at West Point. What does exist are ringing endorsements from the faculty of St. John's College in Maryland, where Mullan

Layout of West Point campus in 1867, and location of academy in New York (inset) —Map by Philip Mobley

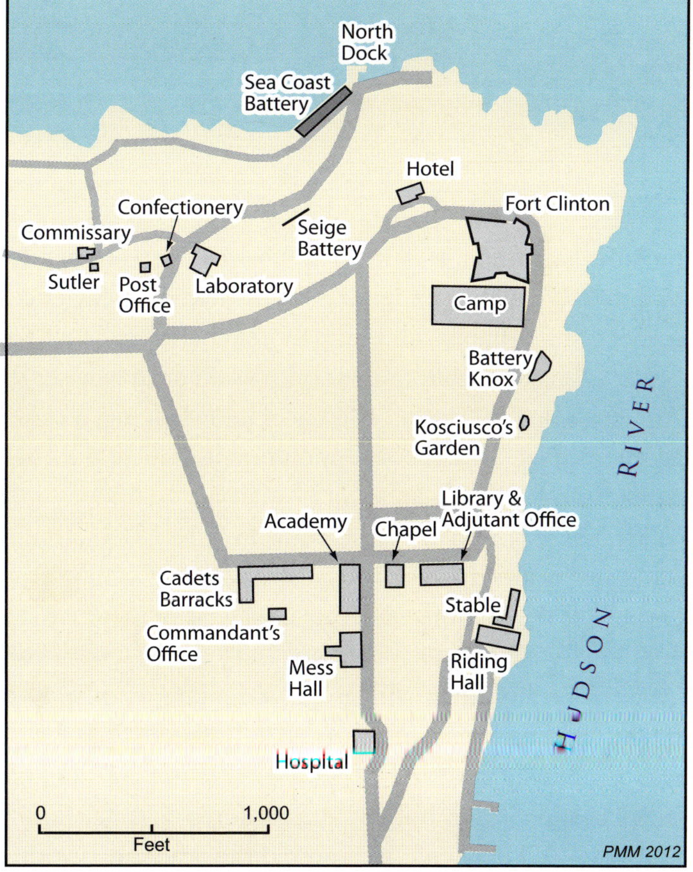

Painting of West Point c. 1870 (artist unknown) –Courtesy West Point Museum Collection, USMA

previously studied, and from his entire congressional delegation, both bodies commenting on his native intelligence, disciplined study habits, and high moral character, as well as his father's long and honorable service as an ordnance sergeant.[2] Also included is Mullan's letter acknowledging receipt and acceptance of an at-large appointment–which did, indeed, come from President Polk.

Mullan reported to West Point in July 1848, along with the other candidates for the class of 1852. Coming from all corners of the country and varying levels of prior education, appointees had to pass initial physical and academic exams in order to be admitted. Mullan passed, along with sixty-seven others, and he went on to graduate fifteenth in a class that was eventually attritted to forty-seven–not an outstanding cadet academically, but by all measures a satisfactory one, and very well behaved. He entered the academy a wide-eyed and promising seventeen-year-old boy; he graduated four years later a confident young man, toughened physically by three summers of drilling, acquainted with a broad diversity of people, and exposed to all the elements of a classical education while deeply immersed in the military arts and sciences. Commissioned a brevet second lieutenant in the army, Mullan was eager to put the whole of his West Point experience to use in the service of a rapidly expanding nation whose future seemed as bright as his own.

The intent of Thomas Jefferson in founding the United States Military Academy is something of a historical puzzle, one that has received significant attention in recent years after a long period of neglect. To the extent that the pacific Jefferson is remembered at all for his role as Commander in Chief, he is best known for reducing the size of the military based on his traditional republican fear of "standing armies" as a threat to liberty. In fact, the Military Peace Establishment Act of 1802, by which he founded the academy at West Point, paradoxically also reduced the size of the army by nearly half, both in numbers of regiments and in raw manpower. Even stranger, Jefferson had vociferously opposed—for lack of constitutional authority, he claimed—the creation of a military academy when it was a pet project of his militant Federalist nemesis, Alexander Hamilton, in 1793.[3] In light of these proclaimed ideological and constitutional scruples, generations of Jefferson scholars have treated his founding of the academy as an ironic twist at best and a hypocritical betrayal of his principles at worst.[4]

More recently, however, scholars have posited a number of more careful explanatory theories. One has it that Jefferson sponsored the military academy because he knew it could succeed, and he thought it was as close as he might get to achieving his much-discussed dream of founding a national university. He had a distinct and comprehensive set of ideas about a liberal and modern curriculum, one appropriate to a New World institution serving a new nation an ocean away from the ancient, musty halls of European universities. This vision was more fully realized in 1819 when he founded the University of Virginia, an achievement of which, unlike the founding of West Point, he was proud enough that he had it inscribed on his gravestone. At the time he assumed the presidency, however, Jefferson may have decided to seize the opportunity to "smuggle his national scientific school into the nation under the guise of a military academy."[5]

Another theory posits that the founding of the academy was part of Jefferson's broader plan to "Republicanize" the army—and indeed, the whole federal apparatus—after having endured the Federalist domination of the Adams and Washington administrations. Republican ideology notwithstanding, Jefferson knew that at least a small corps of professional officers was necessary for the security of the nation, and that a military academy was necessary for their education. His earlier resistance was not to establishing a military academy per se, but to putting such a potentially powerful institution at the disposal of dangerous, elitist Federalists like Hamilton. After his "Revolution of 1800," however, he had the opportunity to create an academy on his terms—one that would provide education and opportunity to the Republican sons of America's virtuous heartland, thereby replacing the "aristocracy of wealth and birth in the army . . . with the aristocracy of virtue and talent." Jefferson's founding of West Point was, according to this theory, "a conscious, purposeful act of an eminently and consistently political man."[6]

More interesting, given Mullan's later work in the West, Robert McDonald and Christine Coalwell McDonald offer evidence suggesting that Jefferson's true ambition in founding the academy was to "advance his efforts to expand, explore, secure, and develop the frontier for liberty-loving Americans."[7] They note that Jefferson's initial selections for faculty and curriculum were clearly designed to provide graduates with just the sort of practical scientific and cultural skills that Meriweather Lewis had had to gather through an ad hoc program of tutelage at the American Philisophical Society in Philadelphia (over which Jefferson presided) before he ventured West, on Jefferson's orders, with the Corps of Discovery.

Before the end of his presidency, Jefferson added two major subjects to the curriculum: French and drawing. The former was eminently useful for the exploration and settlement of the newly acquired and heavily francophone Louisiana Purchase; interestingly, fully half of the cadets appointed to the academy from 1803 to 1805 were from the Louisiana Territory. The latter course of study, drawing, proved essential in army efforts not only to map the West but also to depict its flora and fauna, its ethnography, and its geology to easterners in the age before photography. Jefferson also recommended the addition of "natural and experimental philosophy" (a term that encompassed what we would call earth science, physical science, natural history, and scientific theory), which would produce academy graduates with proficiencies far beyond calculating artillery trajectories, constructing fortifications, and commanding troops in battle. Indeed, once the additions were approved by Congress in 1812, the West Point program expressly equipped its graduates for the kind of scientific exploration of the West that the army had been conducting and would continue to conduct for three-quarters of a century, starting with the Corps of Discovery in 1804 and continuing until 1879, when the army's western survey program was finally merged with civilian efforts under the United States Geological Survey.

Of course, these three theories are not mutually exclusive. They could be, and probably are, simultaneously all true. What is almost certainly not true, however, is the older idea that Jefferson sponsored the academy in order to provide the nation with engineers to develop the national infrastructure. Though the academy was founded under the auspices of the Corps of Engineers, the study of engineering beyond its strict battlefield applications was a later addition to the curriculum–and to the mission–of West Point. The addition of nonmilitary engineering was not attained without difficulty, but by the time of Mullan's arrival, West Point was comfortable in its role as the preeminent civil engineering school in America.

Even practical military engineering was a fairly neglected element of the curriculum until 1812, when Congress authorized the creation of the Department of Engineering, with Captain Alden Partridge as its head.[8] But Partridge, who was also superintendent of the academy, had neither the time nor, apparently, the aptitude to develop a worthwhile engineering curriculum. It was not until 1817, when Partridge was succeeded by Professor Claude Crozet, a graduate of the École Polytechnique in France, that the program was fully developed.[9] Also in 1817, Major Sylvanus Thayer took over as superintendent and began the standardization of disciplinary and academic procedures that earned him the enduring sobriquet "Father of the Military Academy." Between the simultaneous reforms of Crozet and Thayer, West Point implemented a military engineering curriculum of which the nation could be proud. Still, civil engineering was not within the compass of Crozet's program, nor, according to Thayer, did the subject fit within the academy's mission.[10] That change was destined to be mandated from the outside.

In the years after the War of 1812, the nation embarked on a program of "internal improvements"–building canals, roads, and railroads, improving river navigability and harbor access–designed to promote a national economy and presumably to improve military mobility. The problem was, there were few trained engineers in the country qualified to lead such projects. In 1819 Secretary of War John C. Calhoun submitted to Congress a plan for a system of roads and waterways with a suggestion that army engineers might survey them. In 1822 the Chief of Ordnance requested a number of academy graduates to serve as topographical surveyors in the western mining districts. The chief envisioned a symbiotic relationship between the army and private interests, wherein officers would acquire competence in geology and mineralogy while uncovering mineral deposits for both military and commercial use.[11] Despite a suggestion from West Point's Board of Visitors in 1822 that the school update its curriculum to meet these emerging demands, the academic board made no effort to do so.[12] But after Crozet resigned his post in 1823, he was succeeded by David B. Douglass, a progressive and ambitious Yale graduate who was open to the idea.[13]

In 1824 Congress passed the General Survey Act, which authorized the president "to cause the necessary surveys, plans, and estimates to be made of the routes of such roads and canals as he may deem of national importance in a commercial or military point of view, or necessary to the transportation of the public mails." The act mandated the use of military engineers, and the army responded by creating a Board of Engineering for Internal Improvements, to which Douglass was immediately named. He set to work developing a program of civil engineering at West Point, and by the time he resigned in 1831, the Board of Visitors declared West Point to be the only institution in the nation capable of training the kind of engineers required by the government and private business.[14]

The idea of academy-trained engineers working for private business was not without controversy. The year after the passage of the General Survey Act, John Quincy Adams became president, naming as his Secretary of State Henry Clay, whose "American System" called for federally funded internal improvements (along with a high tariff and a national bank) to spur growth. Throughout his single term, Adams stretched the already-broad authorizations of the General

Survey Act to bring West Point expertise to bear on projects great and small, public and private, national and local. This fed the resentment of many of those already alienated by the "corrupt bargain" that led to Adams's election, and it contributed to his defeat by Andrew Jackson in 1828.[15] It also led to attacks on West Point itself, by Jackson and such congressional notables as Davy Crockett, as an antidemocratic bastion of privilege. The cadets were educated at taxpayer expense then paid a federal salary (extending federal patronage to a well-connected few), often receiving pay from private corporations at the same time. Worse, many West Pointers quit the army for more lucrative private employment, frequently before the end of their mandatory one year of service.

This controversy outlasted Jackson's two presidential terms; it fell to Martin Van Buren to end it. Democrat Van Buren increased the size of line units (thereby creating a greater demand for officers in troop-leading roles), increased the mandatory service of academy graduates to four years, prohibited officers from working for corporations or on civilian projects while in federal service, and, in 1838, repealed the General Survey Act.[16] It was a triumph for the notions of equality and democracy that defined the Jacksonian era. Yet in spite of the opposition, it was the General Survey Act and the changes it spurred that led, years later and under a regime of improved accountability and professionalism, to the construction of a road across the Rockies connecting the waterways of the Missouri and Columbia Rivers in the Far West.

Although they were not allowed to work for private companies while serving in the army, once mustered out, West Point graduates continued to work on civil improvement projects in their official capacity. Likewise, the engineering program at West Point continued to develop. Indeed, its most rapid improvement came under the tenure of Douglass's successor, Dennis Hart Mahan, who held the post until his death in 1871. The son of Irish immigrants, Mahan was the top graduate of the West Point class of 1824. His aptitude was so apparent that by his third year as a cadet, he was serving as an assistant instructor of mathematics, and Thayer retained him in that position following his graduation. Mahan would spend the entirety of his professional life at West Point except for four years studying public works and military institutions in Europe, including one year at France's School of Application for Artillerists and Engineers at Metz, before his appointment as department head at the academy.[17] His series of lithographic notes, based on his studies abroad, served as instructional texts in the department well into the twentieth century.

As the rest of America's universities struggled to catch up to West Point in engineering instruction, many staffed their departments with West Point graduates and adopted West Point's texts. Mahan supplemented his original lithographic notes with books on engineering, field fortifications, descriptive geometry, and more. More than a mere technician, Mahan was a theoretician. His *Elementary Treatise on Advanced-Guard, Outpost, and Detachment Service of Troops, with the Essential Principles of Strategy and Grand Tactics*, known simply as *Outpost*, is considered among the most significant nineteenth-century American contributions to the study of war, in a league with the better-known *The Influence of Sea Power upon History,* written by his son, Alfred Thayer Mahan. With his frequent and well-received publications, Mahan became the public face of West Point and was the guiding force of its academic board throughout his long tenure.

In his senior (or first-class) year, John Mullan took Mahan's course entitled "Civil and Military Engineering and the Science of War." Cadets were divided into sections of ten or twelve, based on academic standing. It was customary for the department head to teach the highest section personally, although Mullan, having finished 23rd in engineering, did not enjoy the privilege of sitting in Mahan's section every day.[18] Nevertheless, because Mahan made a habit of visiting each section every day during their two-hour block of instruction–correcting exercises, asking questions, augmenting the instructor's remarks, and on occasion, presenting course-wide lectures–as a cadet, Mullan regularly interacted with the brightest mind in engineering education in America.[19]

Something similar can be said for Mullan's education in drawing. Robert Weir taught drawing at West Point for forty-two years, beginning in 1833. While Weir may not have been the foremost American artist of the period, neither was

Engraving of the library and other buildings at West Point by Lossing & Barritt, 1855
–Courtesy West Point Museum Collection, USMA

he without credentials, and he presided over a drawing curriculum more comprehensive than anything else offered in the country at the time. Before his appointment to West Point, Weir had studied art in Italy for three years and had a studio in New York City. Recognized as a member of the Hudson River School of American artists for his landscapes, he is best known for his *Embarkation of the Pilgrims,* which still hangs in the rotunda of the U.S. Capitol. His allegorical *War and Peace* still watches over the dais of the old Cadet Chapel at West Point.

During his second year, Mullan studied under Professor Weir for two hours a day, learning human-figure and topography drawing. In his third year, he studied landscape drawing for two hours every other day.[20] Additionally, his second-year math classes featured increasingly technical training in drawing and drafting; examples include "Shades, Shadows, and Perspectives," "Spherical Projections and Warped Surfaces," and "Surveying and Analytical Geometry."[21] Moreover, in his engineering courses, Mullan took nearly two hundred more hours of drawing before he graduated.[22] Later, as leader of the road–building project, Mullan was fortunate to have Gustavus Sohon and Theodore Kolecki to do the bulk of the illustration and cartography for his reports, and he was certainly qualified to supervise and verify the quality of their product, but he was also qualified to produce accurate representations himself, had he needed to.

In addition to the courses just described, Mullan took his requisite two years of French, in which he performed quite well. One year of "English Studies" included grammar, rhetoric, and curiously, geography. He studied chemistry in his third year and ethics in his fourth, along with a course in infantry tactics and another called "Artillery and Cavalry." Of particular note in light of his later work, Mullan studied natural and experimental philosophy with William H. C. Bartlett. Bartlett was noted and widely published in the fields of mechanics, optics, and astronomy. In this course, Mullan learned how to use the barometer, the sextant, and other important instruments of the explorer.

Additionally, in his first-class (final) year, Mullan took "Mineralogy and Geology." As James Morrison notes, this was among the most beneficial courses at the academy:

> Cadets not only found these courses interesting but useful as well. Graduates exploring the West were able to render accurate reports of the geological formation and the mineral deposits they found; commanders of frontier garrisons could pick the best plots for post gardens and determine the most suitable fruits and vegetables to plant; and even more importantly, West Pointers could draw on their knowledge to help them in selecting the likeliest sites for finding water. Thus, the [course] had an applicability that transcended the narrow limits of military technology.[23]

One would be hard put to conceive a course of instruction more fitting for a future western road builder than the curriculum at West Point.

Mullan was fortunate to have gone to the academy when he did. In its early years, the school struggled just to survive. Its mission was uncertain, its curriculum was not standardized, its admissions and graduation standards were essentially ad hoc, and its administrators fought among themselves and with their superiors in Washington, D.C. From mid-1811 to early 1813, the academy was actually shut down. But after 1817, the sweeping reforms Thayer initiated brought respectability to the institution and stability to its denizens, advancements that lasted far beyond the end of his long tenure in 1833.

Even so, turbulence still surrounded the academy's place in the republic during Thayer's administration, as evidenced by the Jacksonian-era crisis over internal improvements previously described. In fact, Thayer resigned over a dispute with Andrew Jackson regarding the president's authority to reinstate separated (expelled) cadets. Yet within the confines of West Point, life was predictable tranquil, even, in the unique way that Spartan discipline and military routine can be. For forty years, a standard number of cadets occupied a standardized set of barracks with standardized gear organized in a standard way. Through the academic year, cadets attended classes six days a week and marched together to services at the Cadet Chapel on Sundays. They recited daily from the homework they'd prepared the previous night, and they stood for exams every January and June. During the summer, they drilled at their encampment on the school's parade grounds, known as the Plain.

The tranquility of the era was shattered by the coming of the Civil War. As the sectional crisis escalated, fistfights and duels erupted in the barracks; when the conflict broke out, Southern cadets and Southern faculty alike resigned from West Point to fight for their seceding home states.[24] Following the war, the academy faced many of the same challenges of reintegration and rebuilding as the nation did during the Reconstruction period.

Of course, Mullan's class of 1852 was graduated and gone before all that happened, so Mullan was fortunate to have gone to West Point when he did. He enjoyed academy life under the antebellum Thayer system at the peak of its maturation. Doubly lucky for Mullan, he arrived in 1848, just after the war with Mexico, in which West Point graduates had acquitted themselves in the defeat of Santa Anna's army and the occupation of the Mexican capital well enough to silence almost all of the academy's critics. General Winfield Scott, not a West Pointer himself, was so impressed by their performance that he later offered his "fixed opinion," which every new cadet is still required to memorize in basic training:

> I give it as my fixed opinion that, but for our graduated cadets, the war between the United States and Mexico might, and probably would have, lasted some four or five years, with, in its first half, more defeats than victories falling to our share; whereas, in less than two campaigns, we conquered a great country and a peace without the loss of a single battle or skirmish.[25]

Scott himself visited West Point amid much fanfare on New Year's Day in 1849, bringing blood-stained battle flags and captured cannons for permanent display at the academy.[26] The war brought more than trophies to West Point, though—combat-veteran officers began filling appointments on the staff and faculty, adding to the academy's already impressive theoretical and academic foundation a new degree of practical credibility and battle-tested military leadership.

The nation's citizenry seemed to agree with General Scott's assessment of the academy and its alumni, as West Point became a busy tourist destination during this period. Legislators, too, were impressed—with its value to the nation so convincingly demonstrated, Congress opened its coffers to fund a long-overdue construction program. The funds enabled the school to replace two barracks buildings and add a new mess facility. Thus the tranquility of Mullan's time at the academy was continuously punctuated with the bustle of construction, but the cadets probably welcomed the noise as yet further evidence of their special place in a nation just coming into the fullness of its Manifest Destiny.[27]

In addition to educational enlightenment and professional inspiration, the West Point experience exposed Mullan to the broad social fabric of mid-nineteenth-century America. In his speech honoring the academy's centennial in 1902, President Theodore Roosevelt said that "of all the institutions in this country, none is more absolutely American; none, in the proper sense of the word, more absolutely democratic than this."[28] Roosevelt's words were not mere patriotic hyperbole–the sentiment he expressed would have been just as applicable when Mullan graduated in the year of West Point's semicentennial.

The academy was democratic by design, though it had not always been so. Prior to 1815, there had been no clear standard for selecting cadets. As mentioned above, founder Thomas Jefferson appointed a preponderance of his cadets from the Louisiana Territory in order to bind that new region to the nation and equip its sons to push the frontier even farther west. Later, the Secretary of War and the Chief of Engineers controlled the appointment process, basing most of their selections on political and military connections. Thus, before the War of 1812, three-quarters of West Point graduates came from New England–a full 25 percent from Vermont alone. In 1815 Secretary of War John C. Calhoun began soliciting congressional input into cadet selection, a process which became progressively more formalized. By the 1830s, Congress had de facto control over nominations, and in 1843, the legislators systematized the practice in an act that bound the size and geographical composition of the Corps of Cadets to the size and geographical composition of Congress, plus eleven–the president could appoint one cadet from the District of Columbia and ten, like Mullan, "at-large."[29]

In the face of persistent charges of elitism, the academy in 1842 began compiling statistics on the socioeconomic status of the parents of cadets, and it continued to do so through 1910. Consistently, the reports showed that the majority of cadets came from families of moderate or reduced means. The records for Mullan's class listed one cadet's family as affluent, one as indigent, ten as "of reduced means," and the rest as "of moderate means." Mullan's classmates included the sons of a bank president, a stagecoach proprietor, a schoolteacher, a hatmaker, and a sheriff. Five members of his class were from military families, including Mullan himself, who was the only cadet whose father was a noncommissioned officer. Twenty-three of his classmates were sons of farmers; ten, of lawyers; and eight, of physicians. Two of the fathers were unemployed. Twenty-eight cadets were from rural areas, fifteen from urban environs, and the rest from small towns.[30] In short, Mullan's class at West Point was a nearly perfect cross-section of America at the time, both geographically and socioeconomically. Racially, of course, it was not, nor were there any female cadets at the academy–those integrations would come much later. But given the prevailing mores of the era, Roosevelt's assertion of West Point's egalitarianism could not have been far off the mark. Any cadet who arrived at West Point with a provincial outlook–and no doubt many did–would have been disabused of it by the time he graduated.[31] These cadets, from their diverse backgrounds, forged bonds of friendship in the fires of the strict military regime that governed their lives for four years.

In addition to the academic load already described, the cadets participated in an "encampment" every summer. The plebe encampment–the forerunner to today's Cadet Basic Training–was particularly challenging for the newly arrived cadets, who were in so many cases literally fresh off the farm. After their second (third-class, or yearling) year at the academy, cadets were given two months' leave in the summer–the only leave period of their four-year stay. In every year except their second, cadets spent their summers in encampment, a simulated field environment. In the very shadow of the barracks, cadets lived in tents on the Plain, honing their fieldcraft and soldiering skills. The upperclassmen formed a chain of command to practice their leadership skills on the plebes.

West Point encampment on "the Plain," watercolor by Francis Fowke, 1849. Such encampments were part of John Mullan's training at the academy.
–Courtesy West Point Museum Collection, USMA

Drill was the cadets' most common activity during the encampment–they performed drills incessantly in their wool uniforms in the hot summer sun. They also made extended marches, both mounted and on foot, and conducted tactical training of all types. The cadets maintained a round-the-clock sentry to guard against surprise from any imagined enemies. Before physical training became part of the academic-year curriculum, the summer encampment was the most physically demanding period of a cadet's time at the school.[32] As one cadet in the 1840s wrote to his family, after seeing a newspaper editorial accusing academy cadets of being soft, "All I want of those Editors who say that 'lily fingered cadets lounge on their velvet lawns–attend their brilliant balls and take pay for it,' . . . is that they may go through but one . . . encampment."[33]

Encampment was also a time for hazing. Later in the nineteenth century, hazing at the academy would turn demeaning and even violent, becoming the bane of a whole succession of superintendents. But in the antebellum years, it was mostly confined to summers and consisted primarily of practical jokes perpetrated on the plebes by upperclassmen. Perennial favorites included dragging plebes from their tents in their sleeping bags, cutting a tent's lines to make it collapse on its occupants, and stealing a plebe's clothes in the night, forcing him to report to formation in his underwear or not at all.[34]

Perhaps the greatest hardship of cadet life in these years, and the one most likely to render a cadet hopelessly homesick, was the quality of food in the mess hall. The school's food-service contract belonged to one Mr. Cozzens, who also ran the local hotel. His remuneration was independent of the number of mouths he fed, and his instructions were to provide a "plain but substantial" menu for the cadets. Naturally, he did so as cheaply as possible.[35] The menu rotated weekly through various combinations of meat–beef, mutton, or fish, always boiled–with bread, butter, tea, and famously bad coffee. Cadets ate neither vegetables nor desserts except on special occasions, and they frequently reported mice, ants, and roaches in the food. Theirs were not unwarranted complaints: a congressional committee reported that the food at the school "was neither nutritious nor wholesome, neither sufficient nor nicely dressed."[36]

The best that could be said of mealtime at West Point was that it was over with quickly, as one cadet described:

> We then march to the mess hall, & if one speaks, raises his hand, looks to the right or left (which is the case on all parade) we are reported—indeed we are reported for everything. . . . When we arrive at the tables, the command is given "take seats," & then such a scrambling you never saw . . . not more than two thirds of them get a bit. . . . We have to eat as fast as we can, & before we get enough, the command is given "Squad rise."[37]

In light of these unpleasant dining conditions, it is no surprise that cadets often tried to secretly augment their government rations with extracurricular procurements. They were known to appropriate ducks or geese from a local farmer, or vegetables from the nearby town, and roast their bounty over the fireplace in their barracks. This was, of course, strictly forbidden by academy regulations, but it was a common practice. In fact, the camaraderie of these late-night sessions cooking "hash," as it was known, accounts for some of the cadets' fondest memories at the academy.[38]

Cadet discipline was regulated by a longstanding and elaborate system of demerits. All punishable offenses were divided into seven grades, each warranting between one and ten demerits. For example, being late to roll call cost the offender one demerit, socializing during study hours cost five, and mutinous conduct cost ten. Any cadet accumulating more than two hundred demerits in a year was considered deficient in conduct and recommended for dismissal. There were, of course, offenses that could result in immediate dismissal, including drinking alcohol and playing cards, two vices that plagued the academy's early days and which the administration, as custodian of the young cadets' moral development, was determined to stamp out.[39] A cadet's demerit total was, logically enough, thought to indicate something about his character. To cite two of the more famous examples, Robert E. Lee (class of 1829) received not a single demerit in his time as a cadet, whereas George Armstrong Custer (1861) led his class with 726.[40]

In this respect, Cadet Mullan seems to have been much closer to a Lee than a Custer. He did receive a few demerits, accumulating his highest number, twenty-one, as a senior, but getting none in his second year. He graduated with a total of forty-three demerits in a class whose average was 295 per cadet.[41] But these figures are deceiving. According to Post Orders, on Saturday evening, January 19, 1850—midway through Mullan's yearling year, in which he received no demerits—Cadet Mullan and three others—Cadets Fleming, Forney, and Latimer—were placed under arrest for violation of Paragraph 114 of the academic regulations—no gambling.[42] The rule stated that, "No Cadet shall play at cards, or any other game of chance . . . nor shall have in his room or tent, or otherwise in his possession, the cards or other materials used in these games, on pain of being dismissed [from] the service of the United States."[43]

When Mullan was caught playing cards, he ought to have been kicked out for it. However, by an established West Point tradition, cadets caught drinking or gambling could be pardoned if their entire class signed a written pledge to abstain from the vice for the rest of their time at the academy. In this way, the school could preserve in good standing the not insignificant number of cadets who would otherwise have been dismissed; furthermore, the offenders, along with all of their classmates, would thenceforth be bound not just by regulations, but by their personal honor to uphold the moral standards of the Corps. In Mullan's case, the class of 1852 stepped forward to make the pledge and redeem their wayward classmates.

Mullan's second (third-class) year seems to have been a lucky one all the way through—he was cited five times for being late and once for not having his hair cut, yet no demerits were assessed against him, even though similar offenses during his other three years resulted in their due penalty. Grooming standards seem to have been one of Mullan's greatest conduct challenges—he was cited five other times for lack of a haircut and once for not being properly shaved. Perhaps this explains his later attraction to frontier duty, where grooming standards were considerably more liberal, as well as the chest-length beard he maintained in his old age, after his head went bald. Mullan also received demerits for talking, "gazing about," and allowing his section to talk in the ranks. These indiscretions aside, over the four years, Mullan ended up ranking twenty-fifth out of the roughly two hundred cadets on the academy-wide Conduct Roll.

Mullan and others in the West Point class of 1852 did well by the academy and the nation it served.[44] Of its forty-three graduates, twelve fought for the Confederacy in the Civil War while twenty-three fought for the Union, ten of those as general officers. No less than thirty-six served on the frontier, twenty of whom fought in the Indian Wars, including a full thirteen who served with Mullan in the Pacific Northwest wars of the 1850s. George Crook became one of the nation's best-known Indian fighters, eventually commanding the Department of Arizona and the Department of the Platte in addition to his campaigns against the Yakamas, the Snakes, and with Custer at Little Bighorn, the Sioux. Phil Sheridan–who entered the academy with Mullan but did not graduate until 1853, having been suspended for a year for fighting with a classmate–went on to become Commanding General of the U.S. Army after a successful, if highly controversial, period of Indian fighting as commander of the Department of the Missouri.

But of course–and by design–the graduates' contributions to the West went beyond combat. The top graduate of the class, Thomas Lincoln Casey, built a wagon road between Washington and Oregon and went on to serve as the army's Chief of Engineers, building the Washington Monument along with many other notable works in the capital and around the country. George Mendell surveyed a rail line in California and was put in charge of supervising the construction of military roads in Washington and Oregon from 1856 to 1858. Joseph Christmas Ives served on the Pacific Railroad Survey expedition along the 35th parallel and commanded an important exploration of the Colorado River; later, he assisted Casey with the Washington Monument. After winning the Medal of Honor in the Civil War, David S. Stanley commanded the Yellowstone Expedition of 1873. Sylvester Mowry also served on the Pacific Railroad expedition as well as the California boundary survey before becoming a noted explorer of and writer about Arizona Territory. Henry DougAlass and August V. Katz served on the Northwest Boundary Commission along the 49th parallel in 1857; Katz also made the first ascent of Washington's Mount Rainier. Last but hardly least, John Mullan went on to build the first wagon road across the Rocky Mountains, connecting the headwaters of the Missouri with the headwaters of the Columbia in 1860.

Ten years earlier, while Mullan was still a cadet, Brown University president Francis Wayland claimed that "although there are more than 120 colleges in the United States, the West Point Academy has done more to build up the system of internal improvements in the United States than all the [other] colleges combined."[45] As we have seen, this was no accident. The West Point curriculum was carefully calibrated to equip and inspire its graduates to contribute to national development as well as national defense. Mullan's achievement, like those of his classmates, was a natural culmination of his four years at the United States Military Academy.

1 Helen Addison Howard, "Captain John Mullan," *Washington Historical Quarterly* 25:3 (July 1943), p. 187.

2 "Mullan, John," n.d., National Archives Microfilm Publication 688, U.S. Military Cadet Application Papers, 1805–1866.

3 Thomas Jefferson, *The Writings of Thomas Jefferson*, Definitive Edition, Vol. 27, ed. Albert Bergh (Washington, DC: Thomas Jefferson Memorial Association, 1905), p. 425.

4 For a summary of this debate see David Mayer, "'Necessary and Proper': West Point and Jefferson's Constitutionalism," in *Thomas Jefferson's Military Academy: Founding West Point,* ed. Robert M. S. McDonald (Charlottesville: University of Virginia Press, 2004), pp. 54–76 (hereafter cited as *Thomas Jefferson's Military Academy*).

5 Thomas J. Fleming, *West Point: The Men and Times of the United States Military Academy* (New York: Wm. Morrow, 1969), p. 16. See also Jennings L.

Wagoner Jr. and Christine Coalwell McDonald, "Mr. Jefferson's Academy: An Educational Interpretation," in *Thomas Jefferson's Military Academy*, p. 131.

6 Theodore J. Crackel, *West Point: A Bicentennial History* (Lawrence: University Press of Kansas, 2003), p. 51 (hereafter cited as *West Point Bicentennial).*

7 Christine McDonald and Robert McDonald, "West from West Point: Jefferson's Military Academy and the 'Empire of Liberty,'" in *Light and Liberty: Thomas Jefferson and the Power of Knowledge* (Charlottesville: University of Virginia Press, 2012), p. 240.

8 *The Centennial of the United States Military Academy at West Point, New York, 1802–1902* (Washington, DC: Government Printing Office, 1904), vol. 1, p. 275 (hereafter cited as *Centennial*).

9 Ibid., p. 276.

10 Crackel, *West Point Bicentennial*, p. 97.

11 Sidney Forman, *West Point: A History of the United States Military Academy* (New York: Columbia University Press, 1950), p. 76 (hereafter cited as *West Point History*).

12 Crackel, *West Point Bicentennial*, p. 97.

13 Douglass's son was Mullan's classmate at West Point *(Centennial*, vol. 1, p. 277).

14 *Report of the Board of Visitors*, June 21, 1831, Special Collections and Archives, United States Military Academy Library, West Point, NY (hereafter cited as USMA Archives).

15 In the election of 1824, Andrew Jackson received the majority of popular votes, but neither he nor Adams nor Henry Clay had the required majority of electoral votes. The election was therefore sent to the House of Representatives. Since Clay was the Speaker of the House, many believed that he used his influence to give the election to Adams in exchange for Adams naming him Secretary of State.

16 Robert P. Wetterman, "West Point, the Jacksonians, and the Army's Controversial Role in National Improvements," in Lance Betros, ed., *West Point: Two Centuries and Beyond* (Abilene, Texas: McWhiney Foundation Press, 2004), pp. 144–66.

17 *Centennial*, vol. 1, p. 278.

18 *Official Register of the Officers and Cadets of the United States Military Academy, June 1852,* USMA Archives, 7 (hereafter cited as *Official Register*); James L. Morrison, *"The Best School in the World": West Point, the Pre–Civil War Years, 1833–1866* (Kent, Ohio: Kent State University Press, 1986), p. 91.

19 *Centennial,* vol. 1.

20 Morrison, *"The Best School in the World,"* p. 93; *Official Register, June 1850*, p. 23; *Official Register, June 1851*, p. 23.

21 *Official Register, June 1851,* p. 23.

22 Morrison, *"The Best School in the World,"* p. 97.

23 Ibid., p. 94.

24 The documentation of this period at West Point is extensive and universally poignant. See for instance Tom Carhart, *Sacred Ties: From West Point Brothers to Battlefield Rivals, a True Story of the Civil War*, 1st ed. (New York: Berkley Hardcover, 2010).

25 Winfield Scott, July 25, 1860, in *Report of the Commission Appointed under the Eighth Section of the Act of Congress of June 21, 1860, to Examine into the Organization, System of Discipline, and Course of Instruction of the United States Military Academy at West Point*, 36th Cong., 2d sess., Sen. Misc. Doc. 3, p. 176 (quoted in Forman, *West Point History*, p. 61).

26 George Lucas Hartsuff, "Correspondence," January 22, 1849, USMA Archives.

27 George S. Pappas, *To the Point: The United States Military Academy, 1802–1902*, 1st ed. (Santa Barbara, CA: Praeger Publishers, 1993), pp. 269–71 (hereafter cited as *To the Point*).

28 *Centennial*, vol. 1, p. 21.

29 William Skelton, "West Point and Officer Professionalism, 1817–1877," in Betros, ed., *West Point: Two Centuries and Beyond*, p. 25.

30 "Circumstances of the Parents of Cadets, 1842–1879," Cadets of 1848, USMA Archives.

31 As a testament to the academy's cosmopolitanism if not its Americanism, Mullan's class also included Jerome Napoleon Bonaparte, grand-nephew of Napoleon I, who went on to serve in the U.S. Army and then in the French Army, fighting in Algiers and Crimea and in the Franco-Prussian War before moving back to the United States.

32 Stephen E. Ambrose, *Duty, Honor, Country: A History of West Point* (Baltimore: Johns Hopkins Press, 1966), p. 71 (hereafter cited as *Duty, Honor, Country).*

33 William Dutton to sister, February 18, 1843, quoted in Sidney Forman, *Cadet Life Before the Mexican War* (West Point, NY: United States Military Academy Print Office, 1945).

34 Ambrose, *Duty, Honor, Country*, p. 158.

35 Crackel, *West Point Bicentennial*, p. 121.

36 Ambrose, *Duty, Honor, Country*, p. 154; Crackel, *West Point Bicentennial*, p. 122.

37 William Dutton to brother, June 19, 1842, quoted in Forman, *West Point History*, pp. 96–97.

38 See for example Pappas, *To the Point*, pp. 270–71.

39 *Regulations Established for the Organziation and Government of the Military Academy at West Point, New York* (New York: Wiley & Putnam, 1839), p. 32 (hereafter cited as *Regulations*).

40 Custer's demerit total was not, as is often asserted, the academy record. Ironically, that distinction belongs to his much-maligned subordinate at Little Bighorn, Marcus Reno (class of 1857). To be fair, though, Reno had more time to amass his record–the academy at the time was under a five-year system, and Reno was turned back once, so he was at the academy for six years.

41 All demerit statistics compiled by the author based on the *Official Register* of 1849, 1850, 1851, and 1852.

42 *Post Orders,* Special Order No. 12, Jan. 23, 1850, USMA Archives.

43 *Regulations*, p. 32.

44 The summary of the service of the class of 1852 was compiled by the author from George W. Cullum, *Biographical Register of the Officers and Graduates of the U. S. Military Academy, from 1802 to 1867. Rev. Ed., With a Supplement Continuing the Register of Graduates to January 1, 1879* (New York: J. Miller, 1879).

45 Quoted in R. Ernest Dupuy, *Sylvanus Thayer, Father of Technology in the United States* (West Point, NY: Association of Graduates, U.S. Military Academy, 1958); Wayland's "Report to the Corporation of Brown University" can be found in Richard Hofstadter and Wilson Smith, eds., *American Higher Education: A Documentary History* (Chicago: University of Chicago Press, 1961).

Journey to the Flathead Country

John Mullan's April 1854 Expedition

BY PAUL D. MCDERMOTT AND RONALD E. GRIM

As revealed in the previous essay by Ryan L. Shaw, which focused on John Mullan's training at West Point in the late 1840s, the U.S. Military Academy prepared him well to be an explorer. Certainly, among Mullan's major contributions to the history and development of the West were his explorations, especially his surveys of the Missouri and Columbia Rivers and regions throughout the northern Rocky Mountains. The map in Figure 3-1 shows how much territory he covered. In only a few years' time, Mullan explored over 3,953 miles of land.[1]

Most of Mullan's exploration took place between 1853 and 1854, when he worked under Isaac Stevens on the Northern Pacific Railroad Surveys. Stevens's goal was to determine the economic and physical feasibility of the northernmost of the

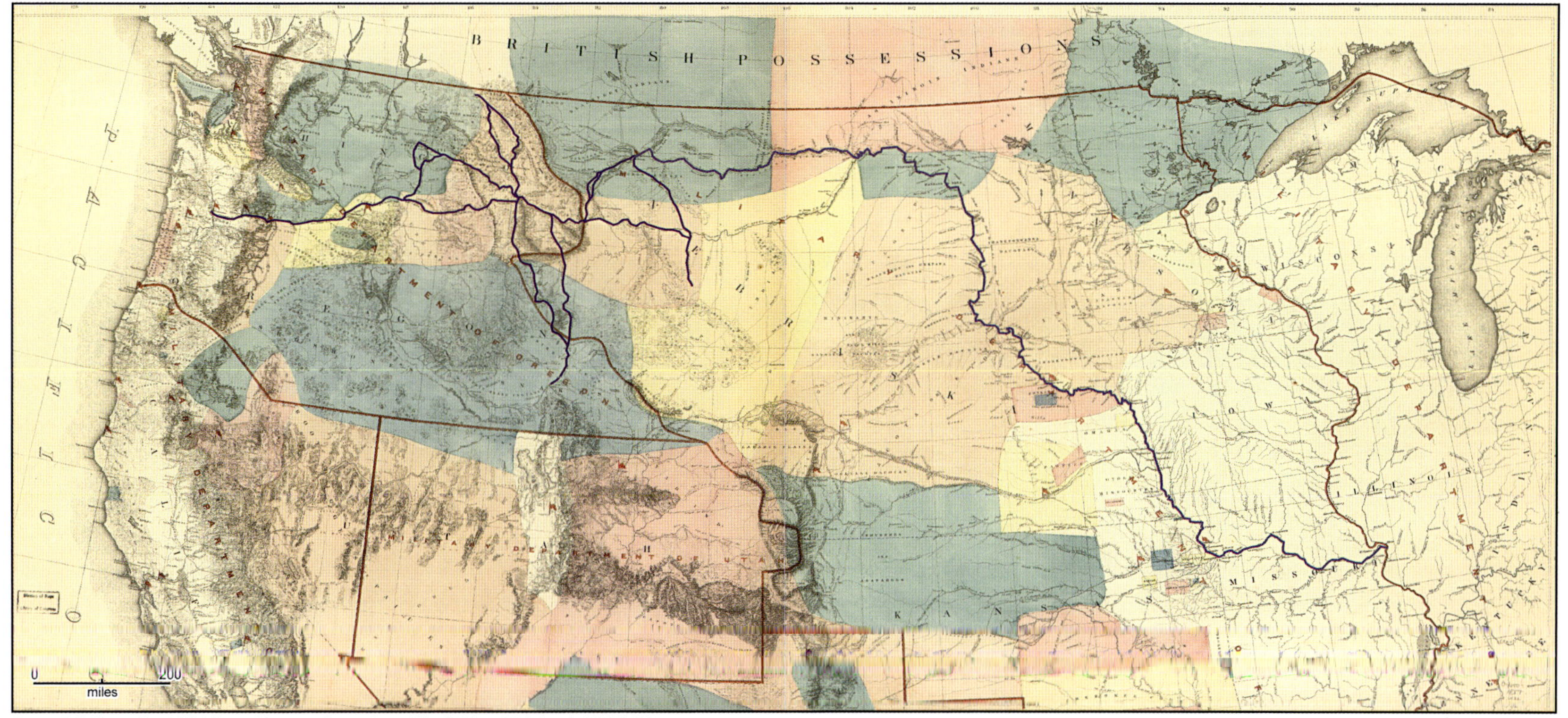

Source: G.K. Warren Map 1858, Library of Congress Geography and Map Division. Original scale 1:3,000,000

Figure 3-1. *Routes of John Mullan's Explorations, 1853–54, as a member of the Isaac Stevens survey party, part of a government effort to select the best route for a transcontinental railroad.*

four potential routes under consideration for the first transcontinental railroad. In Stevens's case, the proposed route would be located, roughly, within the region between the 47th and 49th parallels of latitude.

John Mullan was a key player in Stevens's survey. After the main, east-to-west expedition was completed in the fall of 1853, Stevens gave Mullan the added responsibility of exploring a major portion of the Rocky Mountains during the winter of 1853–54. His primary task was to obtain weather data that could potentially impact railroad and road construction. Secondarily, he was to determine whether there were other mountain passes that would permit passage of a transcontinental railroad. Finally, he was directed to explore the possibility of constructing a wagon road to connect the new Fort Walla Walla with the Oregon Trail.

In only one year, from the fall of 1853 to the fall of 1854, Mullan traveled first from Cantonment Stevens (near today's Stevensville, Montana) to Fort Hall (near present-day Pocatello, Idaho) on the Oregon Trail, then he immediately turned eastward, journeying across the Rockies to Fort Benton before returning to Cantonment Stevens. Mullan's was the first wagon to cross the mountains in this region. In the process, he and his men encountered snow, frigid weather, swollen mountain streams, unfamiliar terrain, and strange people.[2] On January 6, 1854, the party almost lost their horses and their lives crossing the Hell Gate River (now called the Clark Fork River). The difficulty the party encountered in fording the ice-covered Hell Gate is illustrated in Figure 3-2, a painting by Gustavus Sohon.

This eastward expedition was followed in April 1854 by a three-week excursion into Flathead country. After that, Mullan headed back westward to The Dalles, on the Columbia River, using the route created by Lewis and Clark in 1805–6, specifically the Lo Lo (today spelled Lolo) Trail. Mullan explored and took notes describing every mile he covered in this vast region.

Figure 3-2. *"Crossing the Hell Gate, 1854" by Gustavus Sohon* –Courtesy Yale University Art Gallery

This essay focuses on documenting the events and observations of Mullan's April–May 1854 expedition into northwest Montana's Flathead country (see figure 3-3). On this excursion, he explored the vast Flathead Lake region, going north from Cantonment Stevens through the Missoula area and up to the Kootenai River near the Canadian border. Only six people, including Mullan, took part is this foray. Their round-trip horseback journey traversed approximately

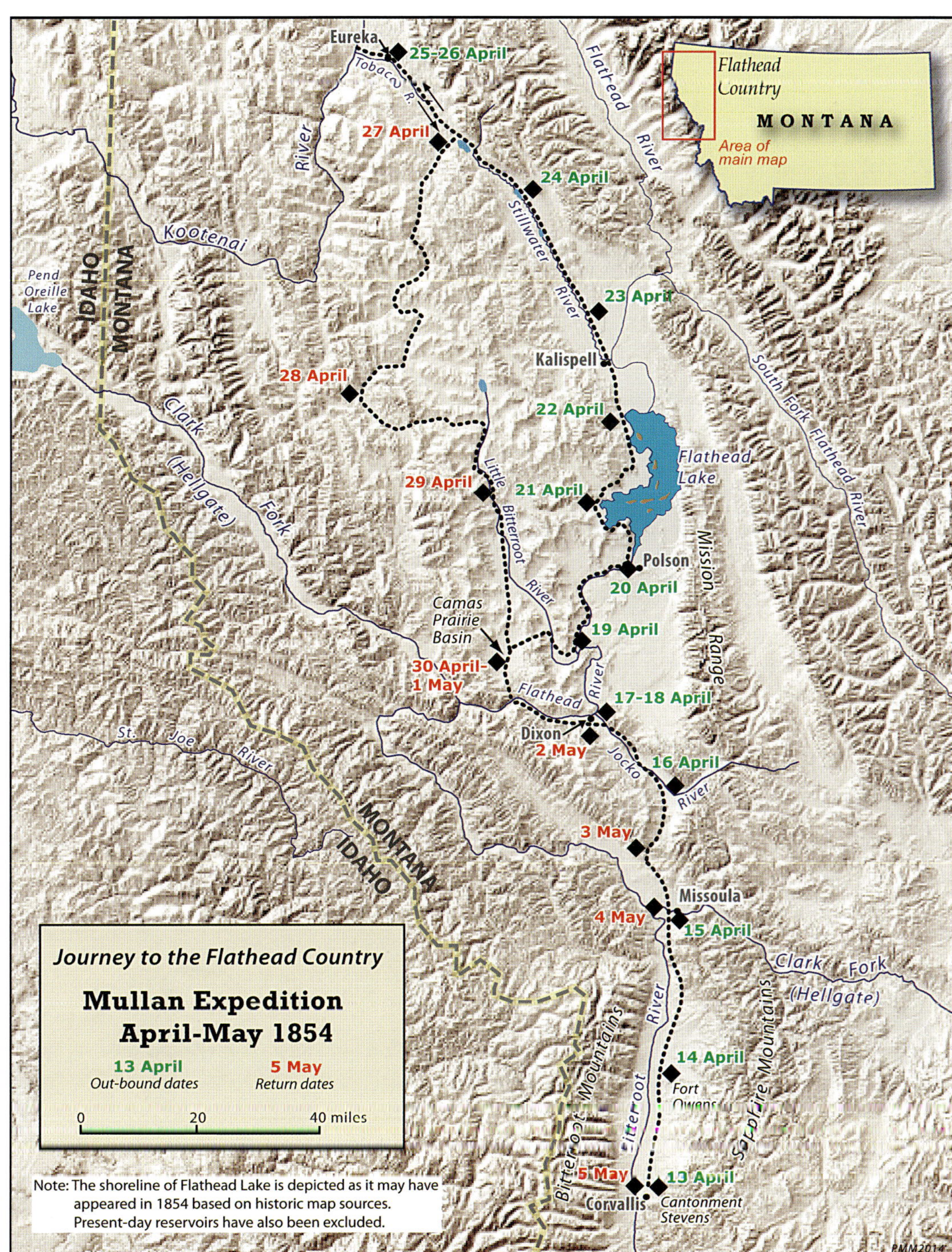

Figure 3-3. *Map of Mullan's exploration route to the Flathead country.* –Map by Phillip Mobley

Figure 3-4. *Mullan's 1854 reconnaissance map of the Flathead Lake region, from Cantonment Stevens to the Kootenay River. This is a small-scale map (1:600,000) depicting 9.6 miles per linear inch. Mountains are symbolized using hachures (starlike shapes).* –Courtesy National Archives and Records Administration, Cartographic Section, College Park, MD

504 miles. The major problem confronting the men in this season was high water in a number of streams, including the Bitterroot, Hell Gate, Clark's Fork, and Kootenai Rivers.

The Flathead Lake trip is memorable for several reasons. First, on May 4, 1854, the entire party nearly died crossing the swollen Hell Gate River. Second, Gustavus Sohon, one of the two primary artists accompanying the survey, began rendering his important portraitures of local Native Americans on this journey. These images are most certainly among the expedition's most significant contributions to American history and the history of the Pacific Northwest. Finally, the trip led Mullan to conclude that the best route for the railroad had been established. Why look any further for an alternate path? The most suitable route had been found!

Reconnaissance Map

Our understanding of this short exploration is documented in the 1855 Congressional Report and the one reconnaissance map drawn by Mullan late in 1854 for his report (Figure 3-4), though this map was not published in the actual report. This simple map, now located in the collections of the National Archives at College Park, Maryland, shows the basic route and periodically specifies where and on what date the group made camp. The exact path followed by the party is difficult to interpret, however. First, multiple dashed routes are etched onto the plot, creating some confusion. Assumedly, these dashed routes represent alternative Indian trails used during that time period. A second problem arises from the map's generalization. The positions are approximations drawn at a small scale and only roughly depict the route followed. The map was drawn in ink at a scale of 1:1,200,000, representing 18.9 miles to the inch. Mountains are indicated with crudely drawn hachures, or dashed lines, arranged in a starlike or molehill pattern. The larger-scale inset (right) shows the geographic character of Flathead Lake.

Cantonment Stevens

The base of operations for these explorations of the intermontane region was Cantonment Stevens. It was built in late 1854 in the Bitterroot Valley, near the present site of Corvallis, Montana, assumedly at the mouth of Willow Creek. John Mullan identified this location as the campsite in his report to Stevens in 1854–55. Like other streams, Willow Creek changed its course over the decades; today, it enters the Bitterroot 2.25 miles northwest of Corvallis. The authors' current thinking is that the cantonment's position was southeast of Corvallis, where the valley of Willow Creek meets the Bitterroot Valley (see Figure 3-5). This location, however, is only an educated guess. To this day, the exact site of Cantonment Stevens has not been definitively determined. One reason for this failure is cultural development. In more than a century and a half, the region has undergone conversion to agricultural uses and substantial residential development. These changes have obliterated any evidence of the early cantonment.

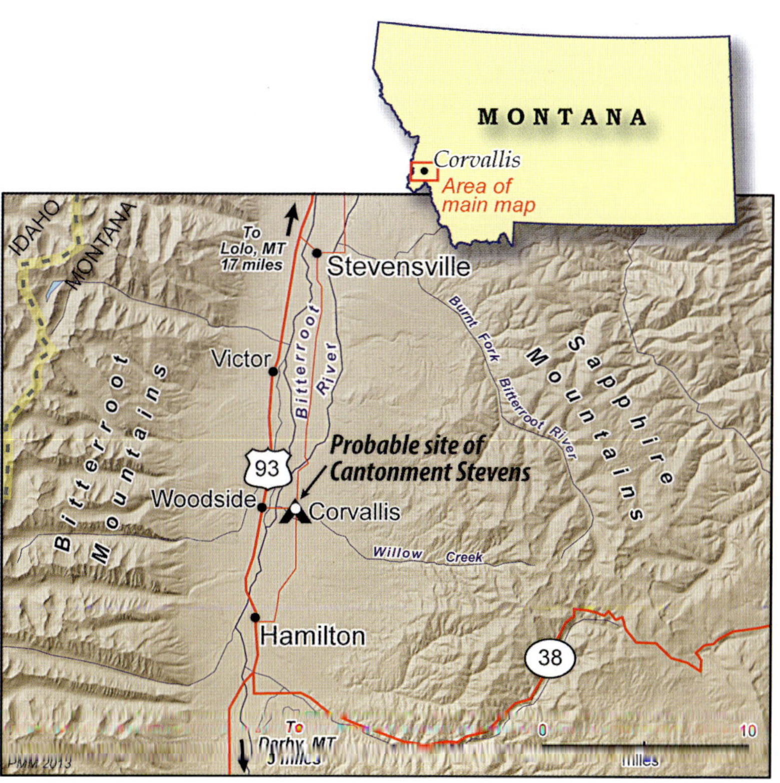

Figure 3-5. *Location of Cantonment Stevens, thought to be southeast of Corvallis, Montana, adjacent to Willow Creek.* –Map by Philip Mobley

What was Cantonment Stevens like? The complex consisted of four hastily constructed log cabins, one positioned at each corner of a large rectangle, surrounded by a split-rail fence. The interior was dominated by a tall flagpole. The entrance to the complex was on the east side, facing the Sapphire Mountains. To the north, nearby Native Americans had erected numerous teepee lodges and several small log buildings around the cantonment, concentrated in a rough cluster; each of these twenty-seven lodges would have housed perhaps 135 to 270 people. Gustavus Sohon made several illustrations of Cantonment Stevens and the surrounding area, including the one in Figure 3-6.[3] In the original sketch on which the final plate was based, Sohon identified the use of each cabin: a storeroom, two men's quarters, and Mullan's cabin.

One interesting component seen in Figure 3-6 is Sohon's depiction of cultural elements. We see several Native Americans practicing their hunting skills, shooting bows and arrows at a target positioned in a dry creek bed. John Mix Stanley, chief artist on the Stevens survey, had directed the

Figure 3-6. *Sohon's depiction of Cantonment Stevens, Mullan's base camp for his explorations with Isaac Stevens's Pacific Railroad Survey (northern route), in 1853* –Courtesy Yale University Art Gallery

Figure 3-6a. *Detail from Figure 3-6 (p. 44), showing a closer view of the encampment* –Courtesy Yale University Art Gallery

expedition's artists to particularly capture Native cultural activity. When the expedition was initiated in May 1853 near St. Paul, Minnesota, Stanley wrote:

> Sketches of Indians should be made and colored from life, with care to fidelity in complexion as well as feature. In their games and ceremonies, it is only necessary to give their characteristic attitudes, with drawings of the implements and weapons used, and notes in detail of each ceremony represented. It is desirable that drawings of their lodges, with their historical devices, carving, etc., be made with care.[4]

The Trip Begins

On April 14, 1854, John Mullan began his expedition from Cantonment Stevens, accompanied by artist Gustavus Sohon; Thomas Adams, a topographer; an interpreter named Gabriel; Corporal W. Gates; and a Native boy.[5] The party began traveling northward along the east bank of the Bitter root River. This outbound route took the party to Flathead Lake and beyond to the Kootenay (today spelled Kootenai) River, near the Canadian border. He chose the first segment

Figure 3-7. *The Clark Fork River near Missoula, Montana, where it joins the Bitterroot River. The Clark Fork is a braided stream with numerous channels. This section of the Clark Fork was originally identified as the Hell Gate River.* –Photo by Paul D. McDermott

of this route to avoid the floodwaters that were overflowing the Bitterroot's banks. They returned using a more westwardly route down to the Jocko River, where they picked up their original trail along the Clark Fork back to Cantonment Stevens. The round trip covered 504 miles.

As the party traveled northward, the Sapphire Mountains were to their right and the higher, snowcapped Bitterroots were across the river to their left. On April 15, the group camped on the south bank of the Hell Gate River near present-day Missoula. In this location, the Hell Gate (now called the Clark Fork) splits up into numerous channels, helping to disburse the river's discharge; otherwise it would have been difficult–if not impossible–for the explorers to cross. Figure 3-7 shows the Clark Fork today as seen from Council Hill, west of Missoula, Montana. The river remains formidable and periodically cantankerous.

The next day, April 16, the party proceeded northward, trudging slowly through a heavy downpour, along a route that roughly coincides with today's U.S. Highway 93. They passed through the Hell Gate Ronde, a huge valley that was formed in part by ancient Glacial Lake Missoula, whose waters were impounded behind glacial ice that dammed the Clark Fork River downstream, near the Canadian border, and filled the valley. As they traveled, the men certainly observed the parallel shorelines etched like stairsteps into surrounding mountain slopes. These lines are relics of the ancient lake's wave action. Many can still be seen today, especially after a light snowfall. Mount Sentinel in Missoula is an especially good place to view these wave-cut landforms (see Figure 3-8).

In due course they reached a small pass, then referred to as Conacan's defile (now spelled Coriacan defile). Upon crossing the defile they entered a portion of the Jocko River

Figure 3-8. *Shorelines on Mount Sentinel in Missoula, Montana, left from the wave action of ancient Glacial Lake Missoula, may be seen clearly after a light snowfall.* –Photo by D. W. Hyndman

drainage basin. The small party was impressed by the basin's camas (*Camassia quamash*) prairie, which provided the succulent camas bulbs that supplemented the local Pend d'Oreilles' diet in the summer. These bulbs are a rich source of dextrose. During the early summer, the camas plants' beautiful blue blossoms blanket the landscape wherever they are found. Mullan noted this camas prairie in his report, and Sohon made a simple sketch of the plant (Figure 3-9).[6]

Later that day, as they crossed the swollen waters of the Jocko River, the explorers encountered some difficulties, having to ford it five times at different spots. In the late afternoon, the party reached the Clark's Fork (the section that is today's Flathead River) near the mouth of the Jocko, around future Dixon, Montana. Crossing this river here proved to be too hazardous–its swollen waters were too deep for the men to safely swim their animals. Instead, Mullan elected to move downstream, near Thompson's Prairie, and construct several rafts to cross the river. The men gathered wood for the rafts before bedding down at their camp that night. The next morning, April 17, they finished building the rafts and safely crossed the river, a distance estimated to be 250 yards. The party then continued their journey on horseback downstream (west) along the river's northern shore. Later the explorers crossed to a better trail on the opposite shore.

Turning northward and following Camas Creek, the explorers entered the main basin where Native peoples harvested camas bulbs. Mullan described the place:

> The Camash prairie referred to is nearly a circular prairie in the mountains, and perfectly level. The grass this season growing in it is exceedlly green and verdant. It is perfectly sheltered on every side by high hills and mountains. . . . [T]here is no better spot in the mountains either for agricultural or grazing purposes than this beautiful valley.[7]

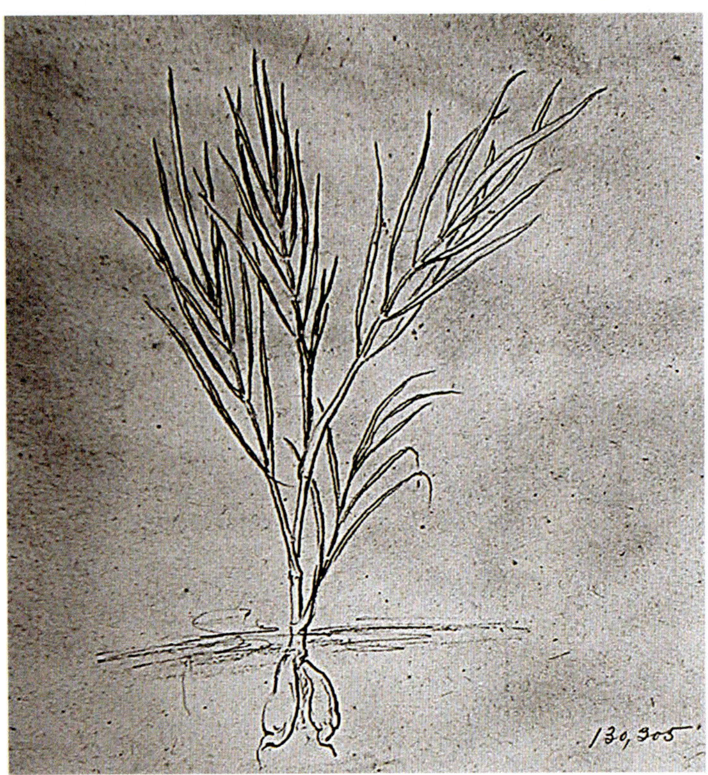

Figure 3-9. *Sohon's drawing of a camas plant, a popular food plant used by Rocky Mountain tribes as a source of carbohydrates.* –Courtesy Smithsonian National Museum of Natural History, National Anthropological Archives

Figure 3-10. *Sohon's image of Camas Prairie, in a large basin southwest of the Mission Mountains* –From Isaac I. Stevens, *Narrative and Final Report of Explorations for a Route for a Pacific Railroad* (1860)

Sohon captured a view of the fecund Camas Prairie, shown in Figure 3-10. In the prairie, they met Michel Ogden, a Hudson Bay factor in charge of Fort Connah, on the Flathead River to the north, and Mullan's party camped near Ogden's camp, where they stayed an extra day and night to rest. On the first night, April 17, the local Pend d'Oreilles women helped them set up camp. Apparently, the women also offered Mullan's men sexual courtesies, but they refrained from taking advantage of the kind offer: "[The women] would probably have extended the limit of their kindness much further had we not requested them to desist."[8] This is not surprising, as Mullan was a very principled and religious individual.

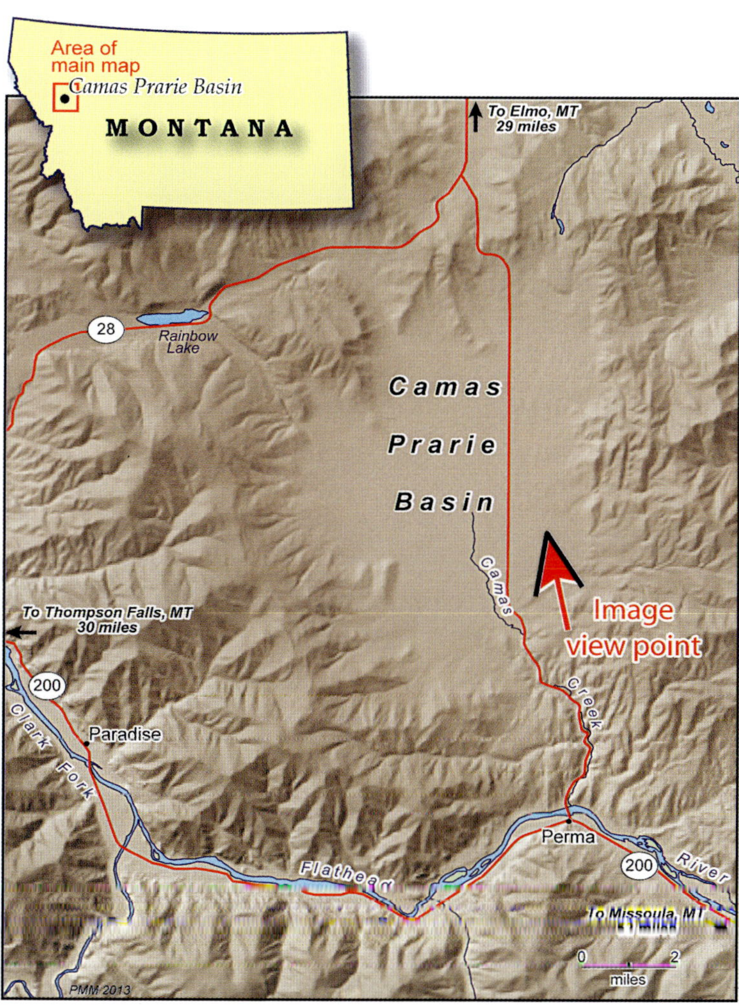

Figure 3-10a. *Location of Camas Prairie* –Map by Philip Mobley

The next day, the women provided them with milk–and lots of it, as Mullan notes: "Having here a number of cows, they brought us milk in such abundance that our lodge might have been taken for a dairy, more than the shelter of a small party of mountain travelers."[9]

Hot Spring Creek (Little Bitterroot River)

On April 19, Mullan and his party resumed their march, heading east through the prairie to the mouth of Hot Spring Creek (today's Little Bitterroot River), where they camped. From there the travelers visited the nearby camp of Pend d'Oreille chief Alexander, which contained forty-seven lodges and included members of other tribes.

It was at Alexander's camp that Gustavus Sohon created the first of his historic portraitures. Figure 3-11 is his portrait of Alexander (native name Tum-cle-hot-cute-se), dated April

Figure 3-11. *Sohon's sketch of Alexander, a chief of the Pend d'Oreille people.*
–Courtesy Smithsonian National Museum of Natural History, National Anthropological Archives

21, 1854. Sohon, an astute observer, wrote this caption for the portrait, following J. M. Stanley's directive as described earlier:

> Alexander. The principal chief of the Pend d'Oreilles is not a Pend d'Oreille proper. But descended on the Fathers side from the Snake Indians and on the Mother's from the the Pend d'Oreilles. He was made first chief by the Pend d'Oreilles themselves and the Jesuit Priests in 1848. He is noted for his high toned[?] noble traits of character. He is a brave man. When a party of his tribe had stolen horses from Fort Benton on the Missouri in 1853, he started with only five men and took them back passing through the whole camp of Blackfeet Indians. Then most deadly enemies. He still rules the Pend d'Oreille and is only 45 years old.

Interestingly, the famous Yakama chief Owhi was also at Alexander's camp at this time, recuperating from a broken leg. On Isaac Stevens's 1855 treaty expedition, Sohon would create a portrait of Owhi, who was later involved in the famous 1858 Yakama uprising (also called the Yakima War).

As John Ewers points out in his 1948 article "Gustavus Sohon's Portraits of Flathead and Pend d'Oreille Indians," published in *Smithsonian Miscellaneous Collections*, Sohon is noted for his simple but detailed pencil images of Native people.[10] By the time Sohon finished his contributions to western art in 1865, he had rendered some eighty significant people, mostly Natives, who were intimately involved in the region's development in the 1850s and 1860s. These images identify

Figure 3-12. *Sohon's view of the Clark Fork River (now called the Flathead River in this area) with the Mission Mountains in the distance* –Courtesy Yale University Art Gallery

Sohon's most important contribution to history. On this short expedition with Mullan, Sohon made six portraits, all of them now in the possession of the Smithsonian Anthropological Collection. Moreover, in the mid-1860s, Sohon took an ambrotype of Father Pierre de Smet, which is now owned by the Montana Historical Society. Altogether, Sohon was responsible for creating 219 known images. Most of these (59 percent) were essentially landscapes, and the remainder (41 percent) were individual portraitures.[11]

On April 20, from their camp near Alexander's village, Mullan and his men proceeded generally northward, following the course of the Clark's Fork. It was a difficult, meandering route. Along the way, the men's attention was drawn to the Mission Mountains. Mullan states: "We had an excellent view of the country in every direction, save to the east, our view here being limited by the high range of snow-clad mountains. . . . This range has a direction nearly north and south, and is exceedingly high."[12] Sohon's illustrations of the Mission Mountains, Figures 3-12 and 3-12a, were done from a ridge overlooking the Clark's Fork as it flows south, meandering along the western flank of the Flathead Valley. The latter (Figure 3-12a) is a lithograph based on the former (Figure 3-12), a sketch.

On April 20, the party camped at the south end of Flathead Lake, the mouth of the Clark's Fork. Here they encountered four lodges of Pend d'Oreille people, who constantly fished near this outlet, even in the winter, when the water was covered by ice. Upon Mullan's arrival at their fishing camp,

Figure 3-12a. *Sohon's lithograph of the same scene, the Clark Fork River with Mission Mountains in the background, created when Sohon accompanied John Mullan on his exploration of the Flathead Region in April 1854* –From Isaac I. Stevens, *Narrative and Final Report of Explorations for a Route for a Pacific Railroad* (1860)

the Natives presented him and his men with both fresh and dried salmon trout. Mullan notes that many of these fish were quite large–some three feet long.[13]

Flathead Lake

Flathead Lake is a magnificent sight. This glacier lake is the largest freshwater lake in the United States after the Great Lakes–covering 194 square miles–and is very deep. At its maximum, it is about twenty-seven miles long and sixteen miles wide; the deepest part of the lake is about 371 feet.

From their camp at the southern end of the lake, the Mullan party traveled north along its western shore. It took them two days to traverse the lake's entire length. According to Mullan's journal, much of the route passed through high open prairie, occasionally punctuated by patches of pine forest. Mullan drew a map of the lake, seen in Figure 3-13; Figure 3-14 is a modern map of the lake.

As they made their way along the lakeshore, the men noticed numerous islands protruding from the lake's surface. Mullan described them: "Several large and beautiful islands lay near the middle of the lake, all covered with an excellent growth of pine; many of these islands are several

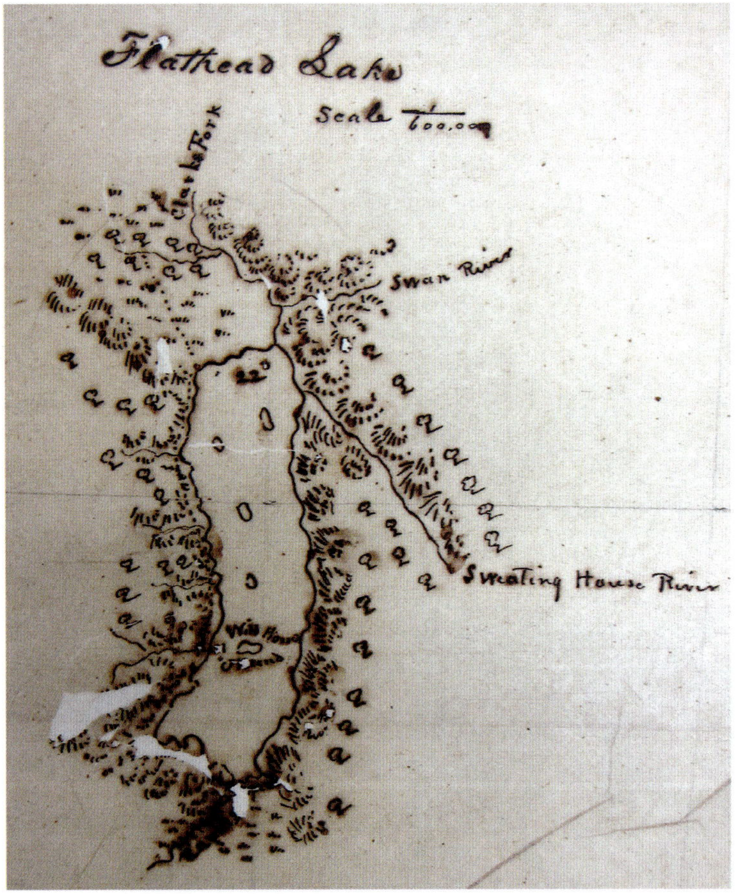

Figure 3-13. *Closer view of Flathead Lake, from inset on Mullan's 1854 reconnaissance map. This shows a different shape for the lake than now exists; the change is due to a rise of water level created by a dam near the lake's southern outlet.* –Courtesy National Archives and Records Administration, Cartographic Section, College Park, MD

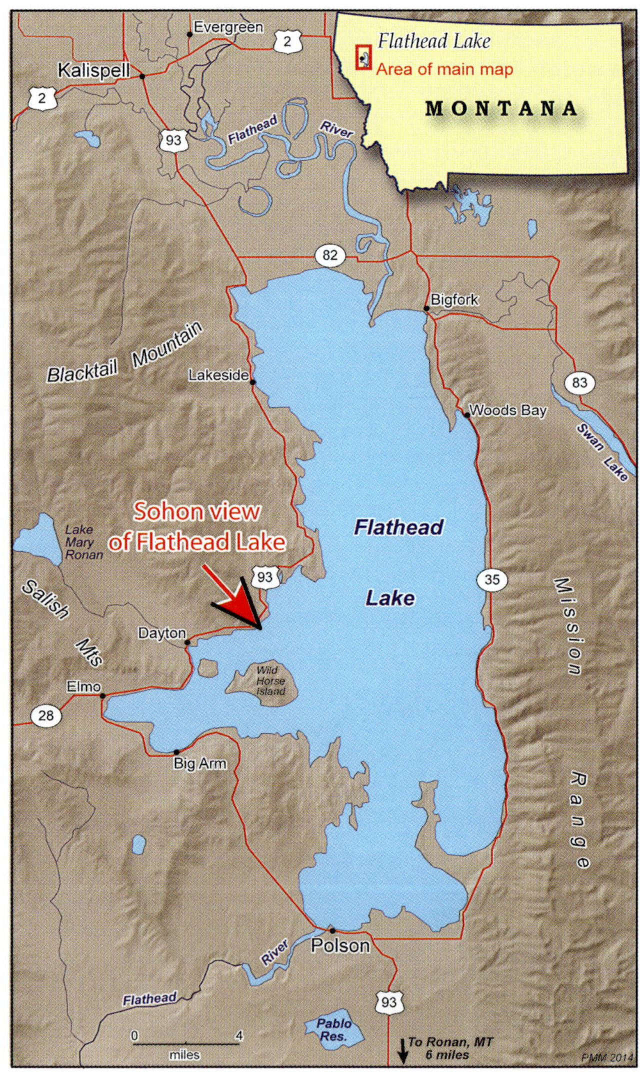

Figure 3-14. *Modern map of Flathead Lake* –Map by Philip Mobley

miles long."[14] One of largest was called Wild Horse Island, so named for its herd of horses, which had been placed there years before by a Pend d'Oreille man. After a band of Blackfeet stole his horses, the man captured a number of Blackfeet horses to replace his loss, then placed the ponies on the island to protect them. In subsequent years, the herd proliferated to some sixty or more animals.

After a cold and snowy day of traveling on April 22, the men established camp at the lake's northern end. Mullan expressed the prediction that when this region becomes settled, its fisheries and lumber will constitute one of the chief activities of trade from this location. As a man of the nineteenth century, he does not envision the role that recreation and tourism would play in the area's economy in the future.[15]

As he continued his exploration, the goal of creating and extending a northwest passage was on his mind. He posed three questions to be investigated. First, could navigation be extended from the Pacific to the base of the Rocky Mountains? Second, what might prevent this extension? Third, were there any insurmountable obstacles to this goal?[16]

Gustavus Sohon sketched in pencil on multicolored paper one of the first known images of the lake, shown in Figure 3-15. Typically, Sohon composed his images on three-toned paper. Blue helped to portray the sky, a neutral cream-colored center dominated the middle ground, and a deeper tan provided an earthy foreground. The chromo-lithograph of the image included in Isaac Stevens's official congressional report in 1860 appears in Figure 3-15a; the original pencil sketch on colored paper now belongs to the Yale University Art Gallery. For comparison, we have included a modern photograph (Figure 3-16), which was taken at the approximate site where Sohon's drawing was rendered in April 1854.

Figure 3-15. *This sketch by Sohon is probably one of the first artistic renditions of Flathead Lake, drawn in pencil on three-toned paper. The white lines were created using a needlepoint tool. The view is toward the southeast.* –Courtesy Yale University Art Gallery

Figure 3-15a. *Sohon rendered this lithograph of Flathead Lake based on the color sketch (Figure 3-15)* –From Isaac I. Stevens, *Narrative and Final Report of Explorations for a Route for a Pacific Railroad* (1860)

Figure 3-16. *Flathead Lake today. With a surface area of 194 square miles and a depth exceeding 365 feet, Flathead is the largest freshwater lake in the western United States.* –Photo by Paul D. McDermott

Journey to the Kootenay River

From the north shore of Flathead Lake, the explorers continued northward across the prairie to the Maple (now called the Stillwater) River, which they followed northwest to the drainage of the Kootenay (now spelled Kootenai) River. This leg of the journey took place from April 23 to 25. The men found the Maple difficult to ford; as Mullan stated, "The Maple River I found to be the swiftest stream that I have ever crossed in the mountains." Considering how many streams Mullan crossed during his ten years in the West, this description is exceptional.[17] Similarly exceptional was his description of the condition of the trail in this section: "Truly, I consider this one of the worst roads, if not the worst, ever traveled by whites or Indians."[18]

Mullan's party encountered several Kootenai lodges as they moved through the corridor. The lieutenant noted that in comparison to other Native peoples, the Kootenai were impoverished:

> The lower Kootenay Indians are represented . . . as being an exceedingly poor, improvident, and miserable tribe of Indians. They are poor in horses, have few or no skin lodges, make but little meat for their sustenance, and, in a word, live a miserable existence. We found them poorly and thinly clad, travelling with few horses, each horse carrying two and sometimes three persons; their lodges were made of mats formed of a tall rush growing in the marshes. Their chief article of food when travelling is roots, and fish when at home.[19]

On April 25 the party made camp a few miles south of the Kootenai River, where they would spend two nights. The next day, Mullan set out to investigate that river, noting that it was one hundred yards wide with a gentle flow. The fertile soil along its banks supported a diverse array of plants. He wrote, "I made a rich botanical collection, a description of which, from my limited knowledge of that science, does not here find a place, but is left for more able hands."[20]

Homeward Bound

On April 27, having explored the region up to the vicinity of the 49th parallel, almost to the Canadian border, the Mullan expedition prepared for their trip back to Cantonment Stevens. Instead of retracing his outbound route, Mullan decided to explore a different path home, hoping to reach the Clark's Fork using a more westerly route. After traveling southeast about ten miles along their previous trail, the men turned southwest. Reaching Tobacco Creek (now called the Tobacco River), the party found it too deep and fast to swim across, so they cut down some trees and built a crude bridge to span the sixty-foot-wide stream.

Tracing Fontine Creek, the expedition traveled southward for about seventeen arduous miles, then moved southeast along Wolf Creek, at the western base of Elk Mountain, and followed this rugged and meandering route to the junction of Wolf and Little Wolf Creeks. Proceeding southward toward the Bitterroot Valley on April 28, the men encountered forty miles of "indescribably horrid" road; Mullan detailed the conditions: "fallen timber piled up for many feet, over which our animals had to jump, innumerable mud-holes and quagmires, rocks, under-brush–in a word, everything to make our road difficult in the extreme, and endangering the lives of both men and animals."[21] Once in a while, however, the men are struck by a beautiful scene–for instance, a "magnificent" sixty-foot waterfall flanked by vertical black rock walls, its sound resonating for miles. Mullan also mentions coming across numerous beautiful lakes in the area. Otherwise, he found the landscape "dreary and dismal."

On April 29, the men reached Hot Spring Creek (the Little Bitterroot River). Moving southward along its east bank, Mullan and his party found their way back to the Camas Prairie, where they again met Michel Ogden, on April 30. The trader provided the explorers with much-appreciated hot coffee and bread; he also gave them provisions he had obtained at Fort Vancouver, on the Columbia River, adequate to get them back to Cantonment Stevens.

Departing the Camas Prairie on May 2, the travelers followed Camas Creek to the Clark's Fork (Flathead River). On their return route from the Kootenai to the Flathead River, the party had explored an additional 140 miles of new terrain. On May 3, the travelers again crossed the Clark's Fork, this time on a birch-bark ferry built by the Pend d'Oreilles. From there they followed the river east to the drainage of the Jocko, camping near today's Evaro, Montana.

The explorers' travel had been rapid, and by the early afternoon of May 4, they were again on the banks of the Hell Gate River. The river, swollen from snowmelt, was deep and flowing fast. The men knew it would be very dangerous to cross under these conditions, but they had little choice except to try. Mullan ordered his men to build two rafts, a task that took them three hours to complete. He then split the party into two groups. On one raft, Mullan took Thomas Adams and another, unidentified man. Gabriel and another man, accompanying a Native woman and her several children, were assigned to the other raft. Mullan directed both parties to move their craft toward a specific destination on the opposite shore, but both missed the location as the rafts were thrown rapidly downstream and collided with protruding and submerged rocks, sawyers (large tree stumps with roots attached), and various small islands. Approximately half a mile from their starting point, Mullan's group was engulfed in the surging rapids, and their raft was dashed against a sawyer that was jutting out from the riverbed on the opposite shore. The men feverishly scrambled to hold onto the stump, clinging to it for dear life. In the process, Mullan was dumped overboard.

Mullan managed to swim back to the raft, and one of his men pulled him back aboard. By this time, however, the travelers had lost their poles, so they had only ropes to hold onto the rafts. Making their way to a rocky island, the men, still holding the ropes, jumped into the water and swam to the island. Through sheer perseverance, they were able to move their belongings and themselves onto the island. In the meantime, Gabriel had managed to get the other raft to the shore and began looking for the lost party.

According to Mullan, the key man responsible for saving their lives and retrieving their property was Thomas Adams.[22] After he, Mullan, and the other man became stranded on the island, Adams dove in, swam to shore, and, barefoot and naked, walked "through bushes, briars, and fallen timber" to the camp, a mile away. He returned with men and horses, and soon Mullan, the other man, and most of the cargo were safe and sound on the shore. "We were rejoiced at finding the whole party thus saved from an untimely end," Mullan wrote.[23]

Adams, a topographer and an artist, had also been with Mullan on his winter trek to Fort Hall. It's possible that Adams made the preliminary sketch for Gustavus Sohon's dramatic image of the Hell Gate crossing, shown in Figure 3-17. This lithograph, reconstructing what was perhaps the most harrowing event on any of Mullan's explorations, was

Figure 3-17. *Crossing the Hell Gate River, May 5, 1854. This lithograph is a re-creation of the incident in which several men escaped death after their raft capsized crossing the river.* –From Isaac I. Stevens, *Narrative and Final Report of Explorations for a Route for a Pacific Railroad* (1860)

published in Mullan's final report. Note the primitive log raft submerged beneath the water and the men struggling to climb onto the tree stump with their belongings. Adams is the man in the middle, and Mullan is the one being helped up. It's possible that the third man was Sohon himself, though Mullan never specifically identified him.

Mullan's long journey to the Canadian border and back was concluded on May 5, 1854, upon his reaching Cantonment Stevens. Having examined the mountains from the 43d to the 49th parallels of latitude, extending from Fort Hall to the nation's northern boundary, Mullan concluded, "I can most unhesitatingly affirm, that the Hell Gate defile is the only one in this section that leads to the passes in the main chain of the mountains that are practicable either for a railroad or wagon route."[24] In short, the expedition was a success. But Mullan's short foray to the north is significant for another reason—Sohon's illustrations. The artist created a magnificent visual record of landscapes and people that forever captured that moment in the settlement of the West.

Mullan's last major exploration took place later that year, when he headed westward over the Lolo Pass, following a well-used Native trail to the Snake River. Eventually, he reached Fort Dalles and later the military post at Vancouver, known at that time as Columbia Barracks. From there he went back to the East Coast for other military assignments. He returned again to the Pacific Northwest in 1858 and remained in the region to build the military wagon road. He later returned to Walla Walla with his bride, Rebecca, to operate a small ranch.

---

1 It is difficult to determine a hard number for how much territory Mullan explored. The numbers here approximate the linear miles he covered; they should be considered minimal calculations—the real distance he traveled was certainly greater. Keep in mind that this calculation does not include repeated travel over parts of the route explored. If one adds Mullan's travels with Col. George Wright's expedition in 1858 and his travels during the construction of the military road between Fort Walla Walla and Fort Benton, the total is even greater—at least 5,480 miles.

2 During a unique climatic period between A.D. 1200 and 1850, temperatures in the northern hemisphere were notably lower than they are today, and they did not rise significantly until after 1850. Consequently, Mullan and his men experienced below-zero temperatures at two major cantonments—Jordan (1859–60) and Wright (1861–62). During these years, winters were characterized by ice cover on streams and rivers with snow depths greater than those of the present. See Brian Fagan, *The Little Ice Age: How Climate Made History*, 1300–1850 (New York: Basic Books, 2001).

3 It's interesting to note that Sohon made more than one image of Cantonment Stevens, each in a different season. Artists frequently make several studies of the same subject.

4 John Mix Stanley, "Memoranda in Relation to Sketches in Natural History, Geology, Botany, and to Views of Scenery and Natural Objects," in Isaac Stevens, *Report of Explorations for a Route for the Pacific Railroad, near the Forty-Seventh and Forty-Ninth Parallels of North Latitude from St. Paul to Puget Sound,* supplement to U.S. War Department, *Reports of Explorations and Surveys to Ascertain the most Practicable and Economic Route for a Railroad from the Mississippi River to the Pacific Ocean*, vol. 1 (Washington, DC: Government Printing Office, 1855), p. 8.

5 Thomas Adams was part of Stevens's division that started from St. Paul in May 1853. Later he was assigned to work with Mullan at Cantonment Stevens.

6 More information about the camas plant can be found in Michelle L. Stevens and Dale C. Darris, "*Ethnobotany, Culture and Use of Great Camas,*" Plant Materials Technical Notes No. 23, in *Technical Notes,* newsletter of the U.S. Department of Agriculture, Natural Resources Conservation Service (September 1999); available in pdf format.

7 John Mullan, "Report of Lieutenant John Mullan, USA, of his Examination of the country from the Bitter Root Valley to the Flathead Lake and Kootenay River" in Stevens, *Report of Explorations*, p. 517. Mullan wrote this report at Cantonment Stevens in the Bitterroot Valley.

8 Ibid.

9 Ibid., p. 518.

10 John C. Ewers, "Gustavus Sohon's Portraits of Flathead and Pend d'Oreille Indians," *Smithsonian Miscellaneous Collections*, vol. 110, no. 7 (Washington, DC: Smithsonian Institution, 1948), p. 62. Ewers was given these images by Sohon's daughter Elizabeth in 1947.

11 Ronald E. Grim and Paul D. McDermott, "Gustavus Sohon's Cartography and Artistic Works: An Annotated Bibliography of His Images," *Phillip Lee Phillips Society Occasional Papers Series*, no. 4 (Washington DC: Library of Congress, Geography and Map Division, 2002), p. 4. A copy of this paper may be obtained by contacting Professor McDermott.

12 Mullan, "Report," p. 519.

13 Ibid.

14 Ibid., pp. 519–20.

15 Ibid., p. 521.

16 Ibid.

17 Ibid.

18 Ibid.

19 Ibid., p. 522.

20 Ibid., p. 520.

21 Ibid., p. 523–24.

22 Ibid., p. 525.

23 Ibid., p. 526.

24 Ibid.

Part Two

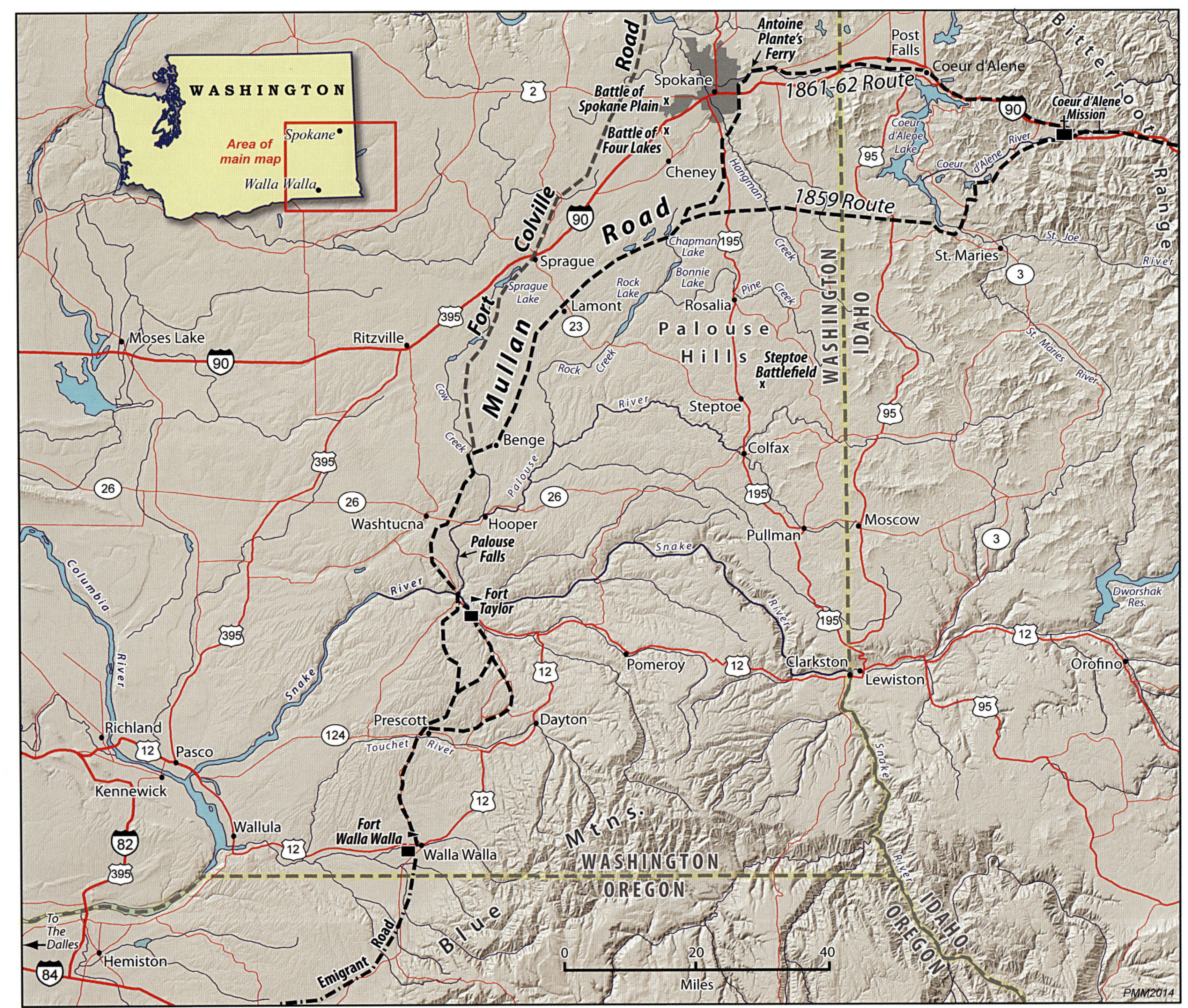

The western section of the Mullan Road –Map by Philip Mobley

Across the Plateau and into the Mountains

The Western Section of the Mullan Road

BY DON POPEJOY

In the spring of 1858 Lieutenant John Mullan was sent by the War Department to build a military road in the Northwest, from Fort Walla Walla, Washington Territory, on the Columbia River, to Fort Benton, Montana Territory, along the Missouri River. Mullan arrived at Fort Dalles, Oregon Territory, along the banks of the Columbia, in May of that year and spent several days gathering men and supplies for the adventure ahead. While there, he suddenly received news of the recent defeat of Colonel Edward Steptoe, commander at Fort Walla Walla, at a battle along Pine Creek (about forty miles south of today's Spokane, Washington) by the united tribes of the Spokane, Coeur d'Alene, Cayuse, Walla Walla, and Nez Perce Indians. Realizing that his surveying assignment would now be delayed, Mullan dismissed all his men except for his two assistants, Theodore Kolecki and Gustavus Sohon.

A few weeks later, General Newman S. Clarke, commander of the Department of the Pacific, sent Colonel George Wright to Fort Walla Walla to put down the uprising. Mullan, seeing a chance to reconnoiter the area for his road project, volunteered to join Wright's staff as a topographical officer. He left Fort Dalles for Fort Walla Walla on July 16, 1858, with his two assistants, three other employees, and an Indian boy.

While Mullan's service with Wright is often considered an interruption in the road-building expedition, I submit that it could be viewed as the first phase of it. The information Mullan and his men gathered about the far-western region during the campaign contributed to the road project before the actual construction began in the spring of 1859. On the fly leaf of his leather-bound journal, Mullan wrote this inscription: "Journal of routes, roads etc kept while surveying Military Road from Fort Dalles to Fort Wallah Wallah and while attached to Staff of Col. Wright on expedition against hostile Northern Indians in 1858, by Lieut. John Mullan USA."

Later we will examine phase two, Mullan's experiences when the construction proper began, from Fort Walla Walla in May 1859 to Montana's St. Regis River, where the men made their winter camp, Cantonment Jordan, in December. The extention of the road to Fort Benton, the eastern sections of the road, in 1860–61 will be the focus of the next several essays in this book. In the spring of 1861, due to the considerable weather damage the road had sustained during its first year, Mullan retraced the road back into Idaho to repair, improve, and modify it. Also during that phase, he decided to reroute the road around Lake Coeur d'Alene. This portion of the construction project, from May 1861 to August 1862, was the final phase.

Phase One: Fort Dalles to Fort Walla Walla, 1858

Mullan's small group left Fort Dalles early in the morning on July 16, 1858, with "our object being to overtake command of 3rd Art. en route to Wallah Wallah under Capt. Keyes &c." On July 17 the team traveled through dry, dusty country, with the Columbia River to the north and high, basaltic bluffs to the south, to reach Mud Springs, twenty-seven miles east of Fort Dalles. Mullan wrote that they traveled "a good road well grassed, but no wood. Fine Bunch grass for countless herds." After crossing the Deschutes River, a "stream 100 yds wide,

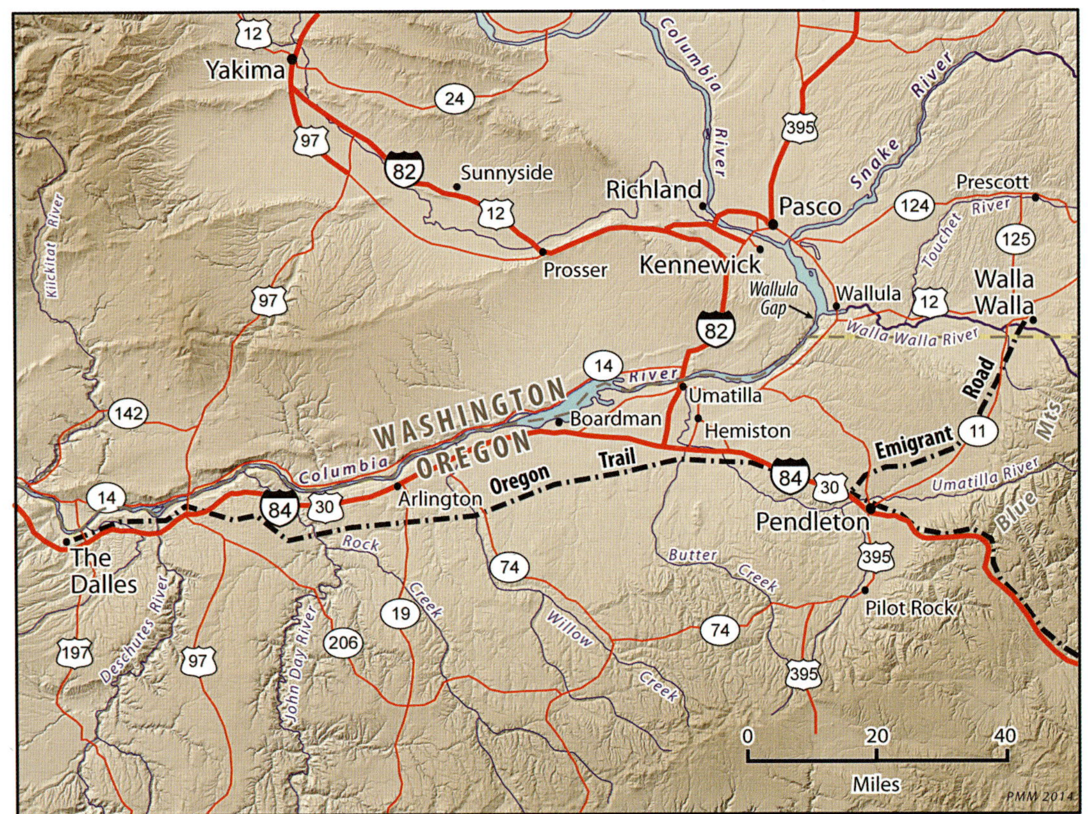

Mullan officially started his road project from The Dalles, but his construction efforts were curtailed by an outbreak of Indian hostilities in May 1858.
–Map by Philip Mobley

rapid," Mullan noted a bridge under construction and estimated that it would be completed within three months. Once across the river by "flying ferry," he visited a friend named Mr. Kingsberry and "after a cordial welcome drank a long life, pledged in a most delightful draught of egg-nogg made by his own hands." That evening, Mullan and his men were treated to a festive musical put on by a local farmer, Mr. Fallin, his two daughters, and a Miss Weatherford, "a relief that we truly appreciated in this desolate region, to us the last mark of a refined civilization that we shall meet with for many months."[1]

The established road that Mullan was using was part of the "Emigrant Road" used since the early 1840s by emigrants on the Oregon Trail. This segment, also referred to as the Overland Trail, followed the Umatilla River, near Pendleton, Oregon, out of the treacherous Blue Mountains to the first sighting of the Columbia River at Biggs Junction. From there the Oregon Trail followed the Columbia to Fort Dalles and then passed around the south side of Mount Hood on the Barlow Road; alternatively, emigrants "took on water" and floated the Columbia River to the Willamette Valley where Portland, Oregon, is today.

As Mullan's crew continued toward Fort Walla Walla, they traversed the "Columbia River Trail" and noted their bearings, mentioning "the snow-capped & prominent Mt peaks of the Cascade Range, that is, Hood, Adams, Jefferson, Rainier and St. Helens, all being in view today." Mullan also mentioned the "low spurs" of the Blue Mountains, not more than fifty miles distant.[2]

As the group reached the John Day River, the road, in need of repairs, became extremely difficult. Near here, Mullan noted the fresh grave of an Oregon Trail emigrant by the name of Fresel who died at age thirty-seven. Mullan stopped to read the headstone, which "some kind friend had erected to point to the traveler the resting place of a man who had left friends & relatives in a far distant land East and who having braved the elements & dangers of both mountain & prairie had reached this distant point to repose his bones, just on the eve of reaching hospitable roofs so near at hand."[3]

Mullan and Kolecki reconnoitered another canyon just south of the John Day Canyon as a possible wagon route, but they found that the trail was not only difficult and very rocky, but it also led them in the wrong direction. At their campsite along John Day's river, the men caught some "fine fish" and reconnoitered the area. After a long, dusty day, Mullan reflected on the demise of the Indians:

> Met no Indians at all in the road from the Dalles to this point. For them what a sad infliction it might induce, that in a region where large & numerous tribes might have been found before the event of the white man & his civilization, and that too at so recent a date–is not found one. The streams that were once their haunts from whence they drew at the cost of little labor their yearly supply of salmon & their berries along the banks, but now the streams flow noiselessly on, their rich and abundant supplies of salmon running undisturbed, and the growths of the new fruit falling from year to year on the ground from whence it sprang, ungathered and uncared for, save when some party of travelers camp near these wild orchards of berries.[4]

The next day, July 19, was a leisurely day of travel. Mullan's party entered the valley of Rock Creek, which they ascended, crossing the creek four times; Mullan noted that the route they took was a "very good road." Rock Creek, lined with thick growths of willow and alder amid grassy hillsides, ran through a valley composed of rich loess soil. In his journal, Mullan recorded that they had traveled seven miles and encamped, "making our total distance from the Dalles 54 miles."[5] Kolecki, Mullan's cartographer, took recordings of latitude and longitude and made other observations.

The explorers' travel over the next couple of days was slow and arduous, but they found an abundance of wood, water, and grass along their route. As they neared the Umatilla Valley, the Blue Mountains came within view, and Mullan commented that the caravan had traveled sixty-one miles and the men "were sorely fatigued and worn out." July 21 would prove to be a very eventful day in several ways. First, their odometer broke near Well Springs, on property known today as the Boardman Bombing Range, leaving the men to figure out their own mileage. Second, during their otherwise "dreary and monotonous march," Mullan's group came upon one of the region's most remarkable land formations: "We had this day a fine view of the Gap of the Columbia near old Wallah Wallah where are situated the well defined Ross's and McKinzies's Peaks."[6] Today we know this landmark as Wallula Gap, a remnant of ice age floods. Third and most important, Mullan was informed by Lieutenant Gibson that while he was at the Willow Creek campsite of July 20, some Snake Indians had stolen five of his oxen. Finally, a prairie fire that had been burning throughout the day threatened their campsite. As night came on, however, a wind from the northwest changed the fire's direction.

The morning of July 22 was bright and clear, and Mullan moved in an easterly direction, soon arriving at the Umatilla River, which he described as "flowing through a bottom rich and fertile & some four or five miles wide. The stream is well wooded with large cottonwood & alder, is from 20 to 30 yds wide, rapid current, gravelly bottom and good banks on either side." Where Mullan's band crossed the Umatilla, he noted the site of the old Indian agency and the remains of Fort Henrietta astride the Oregon Trail. The fort, built in 1855 by Major Mark Chinn to protect settlers from any Indian problems they might encounter, was named in honor of Major Granville Haller's wife, Henrietta. After crossing the Umatilla, the group traveled another five miles following the right bank of the river, coming to a natural corral formed by "a circular ring fringed or bounded by cottonwoods that afford a comfortable & secure camping ground with an abundance of wood & grass."[7]

On July 23, Mullan commented that he could see where "thc Old Emigrant road that crosses the Blue Mts. between the forks of the river. The Blue Mts. today were very distinct making a blue line along the horizon, and we were enabled to mark distinctly the line of the ridges."[8] The Blue Mountains are called "the Blues" for the blue haze that persists along the range's ridge lines, the result of the blue spruce that predominate its slopes and the surrounding forest.

The next evening, the party camped "on the right bank of Wild Horse [Creek] in time to get a good series of observations for latitude & Time."[9] While camped there, Mullan encountered a man making his way to Fort Dalles after his traveling companion was killed by two Walla Walla Indians. A detachment of dragoons lead by Lieutenant Dorsey Pender (a future Confederate officer in the American Civil War) was sent from Fort Walla Walla to rescue the body and find the murderers.

July 25, 1858, would be the last day of a long but interesting journey surveying the road from Fort Dalles to Fort Walla Walla, which would become part of Mullan's road to Fort Benton. "With a good road & an eager desire to reach the end of our journey," Mullan wrote, "we moved rapidly over rolling prairie land."[10] From a bluff overlooking the Walla Walla Valley, Mullan noted:

> To our right lay the range of Blue Mts viewed from the summit to base, in front lay the ocean of rolling prairie that formed the divide of waters of the Wallah Wallah from those of the Snake, while to our left lay the majestic Columbia traced from point to point by high bluffs & buttes that defined its course. While between them lay embosomed the beautiful valley of the Wallah Wallah.[11]

With its mild climate, fertile soil, abundant forests, and relatively easy accessibility, Mullan predicted that this valley could support a population of 30,000 people. He closed this segment of his journal stating, "[W]hat may we not anticipate from this valley as years shall fully develop this region and bring it to our nation's notice!"[12]

Phase Two: South of the Coeur d'Alene Mountains, 1859–60

The Indian wars in the Pacific Northwest finally ended with Colonel Wright's devastating campaign against the "Northern Indians," as he termed them, in September 1858. Mullan spent the next several months putting together his plans to begin the long-awaited road survey in the spring of 1859. He received his survey orders from Captain A. A. Humphreys of the War Department, Office of Exploration and Surveys, on March 15, 1859. The letter stated that Mullan had been appropriated $100,000 for continuing the construction of the road to Fort Benton and that "In conducting this operation, your attention will be first be directed to making those parts of the route where the greatest difficulties and most numerous obstructions exist practicable for the passage of wagons at all seasons of the year."[13]

Humphreys instructed Mullan to estimate the cost of building the road, including the number of assistants he would need and the amount of compensation they would be paid. Humphreys told the young lieutenant that he would be given one hundred men from the Department of Oregon and advised him that construction should not take more then sixteen months. The letter closes with Humphreys' final orders: "Upon completion of the work you will return by such route as the condition of the party may render necessary or desirable, discharge your employees, dispose of your outfit at the most favorable and convenient point, and repair to Washington [D.C.] with such assistants as may be required to complete your office work."[14]

Mullan's estimate of $85,000 included payroll for two engineers, one physician, one wagon master, one clerk, one blacksmith, two carpenters, twelve teamsters, four cooks, and thirty-five laborers. The expedition would start with 3,000 rations, six wagons, sixty yoke of oxen, seventy yoke of cows, plus an unspecified number of chickens. Included in the estimate was the cost of tools, camp equipment, cooking utensils, and incidental expenses.[15]

With the initial railroad survey of 1853–54 behind him, the Fort Dalles to Fort Walla Walla road construction of 1858 finished, and the Indian situation taken care of, Lieutenant John Mullan was ready to begin the next phase of building the Northwest Passage from the Columbia River to the Missouri River, a distance of 624 miles.

John Mullan would rely on several of his men for a successful expedition. His officer staff included Lieutenant James White, Lieutenant H. B. Lyon, and Lieutenant James Howard. His guide, interpreter, and artist was the renowned Gustavus Sohon; David McKay was Mullan's second interpreter; and his meteorologist and astronomer was John Weisner. Mullan also had two topographical engineers with him, P. M. Engle and the indispensible Theodore Kolecki.

Mullan's expedition members were a very diverse group; to handle these rough and tough individuals, the thirty-year-old Mullan would have to draw on his past experiences in the United States Army. The oldest member, at age forty-five, was wagon master John Caldwell, a civilian from Pennsylvania. The youngest was nineteen-year-old John Smith, from Virginia, who was Mullan's clerk. The nineteen Americans in the party represented ten different states, with New York contributing two clerks and four soldiers. Almost half of the men (forty-seven) were from Ireland, with the countries of Mexico, England, Germany, Switzerland, France, and Scotland also represented. A twenty-five-year-old Jamaician, Thomas Towza, Mullan's only printer, rounded out this truly international workforce.[16]

Mullan began his road-building odyssey on June 1, 1859, leaving Fort Dalles for Fort Walla Walla. Meanwhile, he sent P. M. Engle to accompany Pickney Lugenbeel to Fort Colville along the latter's just-established Fort Colville Military Road; Engle was to report on the condition of that road, which would connect with Mullan's road at the Palouse River. Engle then accompanied Captain Kirkham's supply train from Fort Colville back to Fort Walla Walla, where he reunited with Mullan's party just before they departed.

Engle wrote in his journal the first account of this remarkable journey: "Left Fort Walla-Walla early in the morning . . . following a well marked road we reached the Touchet River, and fording it without difficulty, encamped on its north shore, having made 21.4 miles." The next day found the party at the Snake River crossing, which Engle described as about 350 yards wide, "rapid and strong." It would take several days for the group to cross the Snake by means of a local ferry; the crossing site is today known as Lyon's Ferry State Park.

Arriving at Cow Creek, which flows from the Palouse River near present Hooper, Washington, on June 8, Captain Kirkham was directed to take the north fork and left the main group, headed for Fort Coville, per his assignment. As the

Much of the Mullan Road went through the sagebrush-covered Columbia River Plateau. This photograph shows the Colville Road. –Photo by Don Popejoy

Engle party passed through the channeled scablands, which J Harlen Bertz later discovered were carved during the great Missoula floods, Engle wrote in his journal, "travelled over a gently undulating ground, enclosed by isolated prairie ridges, we entered an open pine-timbered forest, which is filled with innumerable little ponds and cut up by pedrigal [basalt] rocks cropping up from the ground in great abundance."[17]

At Cow Creek, the Mullan Road separated from the Fort Colville Road and headed directly northeast to Rock Creek. At Sprague Lake (which Mullan called "Big Lake"), about thirty miles southwest of Spokane, on June 10, 1859, Mullan observed:

> [W]e saw the "Big lake" before us with its pine-clad hills at the northeasterly end. A gentle slope of two miles brought us to the water's edge . . . . The lake is surrounded by low hills; some of them are rocky and precipitous; others again are of gentle grade; both shores of the lake are entirely destitute of timber, and at a few spots only brushwood can be found. . . . Travelling along the easterly shore for two miles we reached a point where, by a succession of rocky points, which extended far into the water, our road turned off from the shore, leading through little ravines and coulees which connected prettily with each other. Striking after one and a half mile the lake again, we reached again after three-quarters of a mile the northeasterly end of it, and turning short to the east we passed a luxuriant cotton-wood grove, and following up a wide bottom, which affords a good road in dry seasons, we crossed a little marshy water-run, and camped near a spring which sends its icy waters to the lake; made thirteen miles. The lake bears the name of "Big lake" in all the Indian languages; it has a length of six miles and a width varying from one-half to three-quarters of a mile. The water is not fresh, and has a swampy taste . . . .[18]

Two days later Mullan, ever the naturalist, described the Grand Coulee area as "high and rich grass intermixed with flowers covers the hills, and groves of luxuriant trees are dotted all over the country. . . . The next four miles of the road . . . led across an undulating prairie country, frequently intercepted by stripes of pine timber."[19]

The expedition reached the St. Joseph's (now called the St. Joe) River valley in mid-July. Approaching Lake Coeur d'Alene, Mullan made a decision that he would later regret: he would go around Lake Coeur d'Alene through the swamps and river bottoms to the south. Thus began the most difficult part of the road building, as the men had to literally cut their way through a dense and rugged landscape, proceeding northeast through the mountains to the Mission of the Sacred Heart. After a solid month of labor, the party reached the mission on August 16. This mission, near present Cataldo, Idaho, was built between 1848 and 1853 by Indian labor, supervised by Jesuit priest Antonio Ravalli.

While camped near Lake Coeur d'Alene, Mullan met with leaders from the local Coeur d'Alene Indians.[20] Still resentful about roads going through their territory, the natives were determined to block Mullan's progress. Mullan met the challenge by stating that he was here on the authority of Colonel George Wright, who would hang any Indian who opposed Mullan's passage. Because the name of Colonel Wright still struck fear in the Indians' hearts, they allowed the road builders to continue their work.

The progress of the road construction was of great interest to the government and the national press. The *Olympia Pioneer and Democrat* reported on September 9, 1859, that "Lt. Mullan is making the most encouraging progress in opening the road connecting the navigable waters of the Missouri and Columbia. The party is exceedingly well organized, is perfectly harmonious, has an adequate escort, is in no danger from the Indians, and is full of hope and confidence in success of it labors. On the 2d. of August, Lieut. Mullan, with his working parties, was eleven miles beyond the Coeur d'Alene Mission."[21]

On August 16, from the Coeur d'Alene Mission, Washington Territory, Lieutenant J. L. White wrote to Captain A. Pleasonton at Fort Vancouver:

> I have the honor to report the arrival on this day of my party at this place. . . . The greater part of the route since reaching the St. Joseph river lies along the river valleys, and is subject to annual overflow. Some places were laid with corduroy before attempting to pass wagons over them, while many others showed by their softness that waters had but recently left them. The road would be utterly impracticable for wagons during the time of high water, but for parties moving westward this would probably be no objection, as before

> they could pass the mountain snows the water would have subsided. The road has required much labor in its construction–cutting in side hills, and through forests, removing fallen trees, marking bridges and corduroys. For a road of its class it may be fairly called a good one thus far.[22]

As Mullan's expedition advanced up the south fork of the Coeur d'Alene River, toward Sohon Pass (now called St. Regis Pass) through the Bitterroot Range of the Rocky Mountains, a critical moment occurred–some of Mullan's workers decided not to cut down any more trees, reducing his work force to about fifty hired hands. As a result of the topography, Mullan's crew had to stay in the narrow valley bottoms, where many river crossings had to be made. By December 4, 1859, Mullan had completed the road over the summit and down into the valley of the St. Regis River on the Montana side. Here, Mullan had his men erect winter quarters, dubbed Cantonment Jordan, at a site near present-day St. Regis, Montana. Over the winter, Mullan realized that he had built his road on the wrong side of Lake Coeur d'Alene! He decided to build a new section along the north side of the lake, in order to detour around the flooding problem. He would start on this project in the summer of 1861.

Final Phase: Rebuilding the Road, 1861–62

Arriving at Fort Walla Walla in April 1861, Mullan turned to his task of improving the military road in the sections that required it. After organizing his team and gathering supplies, Mullan left Walla Walla on May 13, 1861. The new and improved road followed the same route as the original up to the point where it forked north to Fort Colville at Cow Creek. From there, Mullan blazed a new road, traveling north through the channeled scablands of eastern Washington,

This color image by John Madison Alden shows Antoine Plante's Spokane Ferry crossing. The artist's viewpoint was a high lava bluff overlooking the ferry to the east. –Courtesy National Anthropological Archives, Smithsonian Institution

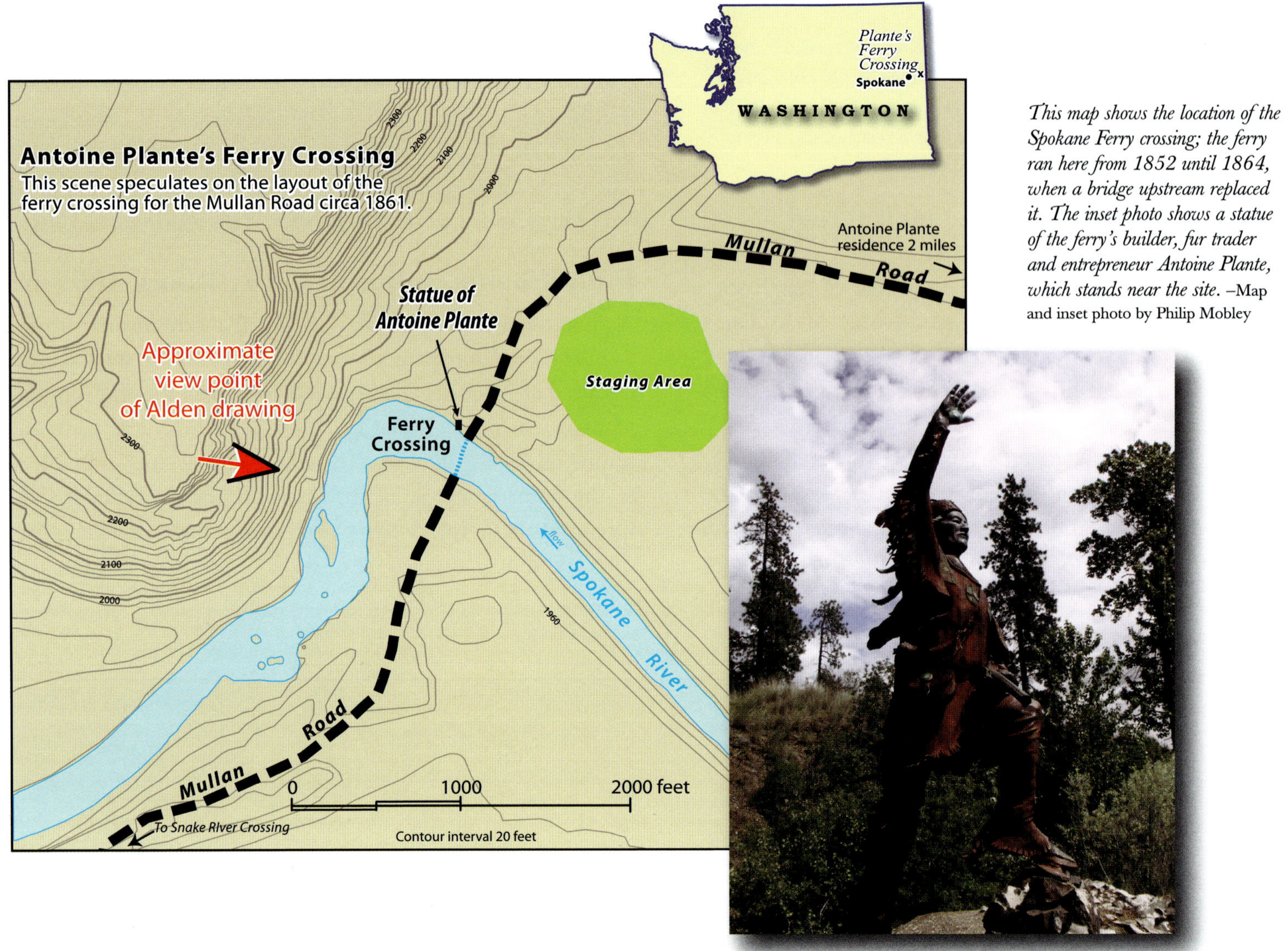

This map shows the location of the Spokane Ferry crossing; the ferry ran here from 1852 until 1864, when a bridge upstream replaced it. The inset photo shows a statue of the ferry's builder, fur trader and entrepreneur Antoine Plante, which stands near the site. –Map and inset photo by Philip Mobley

passing through the present towns of Benge and Lamont. North of Lamont, the road turned east for ten miles, cutting between two pothole lakes, today called Chapman and Bonnie Lakes; from there the route turned suddenly north and a bit east toward today's Spokane, Washington.

Continuing northeast, the road crossed Hangman Creek (a.k.a. Latah Creek) then headed out onto the open spaces of Moran Prairie of present Spokane. Passing through what is now Spokane, the new version of the Mullan Road continued northeast to the Spokane River, crossing at Antoine Plante's ferry, some ten miles east of Spokane Falls, on June 1, 1861. Once across the river, Mullan's road continued east along the north bank of the Spokane River, traversing the broad Spokane Valley for twenty-five miles before reaching the north end of Lake Coeur d'Alene near Post Falls. From there, Mullan's crew crossed the Fourth of July Pass to the Mission of the Sacred Heart at Cataldo, Idaho. En route to the mission, Mullan's men built thirty miles of new road and a bridge across the Palouse River.

From the mission, Mullan followed the Coeur d'Alene River through the present Idaho towns of Smelterville, Kellogg, Osburn, Wallace, and Mullan. Dividing his men into

The Spokane River at Post Falls, looking northwest –Photo by Philip Mobley

several teams, Mullan tasked each to repair the road and build bridges in their respective sections. This construction occupied the workers until mid-September. At today's Idaho-Montana border, the road passed through the Bitterroot Mountains at Sohon Pass. The pass, with an elevation of 4,932 feet, was named after Mullan's talented artist and cartographer. Descending the eastern slope of the Bitterroot Range, Mullan followed the St. Regis River to its confluence with the Clark Fork River, which he would then follow to the town of Hell Gate (today's Missoula, Montana). Here they set up their final winter camp, Cantonment Wright, where they spent the most severe winter yet.

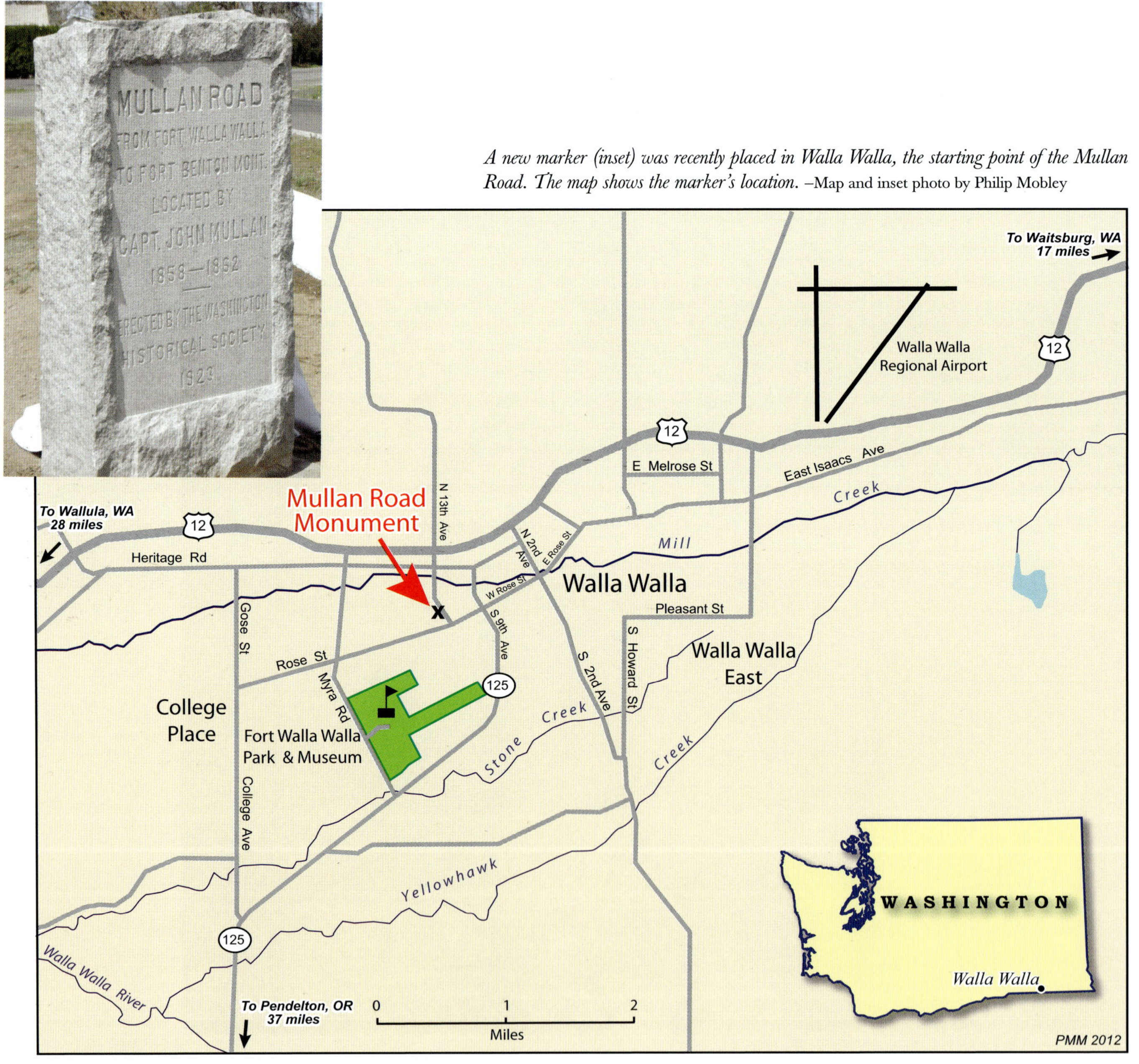

A new marker (inset) was recently placed in Walla Walla, the starting point of the Mullan Road. The map shows the marker's location. –Map and inset photo by Philip Mobley

Most of the original Mullan Road has been obliterated, but markers like this one have been placed on or near the route. –Photo by Don Popejoy

In the spring of 1862, Mullan found that much of his road had already seriously deteriorated. Several bridges along the Coeur d'Alene River washed out during the spring flooding, as did other sections of the road. The army provided no maintenance. Mullan had expected local authorities to care for this "county road" and make seasonal repairs, but no one had done so. Thus it was up to Mullan to go back and repair and refurbish it. His crew spent the summer making all the necessary improvements, finishing the work by early August. Mullan and his men then returned to Fort Walla Walla, where the expedition was finally disbanded. Mullan returned to Washington, D.C., in September to compile his various reports of his historic road-building enterprise.

In less than four years, Lieutenant John Mullan had created America's first interstate road. Much of the credit belongs to his own maturity, organizational skills, and remarkable ability to command men. Afterward, Mullan often said that the happiest years of his life were those he spent in the Northwest, with his saddle as a pillow and pine needles as his mattress.

1 All quotes from Harvey Erickson, "Mullan's 1862 Interstate Road," *Pacific Northwesterner* 18, no. 4 (Fall 1974), pp. 53–66.

2 Ibid., p. 54.

3 Ibid., p. 55.

4 Ibid., p. 56.

5 John Mullan, P. M. Engle, et al, *Military Road from Fort Benton to Fort Walla Walla: Letter from the Secretary of War . . .* H. Ex. Doc. 44, 36th Cong., 2d sess., (Washington, DC: Government Printing Office, 1861), p. 77.

6 All quotes from Erickson, "Mullan's 1862 Interstate Road," pp. 57–58.

7 All quotes from ibid.

8 Ibid., pp. 58–59.

9 Ibid., p. 57.

10 Ibid., p. 59.

11 Ibid.

12 Mullan et al, *Military Road from Fort Benton to Fort Walla Walla*, p. 5.

13 Ibid.

14 Ibid., p. 17A.

15 John Mullan, *Report on the Construction of a Military Road from Fort Walla-Walla to Fort Benton* S. Ex. Doc. 43, 37th Cong., 3rd sess. (Washington, DC: Government Printing Office, 1863), p. 17.

16 Information provided by Professor Richard Scheuerman, Seattle Pacific University, via e-mail correspondence on May 10, 2007.

17 All quotes from Erickson, p. 61.

18 Mullan et al, *Military Road from Fort Benton to Fort Walla Walla*, p. 77–78.

19 Ibid., p. 78.

20 The French fur traders named this tribe the Coeur d'Alene, meaning "the heart of the awl," saying they were the finest traders in the world. The Coeur d'Alenes called themselves the *Schee Chu' Umsch*, which in their native Salishan language means "those who are found here."

21 *Olympia [WA] Pioneer and Democrat*, September 9, 1859.

22 Mullan et al, *Military Road from Fort Benton to Fort Walla Walla*, p. 79.

"A Creditable Piece of Mountain Work"

Building the Mullan Road in the Rockies (Part 1, 1858–59)

BY MAJ. RYAN L. SHAW

> The eastern or western plains, as a general thing, interpose but few physical obstacles to the location of a practical wagon road; but the mountain sections form special problems for solution, which are not always so easily handled.
>
> —John Mullan, 1863[1]

Preliminary work on Lieutenant John Mullan's great road project began in the winter of 1853, which he spent exploring the Rocky Mountains in support of Isaac Stevens's survey for the Northern Pacific Railroad. His work was finished with the submission to the War Department of his *Report on the Construction of a Military Road from Fort Walla-Walla to Fort Benton*, published by the government in 1863. All told, then, the project took a decade. But most of those ten years were but prelude and postscript to the epic battle Mullan waged against the primeval granite, the untamed forests, and the capricious rivers of the northern Rockies, contending with the bitter cold and deep snows of winter and the torrential flooding of spring.

The mountains *captivated* Mullan. He passed the winter of 1853 in Montana's Bitterroot Valley, at Cantonment Stevens, measuring snowfall and exploring passes over the Continental Divide for Stevens's railroad. His writings from and about that time are among his most descriptive. It was then that he conceived of his wagon road plan—it must be remembered that the plan was his from its conception. Mullan was not ordered into the mountains—in fact, after Congress failed to grant an adequate appropriation for Mullan's survey in 1855, he was ordered away from the region, all the way to the swamps of Louisiana and Florida and then to Baltimore with his artillery regiment.[2] But in 1857, when Indian and Mormon disturbances in the West helped persuade the War Department—with encouragement from Washington Territory's delegate in Congress, Isaac Stevens—of the desirability of a northern alternative to the Overland Trail, Mullan was more than ready to spearhead the venture.[3]

The mountains *confounded* Mullan. After finding his route from the Bitterroot Valley to Fort Benton, Mullan spent most of 1854 exploring routes across the Bitterroot Mountains toward the Pacific. Of four possibilities he considered, only two seemed feasible: through the Coeur d'Alene River valley, or farther north along the Clark Fork River. The Coeur d'Alene route seemed more difficult but more direct, and Mullan surmised that the Clark's Fork route, in addition to being longer, would be less trafficable in winter since it was farther north.[4] That assumption, logical as it seemed, proved false, and it ended up costing the crew a great deal of time and money—not to mention misery, trapping them over a very trying winter in the valley of the St. Regis Borgia River (today called the St. Regis River) in 1859–60. The peculiarity of that climactic phenomenon troubled Mullan for years, leading him to lament that "all rules and analogies applicable to the climates of other portions of the country there apparently fail."[5]

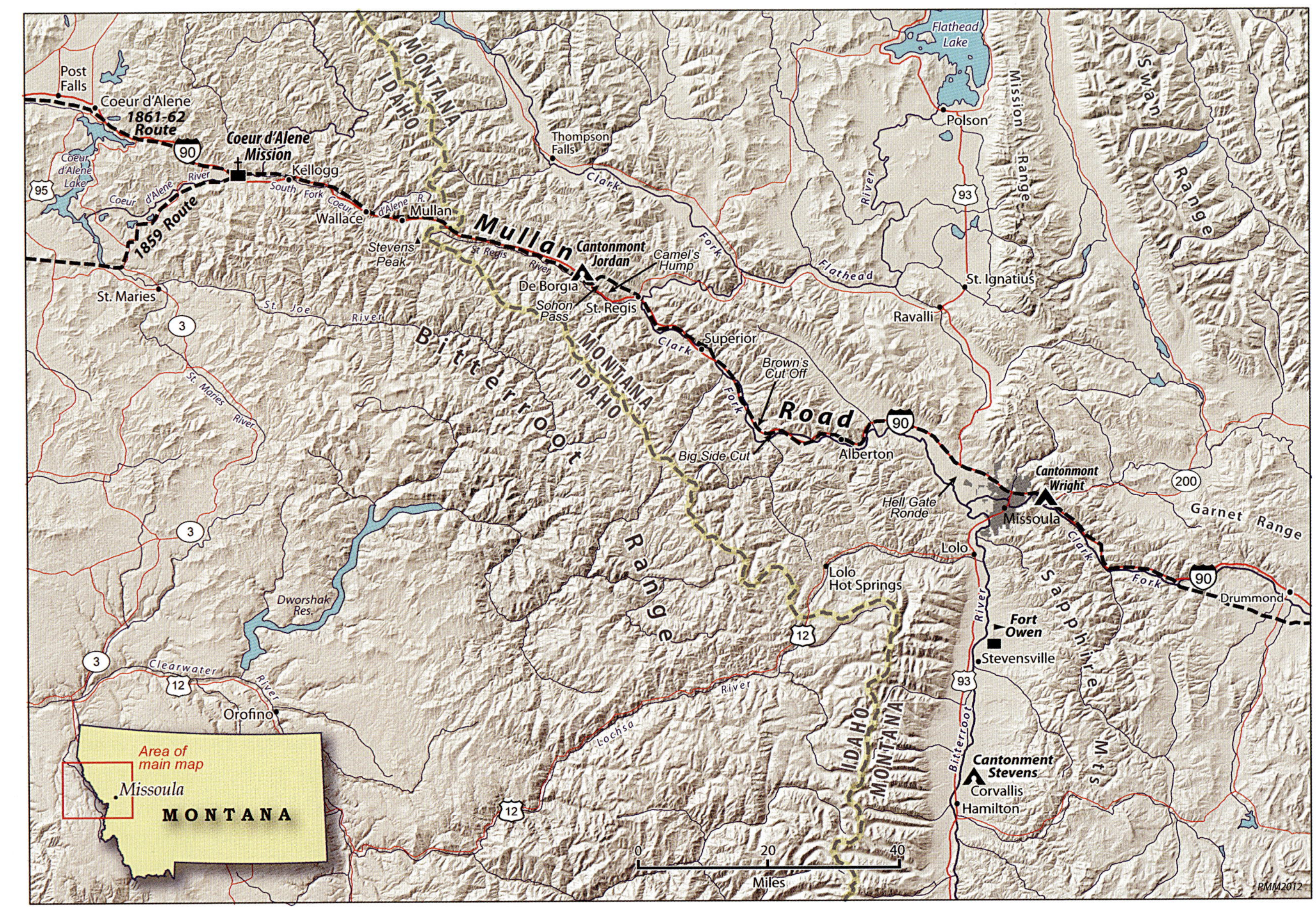

The Mullan Road through the Rocky Mountains –Map by Philip Mobley

The Rocky Mountains region –Map by Philip Mobley

Gustavus Sohon's lithograph of the Rocky Mountain Front for Isaac Stevens's Pacific Railroad Report –From Isaac Stevens, *Report of Explorations for a Route for the Pacific Railroad . . .*, supplement to *Reports of Explorations and Surveys, to Ascertain the Most Practicable and Economic Route . . .*, vol. 1 (1855)

When Mullan finally got his orders to begin the road project in March of 1858, he was en route to Walla Walla with his men by May. But news of Indian attacks and a planned retaliatory campaign directly on the line of march forced Mullan to disband his crew at Fort Dalles. Instead of building a road, he spent the summer of 1858 serving as a topographical officer with the 9th Infantry, collecting further information on the western approaches to the mountain section.

The mountains *surprised* Mullan. By the time the campaign of 1858 was over, Mullan had concluded that in building his road through the Rockies, "the physical difficulties were such that a longer period of time, more ample means, and a larger force of men than either myself or others had imagined were requisite to accomplish our object."[6] He returned to Washington and, along with Stevens, lobbied Congress for an additional appropriation of $100,000, which was granted in March 1859. Even so, by the time Mullan settled into his winter quarters that December, he was forced to admit that "The truth was that the amount of work required was immense, and very much under estimated by both myself and others, for we only truly appreciated it when we came to handle it in detail."[7] His correspondence indicates that the work in the mountains was always slower and costlier than he expected. Some of that underestimation might have been intentional, for the sake of securing appropriation enough to get started without making the project seem too daunting to try. Nevertheless, even after his extensive reconnaissance and planning, the realities of the mountains continued to surprise him.

Construction Begins

When construction of Mullan's road finally commenced on July 1, 1859, Mullan and his crew got off to an auspicious start from Fort Walla Walla. They made rapid progress over what is now eastern Washington state, and by July 16 they were past the plains of the Columbia River and into the timber approaching Lake Coeur d'Alene.[8] To reach the

Gustavus Sohon's view of the St. Joseph Valley, looking to the southeast, c. 1859. The course of the St. Joe River is defined by treed terraces on each side of the river. Poun Lake (today called Chatcolet Lake) is on the right.
–Courtesy Library of Congress, Geography and Map Division

Map of the Poun Lake region by Theodore Kolecki, c. 1859 –National Archives and Records Administration, Old Military and Civil Records Branch, Washington DC

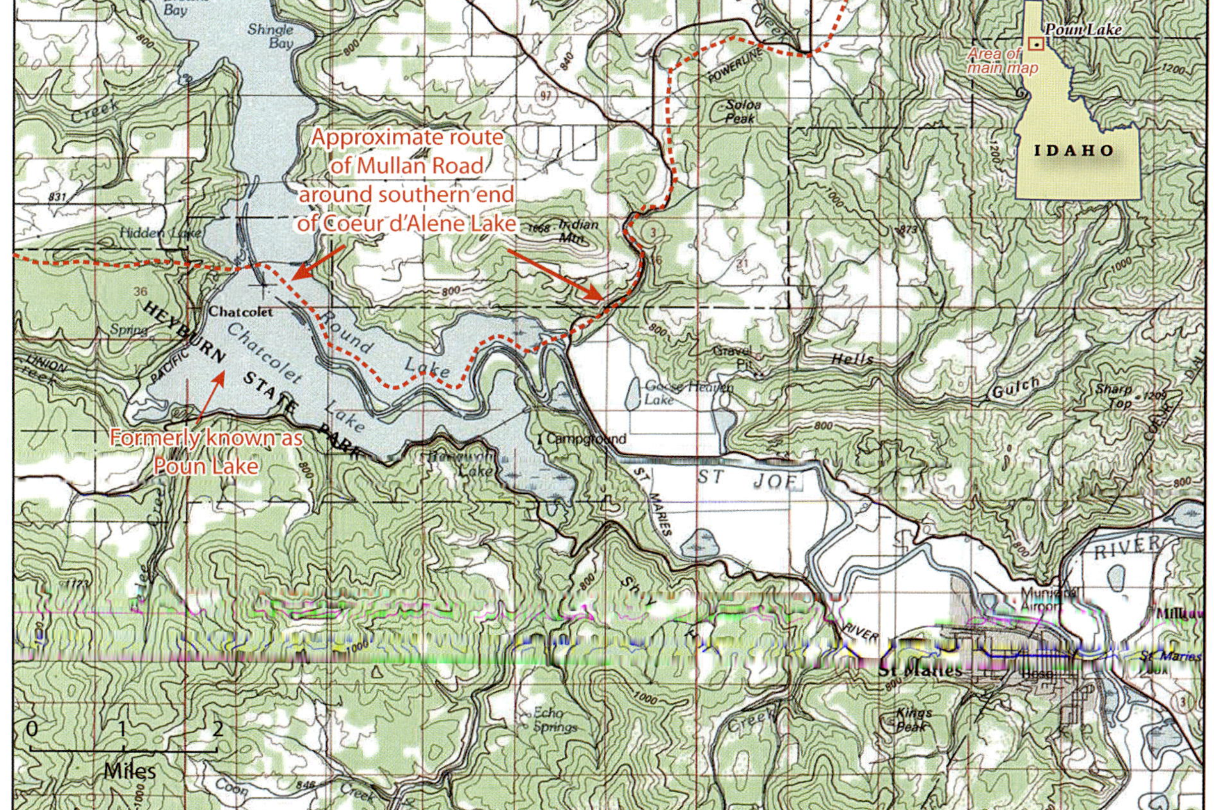

Modern map of the Poun Lake region, near the southern end of Lake Coeur d'Alene, where the St. Joe River enters the lake. The Mullan Road flooded here during the spring, forcing Mullan to relocate the road to the north side of the lake in 1861. –Map by Philip Mobley

lake required some heavy rock work–the expedition's first. Descending the seven hundred feet from the plains to the St. Joseph's (St. Joe) River required felling trees, cutting brush, digging rocks, and blasting boulders. At the bottom, the crew was then obliged to build a sixty-foot bridge across the outlet of what was then called Poun Lake (most likely today's Lake Chatcolet) and a four-hundred-yard corduroy across the marshy Coeur d'Alene Valley.[9]

To cross the lake's tributaries, the St. Joseph's and Coeur d'Alene Rivers, Mullan determined to build flatboat ferries rather than bridges, constructing both on the St. Joseph's. Planning to cross the St. Joe at the southern end of the lake and the Coeur d'Alene River at the east end, he sent one ferry across the St. Joseph's and floated the other into the lake to be rowed to the crossing site on the Coeur d'Alene.[10] From there, to reach their supply point at the Coeur d'Alene Mission (also called Mission of the Sacred Heart or Cataldo Mission), the men had to excavate many more miles of road and construct several more bridges. These labors occupied Mullan's party until mid-August, when they finally arrived at the mission.

Pencil sketch of the Coeur d'Alene Mission, officially called Mission of the Sacred Heart, by Gustavus Sohon, drawn in 1858 when Sohon accompanied the Wright Campaign. –Courtesy Library of Congress, Geography and Map Division

This was all tough work, but a report from Gustavas Sohon, Mullan's longtime interpreter, artist, ethnographer, assistant, and friend, gave ominous indication that the real challenge had not yet begun. Sohon had left Walla Walla in June, in advance of Mullan's main party, to reconnoiter possible routes over the Bitterroots. He tried to secure the services of some Nez Perce guides at Walla Walla, but when he told them what route he intended to examine, they declined, saying that "the whole region was an immense bed of rugged mountains, and over which they never heard of any person having traveled."[11] But Father Joset, visiting Walla Walla from the Coeur d'Alene Mission, provided Sohon with a Coeur d'Alene Indian guide named Augustine, who was said to know the region better than anyone. But Augustine claimed that the streams were still too high and the snows too deep to cross, and they would remain so at least until August. Besides, Father Joset had promised two more guides, who were apparently detained; Augustine suggested they wait.

Nevertheless, Sohon insisted on starting immediately. Augustine, however, "was apparently uneasy and displeased, first, because he disliked to show the country of his people, fearing that they would lose it; nor was he at all anxious to undergo the fatigues and difficulties of the trip."[12] Still, Sohon and Father Joset were able to prevail upon Augustine to find the other guides and then meet Sohon on the camas prairie. When they arrived, "they were evidently excited regarding the construction of the military wagon road, and inquired anxiously as to its probable location."[13] Each of the guides recommended a different route and, failing to agree upon one, the new guides departed, leaving Sohon with only Augustine. Sohon convinced Augustine to go with him at least part of the way, to climb some of the higher peaks in the Bitterroots to get an overview of the country. From there, Augustine said, "you can see all and satisfy yourself."[14]

En route to the mountains, still below any point from which Sohon could "satisfy himself," they were joined by

Augustine's brother, Damass. After conferring with Damass, Augustine made to Sohon the following speech:

> Let us have a better understanding. I told you at Walla Walla that there was no road through this section to the Flathead country; the mountains are too difficult. Those high ridges yonder are small compared to those over which we shall be compelled to pass, and there are many of them. The road passes over high and steep mountains and down in steep ravines and cañons. If all the Americans would work here a thousand years they could never make a road. I think the Pend d'Oreille route is best for you. There you do not pass over difficult mountains. It was Koo-ne-moo-say who told you that a wagon road could be had here, and he says this because he owns land along the Pend d'Oreille route, and he fears that if the soldiers make a road along that route, he will lose his land. I am sorry to see Father Joset and yourself determined to take this route. Those who have told you that a road could be had here have deceived you and made you lose time in exploring. I say, again, that if all the Americans would spend a thousand years here, they could not make a road.[15]

Soon afterward, Sohon reported that Damass and Augustine were "planning mischief against me, and laying for me a trap," although he provided no details, and he abandoned his reconnaissance of the mountain route.[16] Returning to the camas prairie, he determined to see if he could find better luck with a route to the south. He found some Nez Perces there, digging the ubiquitous camas root, and asked them about the potential for a road in the area. They responded matter-of-factly that a road through their country would be impossible. Undaunted, Sohon found an old friend, Three Feathers, who took him to meet their chief, Yah-moh-moh. The aged chief greeted Sohon warmly, and in their first conversation, Yah-moh-moh said that "when young, he hunted in this region and found plenty of game. Water and grass were abundant and travelling good."[17] Encouraged, Sohon promptly asked if he could hire a guide and a few horses to reconnoiter the area for a military road. The chief politely changed the subject. When Sohon repeated his request after dinner, he was "surprised to find that [the Indians'] opinions were now just opposite to what they had expressed in the morning."[18]

> The old man made an energetic speech, and declared his friendship for the whites &c., but described the mountains as formidable, the forests and underbrush impenetrable, and the streams dangerous, if not impassable, and implored me not to think of exploring the route; that if I did, I would perish, and men would say that the Indians had killed me. He thought the route by the Coeur d'Alene Mission the best.[19]

With these dead ends, Sohon returned to Mullan and the main party. Every Indian, it seemed, was pleased to recommend the best route for a road, which was always most certainly through someone else's land. "I satisfy myself of one thing," Sohon reported, which was "the extreme aversion that the Indians have against any wagon road passing through their country; and I have no doubt but that it affords a constant theme for conversation and discussion among them."[20]

The Indians' misgivings about Mullan and his project were understandable. Mullan tried to mitigate the tension and sell the idea of his road to them, as he reported in late July of 1859:

> The Indians have all been restless since we have been in the country, but are now becoming more tranquil. I avail myself of every occasion to impress them with the advantages that must accrue to them on the opening of our road and, indeed, many of them begin to see and appreciate them, while others say that it is "an immense *wedge* that we are planting in their midst, that must rend them and their country asunder."[21]

At the end of that pivotal decade of the 1850s–which started with the organization of Washington Territory, followed by Stevens' treaty councils and the implementation of the reservation system throughout the territory, then the treachery that led to the brutal wars of mid-decade–the Indians' restraint and acquiescence was the truly remarkable fact of the Mullan expedition.

Though the terrain from St. Joseph's River to the Coeur d'Alene Mission was of a different order than that of the prairies–"From the time we first struck the timber to the Coeur d'Alene Mission, the work has been more severe than I had any idea of," wrote Mullan on July 31, 1859–they nevertheless managed to lay their first two hundred miles of road in less than six weeks, arriving at the mission on August 16.[22] However challenging he had found St. Joseph's and the lower Coeur d'Alene Rivers, Mullan recognized that his work

Sohon's lithograph of the Coeur d'Alene Mission, 1859 –From John Mullan, *Report on the Construction of a Military Road from Fort Walla-Walla to Fort Benton* (1863)

would take on an entirely new character as he moved toward the Bitterroot range. As he later wrote, "That this [building through the mountains] proved a difficult task to handle our three years' labor abundantly proves."[23]

Meanwhile, Theodore Kolecki had moved in advance of the work party to the mission, which was "one of the fixed points from which we began our mountain work proper," Mullan noted.[24] Kolecki was there to establish an observatory in order to establish that "fixed point" more accurately. The Coeur d'Alene Mission, constructed by the Jesuits between 1850 and 1853, had been relocated from its original site, which was abandoned in 1846 due to flooding. Despite the supreme isolation of the place, the Jesuit fathers designed and their Indian laborers constructed a strikingly beautiful cathedral in the European style. The Mission of the Sacred Heart, as they called it, featured tin-can chandeliers, wall coverings of painted newspaper, gilded crosses hewn from local pine, and altars and columns of wood painted to look for all the world like antique marble.

The fact that so well-established an outpost of white society existed in the region could lead us to misunderstand the exploratory nature of Mullan's expedition. Although Mullan and his men knew they were at the mission when they arrived, they did not really know where it was. Kolecki's astronomical observations placed it farther west than had Mullan's hasty records of the prior year.[25] This effort to accurately determine absolute, fixed points from which the relative position of the route could be accurately calculated was a continuous focus of Mullan's throughout the construction process, and his ambitions were high: "Be assured," he offered, "that we shall leave nothing undone that shall be necessary to mark and determine every point that will be necessary to refer to at any time in the future."[26]

From the mission onward, Mullan was in a race against the mercury in his thermometer and the funds in his account, both of which were dropping rapidly. He was already over budget by the time he reached the mission, and he wrote to the War Department from there on August 16, 1959: "Although my monthly expenses are greater than originally, roughly estimated, they are not a dollar too much for our work, though it may be for the appropriation."[27]

Leaving the mission, Mullan moved to the valley of the Coeur d'Alene River, which, as Mullan would describe it in his report, "at points verged toward cañons."[28] He also pointed out that "the spurs of the mountains overlapping each other turns or drives the river from bank to bank, rendering it exceedingly tortuous."[29] With his constraints of time and money, Mullan could not afford to spend much of either cutting into the steep banks along the river, so he frequently chose to lay the road across it. He built bridges when absolutely necessary, but in most cases he chose to cross at naturally shallow fords. His regret at leaving a less-than-optimal road in his wake is palpable in his writings but, he reasoned, no one could travel this portion of the route until the grass

The Coeur d'Alene River valley near Lookout Pass (originally called St. Regis Pass), looking west. –Photo by Dan McDermott

was in season, anyway, and by that time the water would be low enough to make for relatively easy fording. This first pass of the road crossed the Coeur d'Alene River forty-one times, and it crossed the St. Regis River, a similar challenge, forty-six times.

When Mullan did choose to build a bridge, he had plenty of timber for the taking: "The standing timber was dense, and the fallen timber that had accumulated for ages formed an intricate jungle well calculated to impress one with the character of impracticability."[30] Indeed, cutting through the thick timber took an even greater effort than grading the road did, and it was certainly more dangerous. By September 19, Mullan reported, two men had been cut with axes and one had been injured by a falling tree. In addition to the construction mishaps, one man was accidentally shot in the knee. In fact, attrition was becoming a problem–Mullan's crew shrunk further when he had to get rid of several unsuitable employees: "Four men, Barstow, Lyne, Brennan, and Irons, were discharged on the 16th instant, as they did not come up to the standard and requisites for laboring men on this expedition."[31]

Despite the difficulties and the urgency of the roadwork, Mullan supervised and reported with great interest on two side projects–one scientific, the other practical, and both wholly incidental to the construction of his road. The first involved the observatory at Coeur d'Alene Mission. Astronomer and meteorologist John Weisner, whom Mullan found "a gentleman of marked talent and ability," gathered observations of sunspots, the results of which Mullan promised to report to Congress at a later date.[32] The other endeavor was determining whether the route in the Coeur d'Alene region could be improved. Mullan sent Kolecki on a reconnaissance of the Lake Coeur d'Alene area to investigate whether it could be made more accommodating to settlement. Specifically, Kolecki was asked to determine if the large rock obstructions at Post Falls, the waterfall at the outlet of the lake, could be removed, and if their removal would prevent the overflow of the Coeur d'Alene and St. Joseph's Rivers each spring. Mullan had conceived this idea as early as 1857, after he first observed the lake. In September 1859, he was pleased to report that the project could be done, asserting that it would be "the means of reclaiming thousands of acres of most excellent agricultural land."[33]

The last week of September and the first three weeks of October saw incessant rain, which slowed progress even further, as did the ever-steepening route up the Coeur d'Alene River, which here gained upwards of eighty feet per mile. By October 20, the wagon trains had crossed the Bitterroots at a place Mullan named, in a nod to his trusted assistant and friend, Sohon Pass. The pass sits 4,932 feet above sea level, 1,353 feet above their last crossing of the river, which, as the crow flies, was just over a mile to the west. The ascent to the pass required a mile and a half of severe grading and "grubbing out the large stumps, the roots of which, at times, would seem to run over the whole mountain side."[34]

Stevens Peak dominates the south side of Sohon Pass (currently called St. Regis Pass), and on one spur of the peak, a reconnaissance team found "a small bowl-like lake five hundred feet in diameter, carved by the hand of nature out of the steep rocky walls."[35] This, they determined, was the source of the St. Regis River (which Mullan referred to variously as the St. Regis Borgia or the St. Francis Borgia). They hoped to follow this river rapidly downstream to reach the Bitterroot Valley before winter set in. Leaving the small alpine lake, Mullan wrote, the St. Regis River "tumbles down three hundred feet in a quarter mile through a narrow rocky channel, after which it flows through a valley a tenth of a mile in average width to the first crossing."[36] Despite the steep drop, the route they found for the eastern descent from the pass was gentle and smooth–clearly a relief to the weary road builders.

When he left the mission, Mullan had expected to reach the headwaters of the Coeur d'Alene River by September 1, 1859, and the Bitterroot Valley by October 15. Though he had reported from the mission in August that "The weather, during the day, is pleasant and well suited for work," he was concerned that already "the nights are cold, the thermometer falling as low as 32°, with ice 1/4-inch thick."[37] From his winter at Cantonment Stevens in 1853, Mullan was familiar with the Bitterroot Valley, which he believed lush and pleasant year-round. As it happened, however, the winter he spent

Post Falls on the Spokane River. Mullan wanted to lower the level of Lake Coeur d'Alene by blasting off a portion of the rock ledge that created this waterfall, but he gave up on the idea. –Photo by Philip Mobley

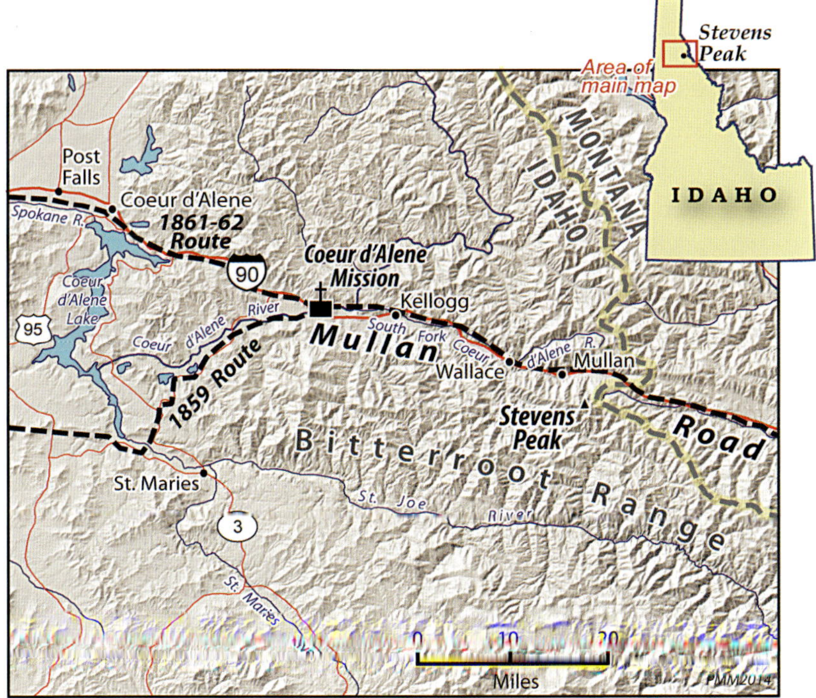

Topographical map of Stevens Peak and the Coeur d'Alene Valley. The terrain along the ridge separating Idaho and Montana is very rugged. –Map by Philip Mobley

there had been uncharacteristically mild, prompting him to refer to it as the Land of Eternal Spring, and to remain attached to it ever after.[38] As late as early October 1859, he still insisted that the Bitterroot Valley was "the *only one* where we can safely winter," and he continued to express optimism that they would beat the snows there.[39] The builders' pace through the Coeur d'Alene and St. Regis River valleys was unexpectedly slow, however, and this became a source of constant concern and frustration for Mullan. By late October, he was forced to concede that "midwinter will, I fear, still find us west of the Bitter Root valley."[40]

None of Mullan's correspondence that season failed to mention his constrained finances, the justification for their depletion, and requests for additional funds. His report of October 26 was typical: "[A]s I before informed the department, my appropriation will be expended by April or May next, and I can only hope the department will press urgently the matter before Congress, and not leave me in the mountains with my hands tied."[41]

Even before winter set in, the pace of Mullan's expenditures exceeded the rate at which the War Department deposited his appropriated funds, and his checks started to bounce.[42] By November, he had begun to make plans for operating with a reduced force in order to avoid overdrafts, yet he still insisted that "I shall push forward with all the energy possible the work to a successful completion, whether it involves a deficiency or not."[43] The primary recipient of all these expenditures was, by an overwhelming margin, the trader Pierre Chouteau, Jr., of St. Louis, who shipped supplies by steamboat to Fort Benton, from whence Mullan had them delivered by pack train.

November found the expedition still in the mountains, many miles from their planned destination in the Bitterroot Valley. The first of the month brought snow—light at first, though continuous through the 5th. After a break of one day, an outright blizzard hit, and by the 8th, the men, the livestock, the wagons, the mountains, the road behind them, and the virgin ground ahead were all sepulchered in eighteen inches of perfect white. When the snow finally stopped, the temperature plummeted, locking the grass under impenetrable sheets of ice, with no hope of another green bite for the animals until spring. Many of them had already died from fatigue, exposure, and want of forage; Mullan had the rest driven ahead toward the Bitterroot Valley. Forced to abandon the supply wagons, Mullan set some men to work fashioning some twenty sleds for moving the loads. The rest kept on building road, though it quickly became apparent that the year's roadwork would soon have to end, and they were destined to winter somewhere on the St. Regis.

The crew could not set up camp where they were–Mullan knew that a frozen winter at the bottom of that gorge would be unpleasant at best, but the spring freshet would be worse still–not only would the camp be awfully wet in all that snowmelt, but the work would not resume until the high water drained out, which could be as late as June. By mid-November the path was clear to within ten miles of the mouth of the St. Regis at the Clark Fork River (which Mullan identified as the Bitterroot), so the crew continued their work in the frigid weather, constructing sixteen footbridges across the St. Regis River for the sleds to pass over.

On December 4, 1859, the men having taken the road about as far as they could, Mullan selected a cantonment site near present-day De Borgia, Montana. It was well-sheltered by the mountains, with convenient water and good timber for fuel and building material. Mullan dubbed their winter camp Cantonment Jordan. The next day, December 5, the mercury fell to 42 degrees below zero before it froze solid inside the thermometer. On that day, Mullan declared, "It was now evident that there was no possibility of making the road any longer to advantage. The men had worked amid snow and cold till it was out of the question to work any longer."[44]

While work on the road was suspended for the season, it was hardly the end of the men's labors. Some of the men set to erecting huts for the winter: first a storehouse for the food and supplies, then a headquarters office, and finally some cabins for lodging. This task required felling trees and trimming the logs, notching their ends and stacking them in the snow, all done using nothing but broadaxes, whipsaws, and their own muscle. Yet the cabin builders probably counted themselves lucky compared to the men who were still struggling to move the supplies on the sleds from upstream. Without the help of the draft animals, the men moved fifty tons of freight, one six-hundred-pound sledload at a time, the eighteen miles to their winter home.

Later that winter, one man left the rear camp to try some hunting, got lost in the snow, and remained out for four days and nights with no food or blankets. He somehow found his way back to camp, crawling frozen and half deranged on his hands and knees. That last frightened step alone in the woods, before he dropped to his knees, was to be his last step ever–his frostbite was so severe that the surgeon amputated both of his legs.[45]

The animals fared even worse than the men. When the expedition arrived at the cantonment, Mullan had his remaining horses and mules driven to the Bitterroot Valley, the "Land of Eternal Spring," some one hundred miles farther east over icy mountain trails. It was a gamble, but, as Mullan wrote, "there was no point for pasturing short of this, and they must either reach it or perish in the snow."[46] Indeed, the majority did die en route; others made it to the valley but broke through the ice on the river and drowned. Upon reaching their destination, the herders found that the forage for the remaining pack animals was hardly better there than on the St. Regis. As for the cattle, rather than move them, Mullan ordered the remaining beeves slaughtered. The meat would keep naturally in deep-freeze until spring, while there was little hope of it lasting that long on the hoof.

With the log huts erected, the crew settled into its winter routine. The officers and clerks completed maps and compiled field notes, keeping strict hours of 10 a.m. to 3 p.m. and 6 to 9 p.m. The soldiers tended the camp, gathered fuel, and prepared tools for the coming work season. Mullan maintained a perimeter guard "merely to preserve discipline, as the snow was an effectual barrier against Indian depredation or Indian surprise."[47]

While vigilance is always commendable in a military command, the truth is, the expedition had experienced very little hostility from the Indians in 1859. Mullan reported that two of Sohon's reconnaissances, always with Indian guides, resulted in danger to Sohon and obstruction to the work, but available records provide no specifics regarding that danger and obstruction.[48] It is worth noting that these reconnoiters were only two among the many excursions expedition members undertook with Indian escorts, most of which concluded amicably. Mullan paid the Indians for their guide services, as well as for delivering mail throughout the winter.

Mullan's overall disposition toward the Indians appears to have been one of general suspicion, though it varied by tribe. The Flatheads, with whom he first established a liaison in 1853, "always treated myself and my parties with a frank generosity and a continuous friendship," and likewise, Mullan was always genial toward them.[49] For the Coeur d'Alenes, however, having fought against them in 1858, he never had a kind word ("The Coeur d'Alene Indians, though whipped last year by Colonel Wright, are still treacherous and faithless").[50] His suspicions even extended to the Jesuits. He reported at one point that though the priests were always kind to him personally, he believed that trouble with the Coeur d'Alenes, dating all the way back to Colonel Steptoe's defeat in 1858, was at least partly the missionaries' fault, as their influence had "rendered the Indians bold, saucy, and impudent."[51] The Blackfeet, meanwhile, were at war with the Pend d'Oreilles and were "disposed to be troublesome," Mullan wrote.[52] For other tribes he had mixed feelings. During Wright's campaign, Mullan had commanded a detachment of friendly Nez Perces, but he nevertheless approached all dealings with the Nez Perces very cautiously.

Throughout the winter, Mullan left snow gauges at key points along the route, particularly the passes, by notching trees with a scale of feet and inches. As mail carriers and others passed by through the winter and reported their findings, he began to develop a general picture of the snowfall on his chosen line.[53] He did not like what he saw. The upper river valleys maintained a snowpack of two and a half to three feet; in the higher elevations, it was up to five feet. And at the summit of Sohon Pass, the snow accumulated to a depth of nine feet, impassable for man or horse, or even for the iron horse, to which object Mullan still looked forward constantly.

Spokane Garry, a friendly Spokane Indian who carried mail for Mullan, arrived at Cantonment Jordan one winter

Sohon's portrait of Spokane Garry (1811–92), a literate man who figured significantly in the history of Indian affairs in the nineteenth century.
–Courtesy Washington State Historical Society, Tacoma

day with mail from posts west. That he arrived on horseback was a total surprise to Mullan. Garry had not taken the new military road but had followed a route along the Clark Fork River to the north. Mullan immediately sent men to investigate, and they "found this fact to hold: that no Indians had ever been known to cross the mountains in winter, via the Coeur d'Alene route, while it was quite an usual thing for them to do so via the Clark's Fork."[54] It turned out that Sohon's treacherous guide Augustine, for all his hyperbole ("If all the Americans would work here a thousand years they could never make a road") had spoken the truth on that summer reconnaissance–the northern route was better.

We can only imagine the frustration that this revelation must have caused Mullan. Had he not guessed wrong in 1854, he would not have lost most of his oxen, mules, and horses to starvation. Had he not guessed wrong, he would have made quicker progress through November and December. And most important, had he not guessed wrong, he would not be stuck on the St. Regis River for the winter, battling frostbite and boredom; instead, he could be clearing road with healthy men and animals up north—*north!*—on the Clark Fork. In his letters to the War Department from Cantonment Jordan, however, he could not, apparently, bring himself to discuss it. In his final report to Congress, composed after a two-year cooling-off period, he admitted that "All our explorations up to the date of our leaving the field in 1854 had been confined to the spring, summer, and autumn months, and it is a self-evident proposition to those familiar with the winter character of the Rocky Mountains, that it is impossible for a man to express a winter view from a summer stand-point."[55]

Based on his various findings, Mullan developed the theory of an *isochimenal* line, a line of approximately equal mean winter temperatures, running from St. Joseph, Missouri, westward to Fort Laramie and then veering, surprisingly, northwest through the valley of the Clark Fork River to Pend d'Oreille Lake before dropping back down in latitude to Walla Walla. In the romantic tradition, he related the isochimenal line to the "kingdoms of natural history, botany, and climatology" and, more practically, to the immediate national purposes of road, rail, and ranch.[56]

Though the work had stopped for the season and the party was locked in winter quarters, the restless Lieutenant Mullan refused to remain idle. In late February he ventured with a small party to the Bitterroot Valley. While he claimed that the trip was to reconnoiter the work of the coming season and coordinate the delivery of supplies from Fort Benton, he may have had personal reasons to go: it was surely no misery for him to be on the trail again, nor to be able to pass some time in his beloved Bitterroot.

While in the Bitterroot Valley, Mullan also met with members of the Flathead tribe at Fort Owen and later with a Blackfeet chief. Mullan's meeting with the Flatheads, with whom he had established great rapport in 1853, was a smashing success. The lieutenant ventured a sizeable request, asking the Indians for 117 horses with pack saddles and a detachment of warriors to escort Sohon to Fort Benton for supplies. After a brief discussion, the Indians left the fort, promising to return the next day with sticks—one for each man and horse they could spare. The next day they returned as promised, with Ambrose, their chief, bearing 137 sticks. Mullan must have been tremendously grateful, but the compliment he left in the record was a somewhat back-handed one: "Such nobleness of character as is found among some of the Flatheads is seldom seen among Indians," he wrote, adding, "I here record to their credit that I never had a want but which, when made known to them, they supplied, . . ."[57] Sohon and his Flathead guides crossed the divide to Fort Benton, loaded the fresh pack string with eleven thousand rations and other supplies, and before March was out, they had delivered the whole load safely to Cantonment Jordan.[58]

After this, Mullan met with Blackfeet chief Little Dog at Fort Owen to address "the unsettled condition of affairs with the Blackfeet Indians, with whom disturbances have again commenced" due to their war with the Pend d'Oreille.[59] He assured the chief of the friendly nature of his mission and of all the advantages to be had from the Indians leaving his road crews alone. The result was an uneasy truce. Mullan concluded that "The Blackfeet are, I believe, if not meddled with, a good people, and might be made better; but the evil elements are so directed that it is evident that it is only a question of time that they shall be included with other miserable and treacherous tribes."[60]

Also at Fort Owen, to further reinforce his stock of draft animals, Mullan dispatched Ned Williamson to Salt Lake City, where, he had heard, mules were selling for between $15 and $50, relative to the $150 price per head in the mountain settlements.[61] Near the head of the Snake River, Williamson got snowed in, separated from his horses, and was rendered snowblind for several days. In a tremendous display of fortitude and resolve, he fashioned makeshift snowshoes from his saddle rigging and covered the rest of the five hundred miles to Salt Lake on foot. He procured fifty mules at Camp Floyd, as directed, and returned on horseback with the string, arriving only fifty days after he'd left.[62]

This drawing shows the layout of Fort Owen, established by John Owen in 1850. The complex originated as a Jesuit mission founded by Father DeSmet. Owen's post lasted until 1871. –Illustration by Philip Mobley

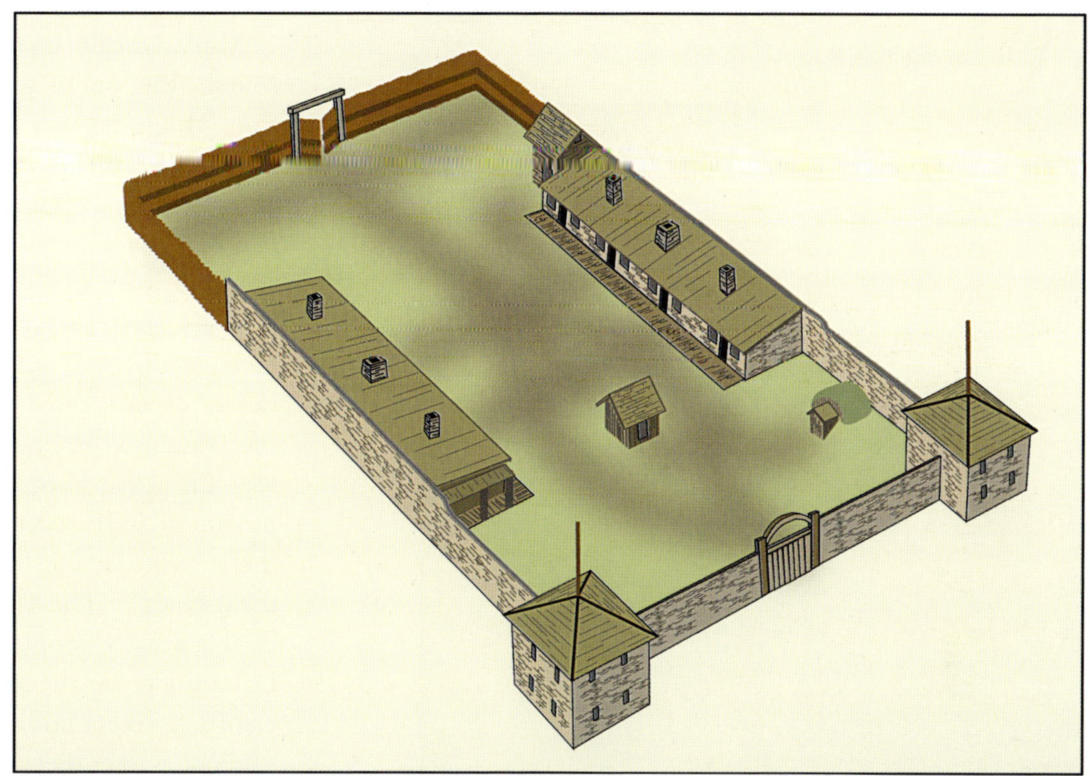

Model of John Owen's office at Fort Owen State Park –Photo by Paul McDermott–

Mullan made one more excursion away from camp in March, this one a grocery run to the Pend d'Oreille Mission. While he was at Fort Owen, scurvy had broken out among the soldiers of his expedition's military escort. Upon investigation, it seemed that the civilian work force was unaffected. The rations of the military and civilian personnel were the same with one exception. Seeing early that the civilian laborers generally worked harder and longer than the soldiers, Mullan had authorized the civilian work force a double ration of desiccated vegetables, which continued through the winter, while the soldiers were fed in accordance with the official army regulations of 1857. At the mission, Mullan procured "fresh succulent vegetables" from the Jesuit fathers there "who let me have them at the expense of depriving themselves."[63] The vitamin-rich vegetables rapidly took care of the scurvy problem, and Mullan reported to the War Department on the value of including more desiccated vegetables (a recent addition to the army's official ration lists) in soldiers' winter diets as an antiscorbutic.[64]

The travails of the winter of 1859–60 only hardened Mullan's resolve. After detailing the severity of the winter, the loss of the animals, the frequent cases of frostbite, and the

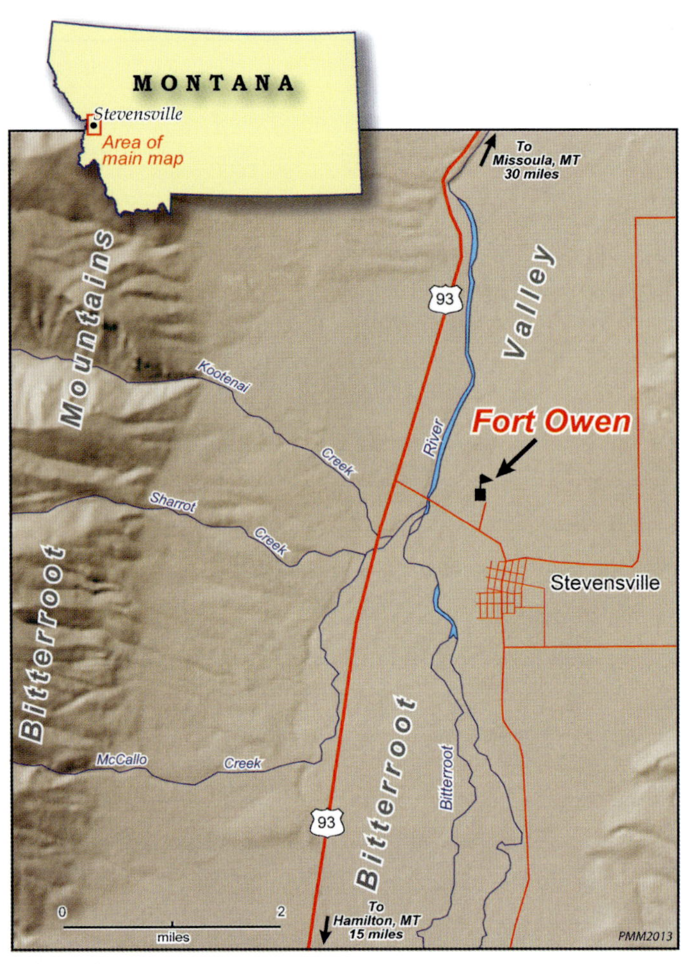

Location of Fort Owen –Map by Philip Mobley

Today, a sign marks the site of Fort Owen –Photo by Philip Mobley

wearing out of all their boots, Mullan wrote from Cantonment Jordan in January that "the department may rely on one thing, as I was sent out to build and construct the road, it *shall* be constructed."[65] He even retained a touch of optimism and good humor: "[T]hough our condition is not such as I could well wish it, still it could be worse."[66] By the time he reported from Fort Owen in the early spring of 1860, however, he had assumed a more realistic attitude than had animated his confident reports of the previous autumn: "Since passing over the Bitter Root [Clark Fork] river section of the line I can see that seven years ago we *much underestimated* the work, and even now I shall not give even probable estimates of progress."[67]

---

1 John Mullan, *Report on the Construction of a Military Road from Fort Walla-Walla to Fort Benton* (1863; reprint, Fairfield, WA: Ye Galleon Press, 1994), p. 3; this source is hereafter cited as Mullan, *Report*, 1863. Note: Research for this paper was funded in part by the Omar Nelson Bradley Fellowship.

2 Helen Addison Howard, "Captain John Mullan," *Washington Historical Quarterly* 25, no. 3 (July 1943), p. 187.

3 Mullan, *Report*, 1863, p. 7.

4 Ibid., p. 6.

5 Ibid., p. 7.

6 Ibid., p. 9.

7 Ibid., p. 19.

8 Mullan to Capt. A. A. Humphreys, July 31, 1859, in John Mullan, P. M. Engle, et al, *Military Road from Fort Benton to Fort Walla Walla: Letter from the Secretary of War . . .* (1861; reprint, Ann Arbor: Michigan Historical Reprint Series, Scholarly Publishing Office, University of Michigan Library, 2005), p. 6; this source is hereafter cited as Mullan, *Report*, 1861.

9 Mullan, *Report*, 1863, p. 15.

10 Ibid.

11 Gustavus Sohon to Mullan, February 15, 1860, in Mullan, *Report*, 1861, p. 129.

12 Ibid., p. 130.

13 Ibid., p. 132.

14 Ibid., p. 130.

15 Ibid., p. 134.

16 Ibid.

17 Ibid.

18 Ibid., p. 135.

19 Ibid.

20 Ibid.

21 Mullan to Humphreys, July 31, 1859, in ibid., p. 8.

22 Ibid.

23 Mullan, *Report*, 1863, p. 17.

24 Ibid.

25 Mullan to Humphreys, August 16, 1859, in ibid., p. 11.

26 Mullan to Humphreys, September 4, 1859, in ibid., p. 15.

27 Mullan to Humphreys, August 16, 1859, p. 14.

28 Mullan, *Report*, 1863, p. 18.

29 Mullan to Humphreys, September 19, 1859, in Mullan, *Report*, 1861, p. 20.

30 Mullan, *Report*, 1863, p. 20.

31 Mullan to Humphreys, September 19, 1859, in Mullan, *Report*, 1861, p. 23.

32 Ibid., p. 22.

33 Ibid., p. 21.

34 Ibid., p. 24.

35 Walter W. Johnson to Mullan, January 2, 1860, in ibid., p. 93.

36 Ibid.

37 Mullan to Humphreys, August 16, 1859, in ibid., p. 13.

38 Helen Addison Howard, *Northwest Trail Blazers* (Caldwell, ID: Caxton Printers, 1963), p. 146.

39 Mullan to Humphreys, October 4, 1859, in Mullan, *Report*, 1861, p. 23.

40 Mullan to Humphreys, October 26, 1859, in ibid., p. 27.

41 Ibid.

42 Ibid.

43 Mullan to Humphreys, November 14, 1859, in ibid., p. 27.

44 Mullan to Humphreys, January 3, 1860, in ibid., p. 28.

45 Ibid., p. 31.

46 Mullan, *Report*, 1863, p. 18.

47 Ibid., p. 19.

48 Mullan to Humphreys, July 3, 1859, in Mullan, *Report*, 1861, p. 2; Mullan to Humphreys, July 31, 1859, in ibid., p. 10. The latter letter refers to two enclosures, neither of which this author has been able to locate, that apparently describe Sohon's misadventures: "But under the most favorable circumstances there are always to be found some *mauvaie sujets* among them. This has been evinced in the recent acts of depredations of the Pelouse, which the enclosed communication to General Harney, headquarters, and marked No. 1, will give you full details. The enclosed paper, marked No. 2, will also show you how guarded we have to be regarding the Nez Percés and their movements." It seems reasonable to assume that the first enclosure refers to Sohon's reconnaissance tale, related above, since Sohon rejoined Mullan on July 7, and that the other, relegated to an enclosure without further comment, could not have been substantial and was probably limited to petty thievery.

49 Mullan, *Report*, 1863, p. 21.

50 Mullan to Humphreys, July 31, 1859, in Mullan, *Report*, 1861, p. 10.

51 Mullan to Humphreys, September 4, 1859, in ibid., p. 19.

52 Mullan to Humphreys, March 11, 1860, in ibid., p. 39.

53 Reports in Mullan, *Report*, 1861: P. E. Toohill to Mullan, January 3, 1860, p. 98; Toohill to Mullan, February 20, 1860, p. 136; and Toohill, "Report of expressman carrying mail across Bitter Root mountains for military road expedition in April and May, 1860," p. 152.

54 Mullan, *Report*, 1863, p. 19.

55 Ibid., p. 6.

56 Ibid., p. 20.

57 Ibid., p. 21.

58 Ibid.

59 Mullan to Humphreys, March 11, 1860, Mullan, *Report*, 1861, p. 39.

60 Ibid.

61 Mullan to Humphreys, March 10, 1860, in ibid., p. 37.

62 Mullan, *Report*, 1863, pp. 21–22.

63 Mullan to Humphreys, April 3, 1860, in Mullan, *Report*, 1861, p. 40.

64 Ibid.; also Mullan, *Report*, 1863, p. 22.

65 Mullan to Humphreys, January 3, 1860, in Mullan, *Report*, 1861, p. 30.

66 Ibid., p. 35.

67 Mullan to Humphreys, March 10, 1860, in ibid., p. 39.

John Mullan on the Hell Gate River

BY KIM BRIGGEMAN

John Mullan's first sweep of construction on his military wagon road through the intermountain West ended in early October of 1860, with a return to his starting point at Fort Walla Walla. He'd been in the field for fifteen months and had built sections of road that were already coming in handy to emigrants. But Mullan knew he didn't have a finished product. He had already sent one proposal to Washington, D.C., seeking to repair and improve the road replicating his original west-to-east route, but after spending more time studying the economics of the situation, Mullan became convinced that the second expedition should work east to west. Specifically, he'd leave from Fort Laramie, on the Oregon Trail, in early 1861 and build a road to meet up with the original one in the Deer Lodge Valley, east of present-day Missoula, Montana. He would spend most of 1862 on the west end of the road and reach Walla Walla by the end of that year. "This would also enable me to test the value of the Laramie and Deer Lodge route, which had always been with me a favorite measure, give me two long working seasons, and subserve all the requisites of economy and efficiency," he reasoned in his official report of 1863.[1] Ultimately, however, Mullan's plans would be readjusted.

In late 1860, nine years before the first transcontinental railroad was completed, Mullan jumped on a stagecoach in San Francisco and rode, in bone-jarring fashion, to St. Louis, then on to Washington to make his pitch. He found matters in the capital in a "somewhat chaotic state" as Abraham Lincoln prepared to take over the presidency in early March, two weeks after the Confederacy inaugurated Jefferson Davis as its president.[2] Interestingly, it was Davis's idea to build a military road linking the Columbia and Missouri Rivers in the first place.[3] Upon arriving in Washington, Mullan learned that his original proposal to improve the road from west to east had already been endorsed, complete with a congressional allocation of $100,000. One can almost imagine the nation's leaders impatiently shooing the nettlesome Mullan back West and out of their hair.

"Seeing so many difficulties interpose, I was forced to relinquish my Laramie scheme and proceed back to Walla-Walla, reaching there early in April," Mullan wrote in his final report.[4] Within days of his return to the western outpost, shots were fired at Fort Sumter, plunging the nation into its bloody, four-year civil war. Mullan, meanwhile, geared up for a May 13, 1861, departure to the mountains.

As critical as Mullan's military wagon road was to the early development of the Northern Rockies, and indeed to the political machinations of western settlement and territory formation, there's no denying that his second sweep in 1861 and 1862 made barely a blip on the national radar. Yet Lieutenant Mullan and his men deserved better. They engineered and completed major construction projects in a river corridor that became the main transportation route in Washington, Idaho, and western Montana. Mullan himself was in the field for nearly thirty months on his two road-building expeditions. Fully one fifth of that time, from November 1861 to May 1862, he spent headquartered at Cantonment Wright, at the mouth of the Big Blackfoot River (where the town of Milltown, Montana, stands today), the longest sustained campsite of the entire project.

It was a damned cold winter.

Soldiers and civilian workers alike shivered through perhaps the most severe January the region had ever known,

The Mullan Road, Hell Gate region –Map by Philip Mobley

followed by a miserable spring of freeze-and-thaw cycles. But mere survival was not their legacy. From November of 1861 to May of 1862, Mullan's men built the first bridge across the Blackfoot River, sinking log-crib piers through holes in the ice. They hewed side cuts in frozen mountainsides to improve the first thirty miles of road along the Hell Gate (now Clark Fork) River above Missoula.

By the time Mullan arrived at Cantonment Wright in the autumn of 1861, minor gold strikes had already drawn a handful of whites to the mountain valleys and canyons of eastern Washington Territory, a region that would later become southwestern Montana. Not until a couple of months after Mullan bid adieu to Cantonment Wright in May 1862 did John White discover gold on Grasshopper Creek, in the Beaverhead River country, setting off a population boom that within two years would fast-track new territory divisions, bouncing the region's territorial jurisdiction first to Idaho and ultimately to Montana on May 26, 1864.

Even before the gold rush, this part of Montana was not an unpeopled wilderness. It was the home or travel corridor of a number of western tribes, as it had been for generations. Moreover, Fort Owen, in the Bitterroot Valley near today's Stevensville, had been around since 1850, when Major John Owen took over St. Mary's Mission and turned it into a trading post. The fort was a decent day's horseback ride from Cantonment Wright. The Jesuits who had abandoned it to John Owen had reestablished a mission at St. Ignatius, a slightly longer trot to the north.

Coming hard on the heels of Mullan's first road crew in 1860 were enterprising traders Francis Worden and Christopher Higgins. Along with their clerk, Frank Woody, Worden and Higgins set up businesses at Hell Gate Village, the precursor of Missoula. From Cantonment Wright, Hell Gate was a quick eleven-mile jog through the mouth of Hell Gate Canyon and over Rattlesnake Creek to Grant Creek. Across the mountains, at the eastern end of the military road, was Fort Benton, the uppermost steamboat port on the Missouri River. To America's fortune seekers and dreamers of the West, Fort Benton was a portal to riches, stamped and certified as U.S.-approved by Mullan's arrival there two years before.

While Mullan was working in what is now Montana, the war was in full swing back East. In March 1862, the *Monitor* and *Merrimac* battled off the coast of Virginia. A month later, more than 100,000 soldiers fought and 24,000 of them died at the horrendously bloody Battle of Shiloh. By early May, David Farragut had won control of New Orleans for the Yanks, and by early June Stonewall Jackson had pushed Union troops in the Shenandoah Valley back across the Potomac.

At Cantonment Wright, the snow began to fall in November and stayed until April. In between was the brutal January cold snap. "The Indians had never before experienced so severe a winter and the poor creatures came in for their share of suffering and loss of stock," Mullan wrote in his 1863 report. He continued:

> This range of cold extended as far east as Fort Benton, the furthest point from which we obtained any data and as far west as the coast. The losses sustained during the winter, together with losses from the heavy freshet that had preceded it, will long be remembered throughout the length of the Pacific coast.[5]

Mullan noted that all travel in the mountains was suspended for a short period in January.

In the fall of 1861, Mullan had established not only his central headquarters at Cantonment Wright, but also four winter camps upriver–Camps Williamson, Campbell, Clark, and Lannon–from which his men worked, despite the cold, on seven miles of side cuts. Mullan denoted these camps on the map that accompanied his official report. Due to significant changes in the landscape over time, the exact location of only the lowest camp, Camp Williamson, has been definitively established. It was on a spring-fed creek between Turah and Kendall Creeks, three miles above Cantonment Wright.[6]

Sohon created the image for this lithograph of Mullan's base camp, Cantonment Wright, in the winter of 1861–62. The cantonment was established at the confluence of the Big Blackfoot and Hell Gate (Clark Fork) Rivers. –From John Mullan, *Report on the Construction of a Military Road from Fort Walla-Walla to Fort Benton* (1863)

A studied guess places the other three, on a modern map, as follows: Camp Campbell, near Starvation Creek, west of Rock Creek and a few miles southeast of Clinton; Camp Clark, near the mouth of Rock Creek; and Camp Lannon, a few miles east of Beavertail Hill. Camp Williamson was supervised by David Williamson, Mullan's top civilian hand, who had been in charge of the advance working party out of Cantonment Jordan in the spring of 1860. Williamson was the brother of Mullan's fiancée, Rebecca Williamson, and therefore Mullan's future brother-in-law. Mullan and Rebecca were married on April 28, 1863.[7]

On Mullan's first pass through these parts in the summer of 1860, he had followed the established trail up the Hell Gate River, which required crossing that stream time and again. To eliminate ten fords, Mullan directed his men to build the road on the right (north) side of the river, finally crossing to the left near the foot of Medicine Tree Hill.[8]

Modern motorists roaring past Medicine Tree Hill on Interstate 90 no doubt find it difficult to imagine its place in history. John Mullan referred to it as "the four-crossing spur," probably because the mile-long toe of ridge has a broad saddle and more than one natural pass.[9] A landmark well into the twentieth century, Medicine Tree Hill is now on private

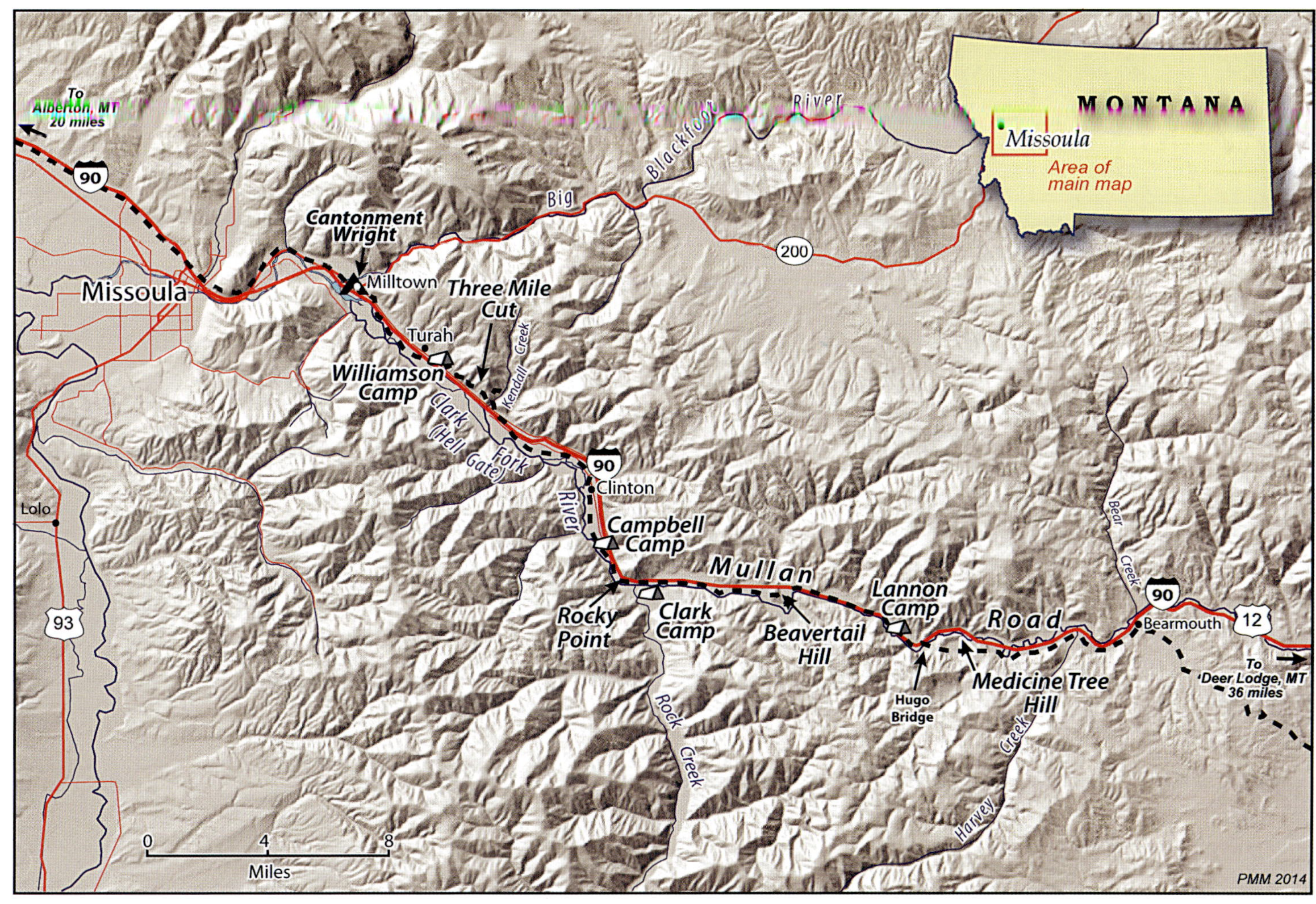

Location of Mullan's work camps along the Hell Gate River. –Map by Philip Mobley

property, conduit to two natural-gas pipelines and a pair of high-voltage power lines, and the tree itself is long gone. Its legends, however, live on.

One story, promoted by Judge Caleb Irvine, an early pioneer, has it that many years ago, a lone Kootenai Indian was being pursued by an enemy band. After running for miles, the young man stopped to rest at the top of Medicine Tree Hill. He hung his medicine talisman on a branch and fell asleep, only to be awakened by a shower of arrows. Not one found its mark, yet when the Kootenai fired back, his aim was always true. As his quiver emptied, it was magically replenished with arrows. The lone warrior had slain a number of his foes before one of them, realizing the power of the medicine bundle, crept up and nabbed it from the tree. Immediately afterward, the Kootenai was killed.

"This legend is still current among the Indians of the western slope," summarized an account of Irvine's story that appeared in *Contributions to the Montana Historical Society* in 1907. "[They]

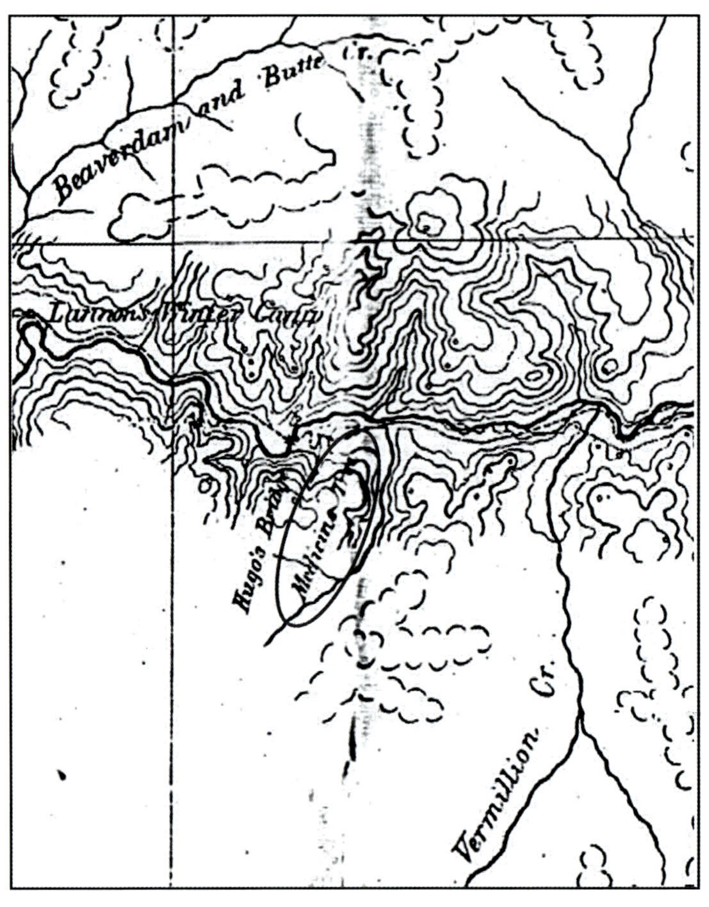

Detail from "Map of the Mountain Section of the Fort Walla Walla and Fort Benton Military Wagon Road from Coeur d'Alene Lake to the Dearborn River" by Theodore Kolecki, 1863, showing the Medicine Tree Hill area –From John Mullan, *Report on the Construction of a Military Road from Fort Walla-Walla to Fort Benton* (1863)

Location of Medicine Tree Hill, a major spur intersecting the Hell Gate River Valley. Such landforms posed obstacles to railroad construction; these were often overcome by building tunnels. –Map by Philip Mobley

never pass the tree without hanging some article from their personal effects upon one of the limbs as a token of awe from their superstitious natures and to keep green the memory of the medicine-wrought tragedy enacted beneath its shade and to the present day the eminence is known as 'Medicine Tree Hill.'"[10]

Irvine, who died in Anaconda in February 1891, had related the legend of Medicine Tree Hill to an interviewer two years before his death, saying he got his information from "several different Indians, all of whom agreed on the material features."[11]

In an article he wrote for the *Missoulian* in 1922, Will Cave, another pioneer and chronicler of western Montana history, called Irvine's story poppycock.[12] Cave, who as a boy of fourteen fought the Nez Perce Indians in the 1877 Battle of the Big Hole, maintained that the white man erroneously bestowed the term "Medicine Tree" on the pine in order to explain the bows, arrows, moccasins, skins, beads, and tobacco that Indians hung on its branches. He was surely referring to Irvine's story when he wrote "I have read another legend concerning the medicine tree, but as it is evidently the product of the vivid imagination of some other than an Indian I will not reproduce it here."[13]

Cave said the real legend was not so creative. His version involved a warrior who was traveling the trail one day and "unaccountably conceived the idea of adopting the tree on top of the hill as his object of personal reverence and worship." The Indian made a mark on the tree, then from a long distance shot an arrow "which with somewhat unexpected accuracy pierced its center." Accepting this as a sign, he adopted the tree as what Cave called "a sponsor for his existence."[14]

This warrior went on to become a great leader among his people, and his tribesmen also began to hang a token on the tree whenever they passed. After the warrior's death, his family and friends continued the tradition, "not however, in the belief that they were making conciliatory offerings to any supernatural being," Cave wrote.[15] They hung the items, he maintained, more as a memorial to the man and his spirit, similar to the decorating of graves on Memorial Day.

Cave even had a personal story about the Medicine Tree, noting that the "old Mullan road" was constructed over the same hill. "Its route lay several hundred feet from trail and tree. Traveling first the road in 1865, and having heard something of the legend, through curiosity, my mother walked the Indian trail to see the tree. I am told my eyes, but two years aged then, also beheld the famous pine. It must be true, yet somehow my memory of the incident appears to play me false," he wrote.[16]

Around 1870, the Medicine Tree was blown over by a fierce gale of wind, rendering it "scarce worthy of further veneration. . . . Even a Medicine Tree may not withstand forever the ravages of time and storm," Cave concluded.[17]

Who's to say, after all these years, why Charlie Schafft started on his fateful walk up the Hell Gate River from Cantonment Wright that January? Various sources, including Schafft's own memoirs, describe the incident but offer little insight. Regardless of his motives for going, after his ordeal in the snow, Charles Schafft was destined to play a remarkably varied role in Montana territorial history. Through his adroit writings he became one of the early champions of John Mullan and especially of the Indians of western Montana. He did most of this without the use of his legs, which, as we'll see, he lost at age twenty-three during the severe winter of 1861–62.

In 1976, *Montana: The Magazine of Western History* published Schafft's memoir, written eighty-nine years earlier, along with his handwritten enlistment record from 1853. His army record described him as blue-eyed, sandy-haired, and five feet, three inches tall.[18] But through a child's eyes, Emma Minesinger Magee had her own impressions of Schafft, who tutored her and her siblings near Missoula in the 1870s. "Tall, blonde, with blue eyes twinkling, he was always cheerful. I think he loved children," Magee wrote in a memoir published as *Montana Memories: The Life of Emma Magee in the Rocky Mountain West, 1866–1950*. She remembered Schafft joking about his prosthetic legs: "On cold days, he never wearied of our thoughtless question, 'Charles, are your feet cold?' If we forgot to ask, he humorously reminded us by stamping his wooden stumps and telling us his feet were frozen."[19]

In January 1862 Schafft was a civilian employee at Cantonment Wright, working as a sutler's clerk with Mullan's main outfit. If it was cold throughout the West, it was absolutely frigid at the cantonment. "The camp was situated upon the high flat in the forks of the Blackfoot and Hell's Gate rivers, where timber was abundant and close; but exposed to the bleak winds that at times came down the valley of the Blackfoot, it was found an abode of not over much comfort," Mullan wrote in his 1863 report.[20]

Perhaps it was due to an argument with another sutler's clerk that prompted Schafft to walk off that day. Schafft's obituary in the *Missoula Gazette* nearly three decades later suggested as much. "During the winter he had a disagreement with the clerk in charge during (William) Terry's absence, and left the encampment," read the front-page story on March 19, 1891.[21]

Or was Schafft on a long-distance mail run to Utah, as one of his rescuers, David O'Keefe, suggested in 1921, not long before he died? "During the winter the snow got so deep and the weather so cold that Mullan offered $500 to the man who would take the mail to Salt Lake," O'Keefe recalled. "A fellow by the name of Charles Shaft [sic] offered to go, but as he did not have a horse he had to go to Deer Lodge to get one."[22]

In his own memoir, written in 1887, Schafft kept the details vague: "On the 8th day of January 1862 I started, for a permanent stay, to Deer Lodge–alone and afoot," he wrote. A few "official" letters had been entrusted to him, he acknowledged, but he wasn't being paid to deliver them.[23] As a civilian, he couldn't be classified as a deserter. Indeed, civilian workers on the Mullan Road seemed to come and go on a regular basis, especially during the winter.

Mullan gave his own account of Schafft's experience in his 1863 report, but his description provided fewer details than Schafft's memoirs did. He wrote of an unnamed "citizen" who arrived at the cantonment in mid-January in great distress:

> He had left one of the camps with the intention of going to the Deer Lodge Valley. Night and severe cold overtaking him before he could reach another camp, he halted to build a fire, and being wet endeavored to slip off his moccasins, when he found them frozen to his feet. He became alarmed, and retracing his steps reached the point he had started from, late at night, but with both feet frozen, and on their being thawed in a tub of water all the flesh fell off. The poor fellow suffered intensely, and his life was only saved by his suffering the amputation of both legs above the knees; the operation was performed by Dr. George Hammond, United States army. A purse of several hundred dollars was raised for him, and he was left to the kind charity of the fathers of the Pend d'Oreille [St. Ignatius] mission, where he remained up to the date of our leaving the mountains."[24]

As previously noted, Mullan's brief account did not mention Schafft's name, nor that he was an employee of Mullan's own expedition. The lieutenant also erred in a number of details. For example, Dr. Hammond, physician for the expedition, amputated Schafft's legs below the knees, not above. In his memoir and elsewhere, Schafft expressed his gratitude for the treatment he received from Mullan and his men, and from Father Urbanus Grassi at the mission.[25] Schafft recalled that when he had recovered enough to get out of bed, some seven months after the amputations, "it did not take very long to learn how to get around on my knees." Thus, he was able to be of "some slight service" to the priests.[26]

At American Fork, sixty miles upriver from Cantonment Wright, brothers Granville and James Stuart had their own camp in 1861–62. The Stuarts' journals reflect a river valley that fairly bustled with Indians passing through and white men grasping for fortune and survival. Granville wrote that on January 1, 1862, "everybody" went to a ball thrown by Johnny Grant at his ranch at the mouth of the Little Blackfoot River, eight miles above American Fork and ten to fifteen miles below settlements in the upper Deer Lodge Valley. A blizzard and temperatures of forty degrees below zero kept them there for two days and two nights. The Grants graciously provided their guests with meals and buffalo robes to sleep on.[27]

The Stuarts were able to save their horses and cattle that winter while many other animals, including most of those of the Mullan expedition, expired. Recalling a tip they'd received from a French mountaineer on the Green River three years earlier, the brothers herded their stock into a sheltered canyon on the north side of the river and kept them from drinking water that would freeze their insides. The result was a surprisingly fat bunch of animals when the chinooks blew through in the spring.

Travel, which had been suspended for several weeks, recommenced in late January 1862, though snows still prevented full movement for at least another month. "No news from the states," James Stuart wrote in late February. "I suppose Bachelder's Express from Walla Walla to Hell Gate and Cantonment Wright is snowed under." He referred, perhaps derisively, to a short-lived mail service instituted by Albert Batchelder of Hell Gate, a civilian who came to the region with Mullan on the 1859–60 expedition.[28]

On February 4, James Stuart noted that a party of Flatheads passed through with horses they'd stolen from the Bannocks in the Beaverhead Valley. That evening, some Bannocks followed in pursuit. The following morning, the Bannocks returned with their horses and two Flathead scalps. A few weeks later, two Flatheads, Little Aeneas and Narcisses, found a Snake Indian camped at Cottonwood Creek and killed him, taking "his lodge and camp fixtures and one of his wives." James Stuart and John Powell, the latter for whom Mount Powell and Powell County would be named, ransomed the wife from the Flatheads, "thinking it just as well not to allow Narcisses to take her down to the Flathead country," Stuart wrote. "It is usual when they take a captive to turn her over to the women of their tribe and they promptly make a slave of her; imposing all of the drudgery of the camp on her, and making life anything but a bed of roses." He described the woman to be "fair with red cheeks and brown hair and eyes and is evidently half white."[29]

Three days later, James Stuart found himself a married man. "Marrying is rapidly becoming an epidemic in our little village," noted his brother Granville, who himself would be wed to a young Snake woman, Aubony, by early May.[30] While James's wedded bliss didn't last long, the union of Granville and Aubony produced nine children and ended only with her death twenty-five years later.

A warm chinook struck the territory in late March. The floods and snow slides that resulted again made travel problematic. Then came more snow and another freeze, making the ground firmer and easier to travel on, and the valley fairly swarmed with activity. Three men from Cottonwood (which would become the town of Deer Lodge) passed American Fork on their way to the Bitterroot River. One was John Jacobs, who the following year would team with John Bozeman to blaze the Bozeman Trail from Wyoming to the gold camps of southwestern Montana. Bozeman himself would make his first appearance in Montana, via Pikes Peak, in June, when he came to the Deer Lodge Valley with a group of prospectors. He was attracted by a description, sent by the Stuart brothers to another brother, Thomas, in Colorado, of encouraging mining prospects in the area.[31]

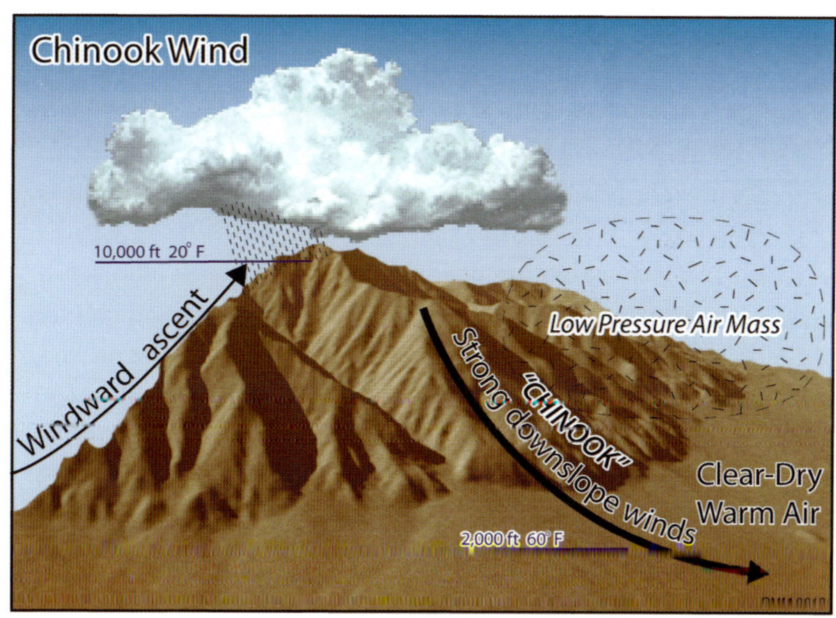

Chinooks are warm winds that heat the air as they descend mountain slopes; at the same time, the air upslope becomes cooler. –Illustration by Philip Mobley

On March 31, 1862, a band of Nez Perces passed by the Stuart brothers' camp at American Fork (later called Gold Creek). Over the next week, bands of Flatheads, Spokanes, and Pend d'Oreilles were seen returning from winter buffalo hunts east of the mountains. Chief Victor and his Bitterroot Salish were passing through American Fork on April 7, and the chief stopped to talk. He warned the Stuarts that the Pend d'Oreilles would cause trouble when they found out that they and other whites had been marrying Snake and Bannock women over the winter. Victor indicated that he would also be visiting Johnny Grant's nearby ranch to caution him to keep the relatives of his Indian wives out of sight or face the consequences. James Stuart dismissed the warning:

> As Deer Lodge Valley and the valleys of the Big Hole, Beaverhead, and Jefferson have been, from times immemorial, a neutral ground for the Snakes, Bannocks, Nez Perces, Pend d'Oreilles, Flatheads, Spokanes, Coeur d'Alenes and Kootenais, it looks like the old chief is too arbitrary in insisting that the Snakes and Bannocks should be forbidden to spend the winter and hunt there the same as all others, because of little squabbles common to all tribes.[32]

In fact, no serious problems came out of the issue Victor raised.

On April 10, James Stuart reported that some men from Fort Benton had camped on the other side of the river. "They have a pack train loaded with supplies and clothing for Lieutenant Mullan's cantonment at the mouth of Big Blackfoot river." While Mullan's men began working in earnest on the mountain cuts downstream, the Stuarts' little settlement began taking on "the lively bustling appearance of a new placer camp," wrote James.[33] (Recall that the landscape-changing gold strike on Grasshopper Creek was still a few months off.)

On April 30, Fred Burr traveled from the Stuart camp down to Cantonment Wright to deliver two horses he'd sold to Mullan for three hundred dollars. He then went on into Hell Gate, returning a few days later with a cache of outdated newspapers. The same day Burr got back to American Fork, Johnny Grant sent a messenger down to that camp to notify the settlers that the dreaded Blackfeet were headed their way. Luckily, the Indians never came.

In the late spring of 1862, around the time that Mullan and his men were putting the finishing touches on their road improvements, Christopher Higgins and some partners found good gold prospects five miles above American Fork. Meanwhile, ten Bannocks stole twenty-five horses from Grant. They were tracked down, but the trackers were bluffed out of returning with the mounts. "They should have gotten those horses," a disgusted Granville Stuart wrote.[34]

On May 20, the Stuarts went up the gulch to a working mine and "sponged" some food from a couple of friendly miners. "They are a jovial set of miners and we had much fun at one another's expense in the way of jokes," wrote Granville.[35] The following day, two days before the Mullan expedition abandoned Cantonment Wright, the Stuarts and their small knot of partners washed out twelve dollars in gold from sluices on Pioneer Creek, a tributary of Gold Creek. Most of it was in the form of a single nugget–the first gold nugget found in Montana.

As ungodly cold as it got during the winter of 1861–62, John Mullan knew what he was doing at Cantonment Wright. It was the third winter he'd spent in the Rocky Mountains. In late 1853 he'd established Cantonment Stevens in the Bitterroot Valley south of Fort Owen. Mullan used Cantonment Stevens as his base for most of a year while, under the direction of Isaac Stevens, he explored potential railroad routes through the mountains.[36]

In late 1859 Mullan was back in Montana, this time supervising the construction of the military road. He and his crew wintered at Cantonment Jordan, near present-day De Borgia, on the St. Regis River. That winter was nearly as cold and severe as the one two years later would be. One of Mullan's men, expressman Ned Williamson, made a potentially disastrous journey from Cantonment Jordan at that time. While caught in the mountains by deep snow, Williamson lost his horses. Crafting a pair of snowshoes from his saddle rigging, he walked some five hundred miles to Camp Floyd, near Salt Lake City. He was back at the Montana cantonment within fifty days.[37] Williamson later served as civilian courier for Colonel John Gibbon, commander of Montana troops during the Sioux War of 1876.[38]

Mullan's crew spent much of the spring of 1860 blasting and digging through the cliffs and rocks of the Clark Fork River between present-day Superior and Alberton. On June 23, the road builders emerged into the Hell Gate Ronde (today's Missoula Valley), a prairie basin twenty miles long that ended at the narrow mouth of Hell Gate Canyon. At this point Mullan deemed their heavy work finished.

On July 2, Mullan's men floated their supplies across the high-running Big Blackfoot on a wagon-boat and a small bateau, once dumping a wagon and retrieving it with some difficulty. That same day, Mullan received word of the arrival at Fort Benton of Major George Blake and three hundred dragoons. They'd come upriver on the first steamboats to the fort and were now waiting for Mullan to escort them to Walla Walla on the new military road. Mullan, who had promised to be at Fort Benton by August 1, was running out of time. In messages he had sent to the Secretary of War and others the previous winter, Mullan had requested such a test run for his road. Now the troops had arrived, and their commander was waiting impatiently for his rendezvous with Mullan.

Speeding up their pace, Mullan and his team would cover the final 250 miles of the road, from the Big Blackfoot River to Fort Benton, in less than four weeks. On July 3, the men moved five miles up the Hell Gate River to a point just east of present-day Turah, and they spent the Fourth of July ferrying wagons and livestock across it.[39]

From Hell Gate Canyon, the expedition crossed the river five more times in the first twenty-three miles, "the valley being narrow (quarter of a mile in width) and the river tortuous in its course," Mullan explained in his 1861 report.[40] At the fourth crossing, a recalcitrant oxen turned downstream and drowned, costing the work crew an extra three hours. Another two days were spent grading a road over Beavertail Hill to eliminate a couple of impracticable river crossings. But for the most part, the wagons and oxen plunged through the river with little trouble, making eleven crossings in thirty miles before crossing Medicine Tree Hill and entering the broader Deer Lodge Valley.[41]

Looking ahead, Mullan realized that this number of river crossings wouldn't do. At high water or in snow and ice, wagons laden with artillery and other military equipment would be unable to make it through. "These fords can be avoided by side hill excavation, part in rock," Mullan noted.[42] He figured it would take fifty men one month to make the side cuts and bridges needed to complete an all-season route. Of course, he did not have the time or resources to do such work that year, so he incorporated it into his plan for the final phase of road construction, which would be done in 1861–62.

Mullan arrived at Fort Benton, as promised, on August 1. There his company was disbanded and his men turned over to Blake's command. A few days later, Blake's troops set out along the new road to Fort Walla Walla, with Mullan and a small crew traveling a couple of days ahead to make minor improvements to the road as they went. Upon reaching Walla Walla, Mullan prepared to return to Washington, D.C., where he would spend the winter.

Mullan returned West in the spring of 1861. For the next several months, he and his men worked on road improvements, changing the route between Walla Walla and the Coeur d'Alene Mission (rerouting via Spokane) as well as building and rebuilding bridges before establishing his winter camp on the Big Blackfoot River.

It was well after dark on a snowy Wednesday, November 20, 1861, that Mullan and his crew, after a busy day, reached their newly built winter camp. "Men engaged in the morning in grading up the slope to gain a high elevation where the road had not been located. . . . Issued supplies for the next ten days," Mullan wrote in his field journal. "Moved forward to the Cantonment at the Big Blackfoot, which we didn't reach until late at night. Found that Lieut. [Salem] Marsh had made most excellent progress in his work and had his quarters nearly completed."[43]

Marsh, who was in charge of Mullan's military escort, had been sent ahead from the previous river crossing at St. Regis, some seventy miles downstream, to build wooden huts at Cantonment Wright near today's Milltown, Montana, a few miles east of Missoula. The camp was named for Brigadier General George Wright, who had registered a decisive victory at the Battle of Four Lakes near Spokane, Washington, during the Yakima War of 1858. Wright had been promoted to commander of the Department of the Pacific in October 1861, and as such was in charge of a record 6,000 soldiers when the Mullan expedition set up the cantonment in his name.[44] No trace of Cantonment Wright remains at its original location, but another post named for the general was built in Spokane in the late 1890s. Today, the Fort George Wright Historic District is listed on the National Register of Historic Places.

Once the crew had settled in at the cantonment, Mullan divided his workforce into five sections. One group stayed at Cantonment Wright to build a bridge over the Big Blackfoot River. The other four were scattered among the winter camps upriver, assigned to make side cuts where the Hell Gate River curled close to steep embankments. "My programme of work from this point forward was to avoid all the crossings of the Hell's Gate except one, which would be bridged, and to put the road thence to Fort Benton in such repair as time and means warranted," Mullan wrote.[45]

Snow had already fallen by the time Mullan arrived at the cantonment, and ice floated on the river. As soon as the camp shelters were finished, construction of the bridge began. Timbers were cut from nearby groves, hauled to the side of the river, and hewn. A log barrier was laid across the river, which was six feet deep at that point, to contain a section of the ice-laden waters. In one night, the dammed river froze solid enough for horses to cross.

The men cut holes in the ice through which they would submerge large, square log frames, or piers. Rocks, moved by hand-sleds across the ice, were placed inside the piers to sink them to the river bottom, where they were leveled and anchored with more rocks. Meanwhile, whip sawyers produced seventeen-foot-long planks for the deck. By March 1 the workers were finished with the four-span, 235-foot-long bridge, whereupon Mullan dispatched them upstream to help on the side cuts.[46]

The hard winter of 1861–62 was followed by a spring of high waters and freshets (floods), the bane of the Mullan Road. These soggy conditions prevented Mullan, who was preparing to leave the mountains and head to Fort Benton in late May, from building the needed bridge across the Hell Gate River. Thus he left the construction of the bridge, which would span the river at Medicine Tree Hill, to a local miner and farmer named Sam Hugo—"old man Hugo," as James Stuart referred to him in his journal. Hugo is thought to have been the first private contractor to work on a federal government project in Montana.[47] Not long after, the bridge was bought and maintained by Charles McCarty; it was referred to as McCarty's Bridge well into the twentieth century.[48]

When Mullan returned to Cantonment Wright later that summer, he found that the new Big Blackfoot bridge had been "thrown out of shape" by flooding. Again, having no time to make repairs, he hired Sam Hugo to fix it. Within a handful of years, the Blackfoot bridge became one of many pay-to-cross spans along Mullan's road. "The mania for toll roads is immense," complained John Henry Bryant, who traveled the road in 1879. He said men with no authority to do so took advantage of "small mountain torrents that an individual could almost leap across" to operate a toll bridge.[49] This in spite of the fact that in 1869, the Montana territorial legislature had passed a law that declared the Mullan Road a public highway "and the collection of tolls from Fort Benton to the summit of the Coeur d'Alenes [was] made a misdemeanor," according to Michael Leeson in *History of Montana, 1739–1885*.[50]

The bridge Mullan built over the Big Blackfoot River was washed out after only a few years, but other bridges soon replaced it. The horses and wagons that first crossed the river by bridge became sheep and cattle en route to and from summer pastures, then trains, electric streetcars, and early automobiles, and eventually modern cars, semi trucks, logging trucks, and recreational vehicles, and even bicyclists and joggers. Today, no fewer than five bridges span the Blackfoot River in the first half-mile above its mouth. Two are spans of Interstate 90; one serves the Montana Rail Link line; and a fourth accommodates traffic on U.S. Highway 10. The fifth and uppermost is a pedestrian bridge, known locally as the Black Bridge. Built in 1921, it was the lone highway bridge at the time. Half a century later, Missoula County closed it to motorized vehicles. In 2008 the Black Bridge was renovated in response to pressure from a community unwilling to see it scuttled.[51]

In addition to the Blackfoot bridge, Mullan and his men cut a narrow road high above the valley floor on the south side of Turah Peak, a few miles east of the Blackfoot. The grading of Three Mile Hill was hardly begun by January 9, 1862, when the cold suspended most work until April. When construction resumed, Gustavus Sohon sketched the progress of the work from a vantage point high on the east end, looking down the valley toward what would later become the Milltown Dam and reservoir. When completed, the Three Mile Hill cut eliminated the first and second fords of the river. This section of the road bore wagons, stagecoaches, horses, foot traffic, and, eventually, automobiles. Though unused for most of a century, the cut remains perhaps the most enduring piece of roadwork that Mullan's crew accomplished east of Missoula; much of it is still visible from Interstate 90, high on the mountainside between Turah and Clinton.

It was cause for celebration in 1911 when Missoula County commissioners approved the money to build a road to bypass the grade. The road there was "narrow and dangerous and so steep in places as to be almost impassable," the *Daily Missoulian* reported on June 25, in anticipation of the state's Good Roads convention in Missoula. "By securing the old roadbed of the Northern Pacific on either side of the highest and most precipitous point . . . and then constructing a low grade about this point over the railway tracks, the whole Three Mile Hill was eliminated," the paper reported.[52]

The new grade, about half a mile long, required heavy rock work. It was on this project that Missoula County experimented, for the first time, with convict labor furnished by the state penitentiary in Deer Lodge. The road was further improved with a bridge across the Hell Gate (Clark Fork) River at the Missoula-Granite county line, where the previous bridge had been wiped out in the historic flood of 1908. "With this finished and other minor repairs made the main

This sketch of the Three Mile Hill Cut by Sohon shows two of Mullan's workers walking along the newly built side cut. —Courtesy National Archives and Records Administration, Cartographic Section, College Park, MD

This modern photo of the Three Mile Hill Cut shows a portion of the Mullan Road along that grade. –Photo by Kim Briggeman

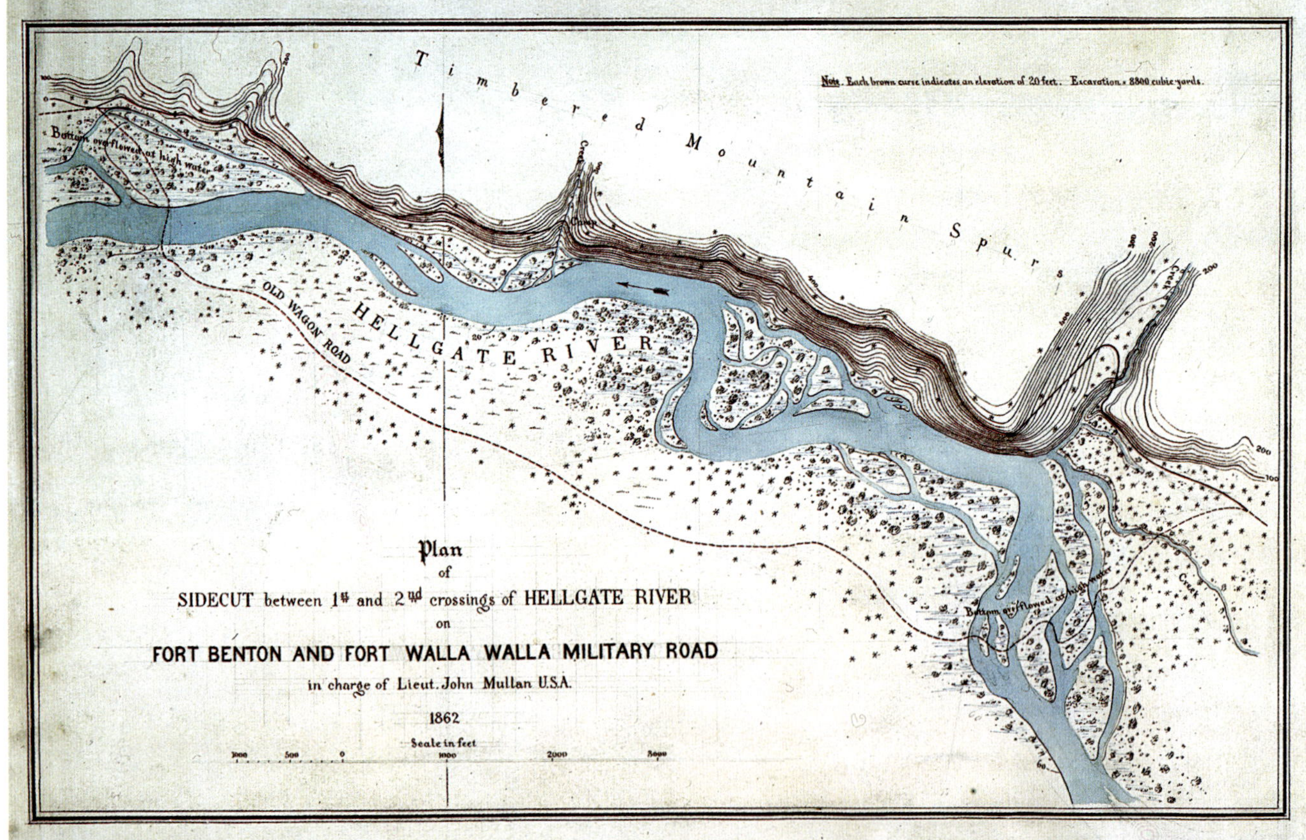

Map of the Three Mile Hill Cut by Sohon, showing the two crossings of the Hell Gate River on the original road, created in 1859-60, and the side cut, completed in April 1862, that moved the road inland from the river plain. Even after the cut was built, wagons and later road traffic found travel difficult in this region. – Courtesy National Archives and Records Administration, Cartographic Section, College Park, MD

road reaching east from Missoula was put in excellent condition," the *Missoulian* concluded.[53]

Charles Schafft, born in Berlin, Germany (then Prussia), in 1838, was a civilian commissary stock herder for John Mullan's first excursion. After Cantonment Jordan was established in December 1859, he and three other men drove what few of the expedition's cattle had survived down to the Bitterroot Valley, then set out on foot back up the road to Walla Walla. Their path was blocked by bad weather and ice on the Spokane River, so Schafft spent the winter at Walker's Prairie, a small settlement and the site of an abandoned mission a few miles north of the river, and Fort Colville, some seventy miles north of Spokane. Come March of 1860, he set out alone and walked first to Walla Walla, then down the Columbia River trail to The Dalles. From there he took a steamer to the Willamette Valley and worked on a farm until July 1861, when he rejoined the Mullan expedition, now in the Coeur d'Alene Mountains, as a sutler's clerk, assistant to William Terry.[54]

As a sutler's clerk, Schafft became part of a military tradition dating back almost to the beginning of warfare. The sutler was a civilian merchant who sold provisions to an army in the field or in camp. In his book *Peddlers and Post Traders: The Army Sutler on the Frontier*, David Michael Delo said the sutler's role was formally recognized in the United States as a quasiofficial service in 1822, and it soon evolved into a profession. There's little doubt that Terry's store at Cantonment Wright carried whiskey. Most did at the time. At least in part because of that, the army ended the era of the independent sutler soon after the Civil War.[55]

From his previous experience, young Charlie was familiar with the route upriver from Cantonment Wright. In fact, he'd made a little history there just a couple of weeks before his fateful trek in the snow. Mullan authorized him and two others to pursue a pair of horse thieves, and their chase bore fruit. According to Schafft, they caught one, named Butler, at Johnny Grant's ranch near the mouth of the Little Blackfoot River, and the other, named Williams, ten miles farther on, at Cottonwood Creek near present-day Deer Lodge. In his 1887 memoirs, Schafft notes that he mentioned the incident only because it was "probably the first arrest within the Territory."[56] According to Granville Stuart's *Forty Years on the Frontier,* James Stuart reported on December 27, 1861, that "Roland and another man passed with Jack Williams, prisoner for stealing horses." In parentheses he noted that it was the "first arrest for this offense here," confirming Schafft's notion.[57]

They hanged horse thieves in those days, but Butler and Williams survived. As Granville Stuart noted, the closest court of law in Washington Territory was in Walla Walla. When the posse delivered the prisoners to Mullan, the lieutenant opted to put them to work. The two were chained together and spent the rest of the winter working on the Blackfoot bridge.[58]

As Schafft recalled, in 1861 he and the other captors shared Christmas dinner with the Stuart brothers and, presumably, the thieves. Perhaps it was this kind of hospitality, and the prospects of a bustling mining region, that drew him back to the Deer Lodge Valley a couple of weeks later.

So when Charlie Schafft left Cantonment Wright on January 8, 1862, he knew where he was going. His walk that morning would have taken him past Williamson's camp, the first winter camp of Mullan's workers, at the Three Mile grade, between the first and second crossings of the Hell Gate River. The next camp upstream was Camp Campbell, about fifteen miles from Cantonment Wright and just downstream from the mouth of Rock Creek, near what Mullan's crew dubbed Rocky Point (which Schafft, in his 1887 memoir, called the Rocky Grade).[59] By evening he had probably reached the third camp, Clark's camp, a couple of miles farther up the valley on the other side of Rocky Point. It was likely here that he became snowed in for a week.

On the morning of January 15, Schafft pressed on, through snow almost three feet deep and too light and feathery to bear snowshoes. Passing the fourth and final work camp, Lannon's camp, he presumably crossed the frozen Hell Gate several times, since the side cuts weren't finished by then.

The last crossing, a couple of miles above Lannon's camp, was at the point where Sam Hugo would build his Hell Gate bridge later that year.

David Morton, a Methodist minister from Kentucky, traveled the Mullan Road from east to west in the summer of 1876, shortly after he arrived in Helena. His description of the section from Deer Lodge to Missoula gives a hint of the geography that Schafft had trod fourteen years earlier:

> From the town [of Deer Lodge] you go down the Deer Lodge River, over the Yam Hills and Flint Creek Hills . . . for about fifty miles, passing several mining camps, attractive only for the gold, which is found in large, remunerative amounts, till at McCarty's Bridge you enter Hellgate Canyon. This canyon severs one of the main ranges of the Rocky Mountains in twain and furnishes an outlet for the Hellgate River. It is nowhere over a mile wide, and in many places the mountains come to the water's edge. Through this pass the United States government has constructed a military road forty miles in length, which in many places is literally hewn out of the mountains and which is the great emigrant route to Oregon and the only post road in all this region–in fact, the only road over which a vehicle of any kind can pass. In many places the road, which averages about ten feet in width, is so steep that the teams ascend it with great difficulty and go down with almost railroad speed. On one side it is walled in by the mountains, and on the other are precipices, almost perpendicular, from ten to three hundred feet high.[60]

Of course, Charlie Schafft's own description of what he encountered on the route, as recounted in his 1887 memoir, is far more agonizing than what Morton observed on his summertime trek:

> Crossing the Hell Gate on the ice near the site of [the future] McCarthy's [sic] bridge, I took a direct course over Medicine Tree Hill, and having descended again to the bottom I broke through the ice into a slough. After this I had hard work (owing to snow sticking to the wet clothes) to reach the river bank where I could obtain wood to make a fire. It was now evening and I had only made twelve miles since daylight.[61]

It was, he learned later, forty degrees below zero that night. Discouraged by the prospect of the timberless Flint Creek hills ahead, Schafft concluded to retrace his path and head back to the worker camps. Back up over Medicine Tree Hill he went, reaching the bank of the Hell Gate crossing at around midnight. Ironically, the ice he'd crossed that morning had weakened, so he was forced to wait, building a small fire and walking around it to keep from freezing, until the ice again froze hard enough to bear his weight. After two or three hours, he finally crossed and plodded to the nearest camp, "really feeling warm, yet not knowing that my feet were freezing all the time," he wrote. "The intense cold acting on the trees made them give reports like pistol shots in all directions," he related. "The timber wolves were howling dismally and altogether it was not a very pleasant situation."[62]

At daylight, Schafft was still four miles from help when he found that both of his feet were frozen solid up to the ankles, where the moccasin strings were tied. The path ahead lay over steep Beavertail Hill, but somehow he made it to Clark's camp. Still, Schafft's problems were far from over. The only treatment for his frozen feet that he and the others knew was cold water and salt. An Indian agreed to carry word to Cantonment Wright, twenty miles away.

Upon receiving word of the episode, Mullan sent David O'Keefe, a civilian worker at the main camp, to go pick up Schafft. Shortly before his death in 1922, O'Keefe recalled the incident to a local reporter: "Mullan sent 'Bill' Hengan and I after [Schafft]. . . . It took us a day to walk 12 miles. After we got there, we rested a day, then started home, with Shaft [sic] on a hand-sled. We made half way that night and arrived there late the next afternoon. . . . We got a big drink of whiskey for bringing Shaft down."[63]

It's unclear whether O'Keefe and Hengan took Schafft back to Cantonment Wright, twenty miles from Clark's camp, or to Williamson's camp, roughly eight miles closer. Whichever the case, Schafft suffered more disappointment when he arrived: the doctor wasn't there. The expedition's physician, Dr. George Hammond, was snowed in at Fort Owen, thirty-five miles south of Cantonment Wright in the Bitterroot Valley. It took him more than a week to get to Schafft, who by then was certainly at Mullan's camp. "When he came in my case was hopeless," Schafft related. "Being too weak to be performed upon at once, the inevitable operation was delayed until the 7th and 8th of March, when both of my legs were successfully amputated within six inches of the knee joints and I was henceforth a cripple."[64]

When Mullan was ready to break camp in May, Schafft was still in no shape to be moved anywhere by wagon. "[I] was about to be left in camp with a guard of soldiers until I should be able to move or die (the latter was expected by the Doctor and the men) when Father Joseph Menetrey came to the rescue . . . ," he wrote. [65] The Swiss-born Menetrey, who had come to the missions of the Pacific Northwest in 1847, talked the situation over with Mullan at Schafft's bedside and agreed to oversee Schafft's recovery if some way could be found to get him to the Jesuit mission at St. Ignatius, nearly fifty miles away. It took three days for six soldiers and an army hospital steward to carry Schafft on a litter through the Missoula Valley, up and over Evaro Hill into the Jocko country, and over another pass to the mission. Schafft termed the trip "severe." It's amazing he survived another twenty-five years to remember it.[66]

A screenwriter could make salad from the scenes of Schafft's life, starting with the day a nine-year-old Charlie and his older brother were stopped by a mob at the walled city of Berlin on March 18, 1848. The March Revolution had broken out, and barricades were being thrown up at the city gates. The Schafft boys were made to pack "a few paving stones" before they were allowed to leave–as a joke, Schafft insisted.[67] At age eleven, he was sent alone to America. At fifteen he joined the army as a drummer boy and was ordered West. En route, he survived a catastrophic shipwreck off Cape Hatteras, South Carolina.

In the twenty-nine years he lived after losing his legs, Schafft made an indelible mark on Montana history. Perhaps it was the Jesuits' influence on him at St. Ignatius, and surely his own resilience and courage had something to do with it. He maintained close ties to the mission at St. Ignatius for many years and supported the mission's work with the Flathead Indians. In October 1863 Schafft went to the Flathead Indian Agency near Arlee, where he clerked and kept books. Starting that winter, he was even put in charge of the agency during the cold months, when head agents removed themselves to more civilized climes. In 1865 he copied a Salish-to-English dictionary for Father Urban Grassi at St. Ignatius and penned the physician's report for the Flathead Nation, signing it "Dr. Charles Shaft" (Schafft spelled his own name at least three different ways over the years).[68] Later, Schafft returned to the agency as a clerk, working there off and on for years. He also clerked occasionally for Jesuit missionary Anthony Ravalli at St. Michael's Church in Hell Gate (Missoula).

Not all of Schafft's jobs were with the Indians. In February 1865 he became the first clerk and recorder and one of the first two justices of the peace of the newly formed Missoula County. Later that year he moved from the Flathead Agency to Hell Gate Village, where the clerk's office was set up in the back of a butcher shop "amongst grease and tallow."[69] In the fall of 1865, he partnered with Frank Woody, pioneer settler and future first mayor of Missoula, to build a house in town.

Because both the clerk and the justice of the peace positions were unpaid, Schafft resigned in early 1866 to make a living, returning to Flathead Agency as clerk as well as becoming postmaster for Missoula. Through the years he also kept books for various hotels, merchants, attorneys, freighting companies, and other outfits.

In 1867 Schafft was working at the Flathead Agency when he received his first artificial limbs, ordered from Philadelphia for three hundred dollars. Two years later, he left Montana to visit his family in Berlin, then he toured some American cities before returning to Missoula in the fall of 1870. Back at the Indian Agency in 1872, Schafft was around when General James Garfield, the future president, came to Montana to negotiate a treaty to remove the Flatheads from their home in the Bitterroot Valley. Schafft "had the honors to make copies of an agreement," he wrote.[70] The following year, he superintended the building of a new Indian agency in the Jocko Valley while agent Daniel Shanahan was on a six-month leave in Washington.

In July 1874 Schafft fled to Alberta, Canada, to avoid testifying during the federal "Indian Ring" investigation of one of the Flathead agents. Indian agents from several western reservations were suspected of misappropriating government flour and other goods from the Indians' subsidized provisions. In

Canada, Schafft ran whiskey at "Fort Whoop-Up." When Colonel James MacLeod and the newly formed Mounties showed up "with siege guns" to drive out the illicit whiskey traders, they found "open gates, a cripple as second in command, and six or seven peaceable looking citizens," Schafft wrote. The fort's liquor supply, he noted wryly, "was cached at the bottom of Belly River."[71]

Schafft himself was no stranger to drink. He made no bones about the fact that the bottle was a major influence in his life, both before and after he lost his legs. "A Jesuit father once told me that every man has his fault or failing," Schafft wrote in 1887. "I have mine–a habit of drinking 'fire water,' a habit adopted in early youth and nourished by frontier life and usage. It has led me into a great many comical adventures and some serious ones. It has also made me some enemies, but none greater than myself."[72]

Nevertheless, as Schafft took pains to point out, "against general opinion," he didn't taste a drop of liquor from January 8, 1862, when he set out on his fateful walk, until March 7, when Hammond performed the first amputation and "it was necessary to stimulate me." Schafft maintained, "I was as sober in the night when I froze my feet as ever I was in my life."[73]

Though alcoholic, Schafft was an accomplished writer and artist. He literally wrote remarkable chapters of the formative years of the territory, from his arrival in the 1850s through statehood in 1889 and a couple of years beyond. After returning from Canada in November 1874, he again returned to the Flathead Agency as a clerk. After knocking around for a few months in 1878, he moved to Fort Benton in December and spent a year and a half working for the *Benton Weekly Record.* There he kept books, wrote short items and articles, and penned a series of "literary contributions," the first one being a reminiscence of the Mullan Road expedition.

After a little more traveling, including a stint in Washington, D.C., in which he helped with the 1880 U.S. Census, Schafft spent the final decade or so of his life in the Missoula area. For a time, he lived with "Baron" Cornelius O'Keefe (brother of David O'Keefe) at his "castle" at Evaro Hill. At this point, Schafft found work hard to come by. By 1887, when at age forty-nine he wrote his memoirs, he was destitute. "The end may come now at any time," he concluded, "but should I discover some strange country, in a possible hereafter, and come there a stranger, flat broke and without recommendation, I will undoubtedly soon meet a friendly spirit who will ask me–to take a drink."[74]

Charles Schafft died from pneumonia in 1891; the *Missoula Gazette* said the deceased "was very popular, and leaves a host of friends."[75]

Far from harboring bitterness for the fate that befell him in the harsh winter of 1862, Schafft remained a staunch champion of John Mullan and his road-building expedition. He expressed this sentiment in his reminiscences of the expedition, which appeared in the January 2, 1880, edition of the *Benton Weekly Record*:

> A part of the road he built will probably long remain in public use, notwithstanding the fault-finders who never saw the principal part of the work, and whose imagination can't picture the hardships endured by those who toiled upon it. . . . Mullan was one of the pioneers of this country. His name became permanent by a public work of peculiar difficulties, and those that are acquainted with the circumstances well know that he rather lost than made a fortune during the time he was employed upon it. There certainly is some credit due him.[76]

1 John Mullan, *Report on the Construction of a Military Road from Fort Walla-Walla to Fort Benton* (Washington, DC: Government Printing Office, 1863), pp. 28–29. This source is hereafter cited as Mullan, *Report*, 1863.

2 Ibid., p. 29.

3 Ibid., p. 20.

4 Ibid., p. 29.

5 Ibid., p. 32.

6 On an 1862 map in Mullan's 1863 report, titled "Plan of Sidecut Between 1st and 2nd Crossings of Hellgate River on Fort Benton and Fort Walla Walla Military Road in Charge of Lieut. John Mullan USA," Camp Williamson is denoted by the word "camp" on this small creek. A private home currently sits on the site.

7 Evidence that David Williamson was Rebecca's brother was recently uncovered by Keith Petersen of the Idaho Historical Society, in the form of letters written by Rebecca to Mullan when he was in the West. See Keith C. Petersen, *John Mullan: The Tumultuous Life of a Western Road Builder* (Pullman: Washington State University Press, 2014), pp. 100-101.

8 Mullan, *Report*, 1863, p. 33.

9 On August 22, 1860, Mullan wrote: "Continued down the valley of the Hell Gate stopping for a few minutes at a small creek near the 400th mile post to fix the crossings hence the road is good to what is termed the *four crossing spur* where much work was done while we were here in July last." From U.S. War Department, *Reports of Explorations and Surveys, to Ascertain the most Practicable and Economic Route for a Railroad from the Mississippi River to the Pacific Ocean . . .* (Washington, DC: Thomas H. Ford, Printer, 1860), p. 57. This source is hereafter cited as War Dept., *Reports*, 1860.

10 Caleb Irvine, "Medicine Tree Hill: A Legend" (Helena: State Historical Society of Montana, 1907), pp. 1–2.

11 Ibid.

12 Will Cave was born in 1863 near Virginia City, Montana. His parents moved the family to Missoula when he was ten, and he remained there the rest of his life. Cave graduated from the College of Montana in Deer Lodge and worked for the Missoula Mercantile Company from 1884 to 1891. He organized what's said to be the nation's first troop of volunteers for the Spanish-American War, where he served as first lieutenant in the Third Cavalry. Later he wrote historical articles for the *Missoulian*. Cave died in Missoula in February 1954. From "Guide to Will Cave Papers, 1892–1950," Maureen and Mike Mansfield Library Archives and Special Collections, University of Montana, Missoula.

13 Will Cave, "This Cave Article Tells of Our Names," *Missoulian*, June 4, 1922, p. 10.

14 Ibid.

15 Ibid.

16 Ibid.

17 Ibid.

18 Schafft's handwritten enlistment record from 1853 was found in Record Group 94, p. 240, in the National Archives by Vivian Paladin, editor of *Montana: The Magazine of Western History*, and published with Schafft's memoir, "Sketch of a Life–Charles Schafft," in *Montana: The Magazine of Western History*, January 1976, p. 28. This source is hereafter cited as Schafft, "Sketch."

19 Quotations from Ida Smith Patterson, *Montana Memories: The Life of Emma Magee in the Rocky Mountain West, 1866–1950* (Pablo, MT: Salish Kootenai College Press, 1981, 2011), pp. 39–40.

20 Mullan, *Report*, 1863, p. 33.

21 *Missoula Gazette*, March 19, 1891.

22 "David O'Keefe Dies Suddenly at Home, Lived Here 60 Years," *Missoulian*, August 24, 1922, pp. 1, 3, and 6.

23 Schafft, "Sketch," p. 30.

24 Mullan, *Report*, 1863, p. 33.

25 Fr. Grassi opened the first school in Montana's Mission Valley at the St. Ignatius Mission in 1863.

26 Schafft, "Sketch," p. 31.

27 Granville Stuart, *Forty Years on the Frontier*, ed. Paul C. Phillips (Cleveland: Arthur H. Clark, 1925), p. 193.

28 Ibid., p. 198.

29 Ibid., p. 196.

30 Ibid., p. 206.

31 Ibid., p. 211.

32 Ibid., p. 204.

33 Ibid., pp. 204–5.

34 Ibid., p. 208.

35 Ibid., p. 209.

36 Cantonment Stevens occupies a place in history as the first army post established in present-day Montana. While Lewis and Clark, who commanded a military expedition, established temporary tent camps in Montana in 1805 and 1806, and subsequent fur traders built nonmilitary forts in the decades after that, Cantonment Stevens was the first government-funded military post constructed in the state. Mullan himself built the log huts at his winter camps Cantonment Jordan and Cantonment Wright in 1859 and 1861, respectively.

37 Mullan, *Report*, 1863, pp. 21–22.

38 George F. Weisel, *Men and Trade on the Northwest Frontier* (Missoula: Montana State University Press, 1955), p. 203.

39 Mullan, *Report*, 1863, p. 24.

40 John Mullan, P. M. Engle, et al, *Military Road from Fort Benton to Fort Walla Walla: Letter from the Secretary of War . . .* (Washington, DC: Government Printing Office, 1861), p. 51. This source is hereafter cited as Mullan, *Military Road*, 1861.

41 Mullan, *Report*, 1863, p. 24.

42 Mullan, *Military Road*, 1861, p. 51.

43 John Mullan, "The Journal of John Mullan, U.S. Army Lt., Fort Walla Walla to Fort Benton Military Road," entry for November 20, 1861, John Mullan Papers, Whitman College and Northwest Archives, Walla Walla, WA. This source is hereafter cited as Mullan, *Journal*.

44 Carl Schlicke, *General George Wright: Guardian of the Pacific Coast* (Norman: University of Oklahoma Press, 1988).

45 Mullan, *Report*, 1863, p. 33.

46 Ibid., p. 33.

47 Jon Axline, *Conveniences Sorely Needed: Montana's Historic Highway Bridges, 1860–1956*. (Helena: Montana Historical Society, 2005), p. 13.

48 Hugo's bridge apparently didn't last long. In 1866 Charles McCarty and J. A. Johnson were authorized by the territorial legislature to operate a bridge or a ferry over the Hell Gate River. McCarty was licensed to establish an inn there in 1867 and a ferry and saloon in 1868, but the license was never renewed. By 1869 the *New North-West,* a newspaper published in Deer Lodge, Montana, reported that McCarty had bridged the river and "built a good road to avoid the Medicine Tree Hill and ford." A toll gate was established west of the bridge on the north bank of the river, just inside the Missoula County line (*New North-west*, October 1, 1869). It was said that McCarty let Indians heading east cross his bridge without paying a toll, but when they returned with their buffalo, McCarty demanded payment with buffalo tongues (*Missoulian*, January 12, 1871). He died in 1876 at age forty-five (*Missoulian*, August 9, 1876). Six years later, the *Missoulian* reported that McCarty's toll bridge and 100 acres were for sale (*Missoulian,* January 20, 1882). The newspapers are archived in the Audra Browman Research Files, Maureen and Mike Mansfield Library, University of Montana, Missoula.

Remnants of a subsequent bridge built in the early twentieth century can still be seen at the west base of Medicine Tree Hill, below the south shoulder of Interstate 90 near Milepost 136.

49 John Henry Bryant, "Letter to family in New York" (1870), Seattle Genealogical Society Bulletin 39, no. 1, (Autumn 1989); reprinted in *Mullan Chronicles* 7, no. 3 (Summer-Fall 1998); the *Mullan Chronicles* was a quarterly published by the Mineral County Museum from 1989 to 2007.

50 Michael Leeson, *History of Montana, 1739–1885* (Chicago: Warner, Beers & Co., 1885), p. 837.

51 "Walkers commemorate new Blackfoot River bridge" *Missoulian*, November 3, 2008 (retrieved March 11, 2012, from missoulian.com).

52 "Meeting of Montana Good Roads Congress Great Boost for State and County Highways," *Daily Missoulian*, June 25, 1911 (editorial section).

53 Ibid.

54 Schafft, "Sketch," p. 23.

55 David Michael Delo, *Peddlers and Post Traders: The Army Sutler on the Frontier* (Salt Lake City, UT: University of Utah Press, 1992), pp. 2–3.

56 Schafft, "Sketch," p. 30.

57 Stuart, *Forty Years,* p. 192.

58 Ibid.

59 Schafft, "Sketch," p. 30.

60 Quoted in Bishop Elijah Embree Hoss, *David Morton: A Biography* (Louisville, Ky.: Methodist Episcopal Church, South, 1916), p. 127.

61 Schafft, "Sketch," p. 30.

62 Ibid.

63 "David O'Keefe Dies Suddenly." David O'Keefe was one of several men who stayed in western Montana after Mullan's road crew disbanded. He joined his brother Cornelius, who had worked on Mullan's first road crew and came to be known as the "Baron," to settle on a creek at the foot of Evaro Hill, west of Missoula.

64 Schafft, "Sketch," p. 31.

65 Ibid.

66 Ibid.

67 Ibid., p. 28.

68 In the *English-Flathead Dictonary*, Schafft's inscription read, "Copied from M.S. of Fr. U. Grassi by Chas. Shaft [sic], 1865." At the end of the 144-page dictionary, all in Schafft's handsome script, he signed it, "Flathead Agency, Hell Gate, M.T., June 20th, 1865. Chas. Shaft." At the top of the following page, Schafft had begun transcribing a translation of the Lord's Prayer: "Kae-Loere (Our Father) lu l'schichemaskat (in heaven) si (who) ku-lzii (art)." The prayer and the dictionary end at this point. A rare copy is in the author's private collection. Regarding the physician's report, see Flathead Indian Agency Records, SC 885, Montana Historical Society, Helena.

69 Schafft, "Sketch," p. 32.

70 Quotation from "General Garfield and the Flatheads," in the *Council Fire and Arbitrator* 5, no. 1 (January 1882).

71 Schafft, "Sketch," p. 34. In a letter to the *New North-West*, published October 24, 1874, Schafft elaborated on the incident: "The Manitoba Mounted Police, a force of 15 men under command of Major McLeod, encamped on Belly River a short distance below here last night. In the evening the Major, some other officers, and a squad of men paid us an official visit. They acted with courtesy toward every one, but all appeared 'dry' which after a 4 months' march on arid plains is perhaps not to be wondered at. They asked for whisky, but when we regretted our inability to give them a drop, they evidently took it as a joke, for several details under command of proper officers were soon engaged in trying to find the 'critter' (whiskey). They searched up stairs and down stairs, peeped into all kinds of holes and crevices, but their search was of no avail and they left for other fields."

72 Schafft, "Sketch," pp. 36–37.

73 Ibid., p. 31.

74 Ibid., p. 37.

75 *Missoula Gazette*, March 19, 1891.

76 *Benton Weekly Record*, January 2, 1880.

"A Creditable Piece of Mountain Work"

Building the Mullan Road in the Rockies (Part 2, 1860–62)

BY MAJ. RYAN L. SHAW

In the spring of 1860, even after a difficult fall and winter full of discouraging developments, Mullan remained, as ever, utterly committed to the completion of his road. Eager to demonstrate its value and to extract a greater level of commitment from the War Department and Congress, he had struck upon an idea in January 1860 and wrote to Washington, D.C., suggesting that the War Department send a detachment of fresh recruits, up to three hundred, to reinforce the regiments of the Department of Oregon (formerly the Department of the Pacific). His plan was to have the troops travel to Fort Benton by steamer from St. Louis, ideally just in time to rendezvous with him there in early August, when he expected the road to be completed to the fort, its eastern terminus. With his then-empty wagons, he could escort the detachment and their supplies in perfect safety and security across his new road, reaching Fort Walla Walla within sixty days.

His plan would save the government $30,000, he figured, "to say nothing of the additional efficiency of the men, the effect on the Indian mind, and the want of the experiment of testing the value of a new line via the Columbia and Missouri rivers, with a land portage of only 600 miles."[1] The recruits would be escorted by some capable officer as far as Fort Benton, Mullan continued, and from there they would be turned over, obviously, to the sole command of Mullan himself. He asked that the War Department notify him of their decisions on his plan as soon as they possibly could.

Word was slow in coming. Fearing his plan would be altered or ignored altogether, Mullan became indignant. In the end, the government did alter Mullan's plan in ways that left him forever bitter. Recounting the process years later, he stated tersely that "It is needless to narrate all the details that followed my recommendation, nor the various difficulties that beset our pathway to the right and left on the part of callous and apathetic persons."[2]

In the spring of 1860, having received no answer regarding either the recruits or a renewed appropriation, Mullan was palpably disgruntled, almost belligerent, in his April 3 correspondence:

> To abandon the work when work is most required, and at its maximum, might be in accordance with the possible action of Congress, but certainly not in accordance with my ambition and desire to see the route practically and successfully opened, nor can it be in accordance with the spirit that explored, projected, planned, and has, thus far, constructed it. And though the non-action of Congress may have the effect to lay my plan of progressing (even without means) open to censure, still I feel that public necessity and public economy dictate that our work should be pushed on to completion with the force now working it.[3]

This remarkable quote reveals much about the U.S. Army in 1860, about the government's attitude toward the evolving frontier at that time, and about Mullan as an individual. The letter amounts, in essence, to an ultimatum from a lieutenant to both the War Department and Congress: *I don't care about your priorities*, it says, *I'm building this road and you can't stop me. If you don't send me the money to complete this project, I will take on debt in your name to do it. No matter what you say or do, I will finish building this road.* Interestingly, the reason he provides for his demands, even before mentioning "public necessity and public economy," is his own "ambition and desire to see the route practically and successfully opened." An insubordinate letter

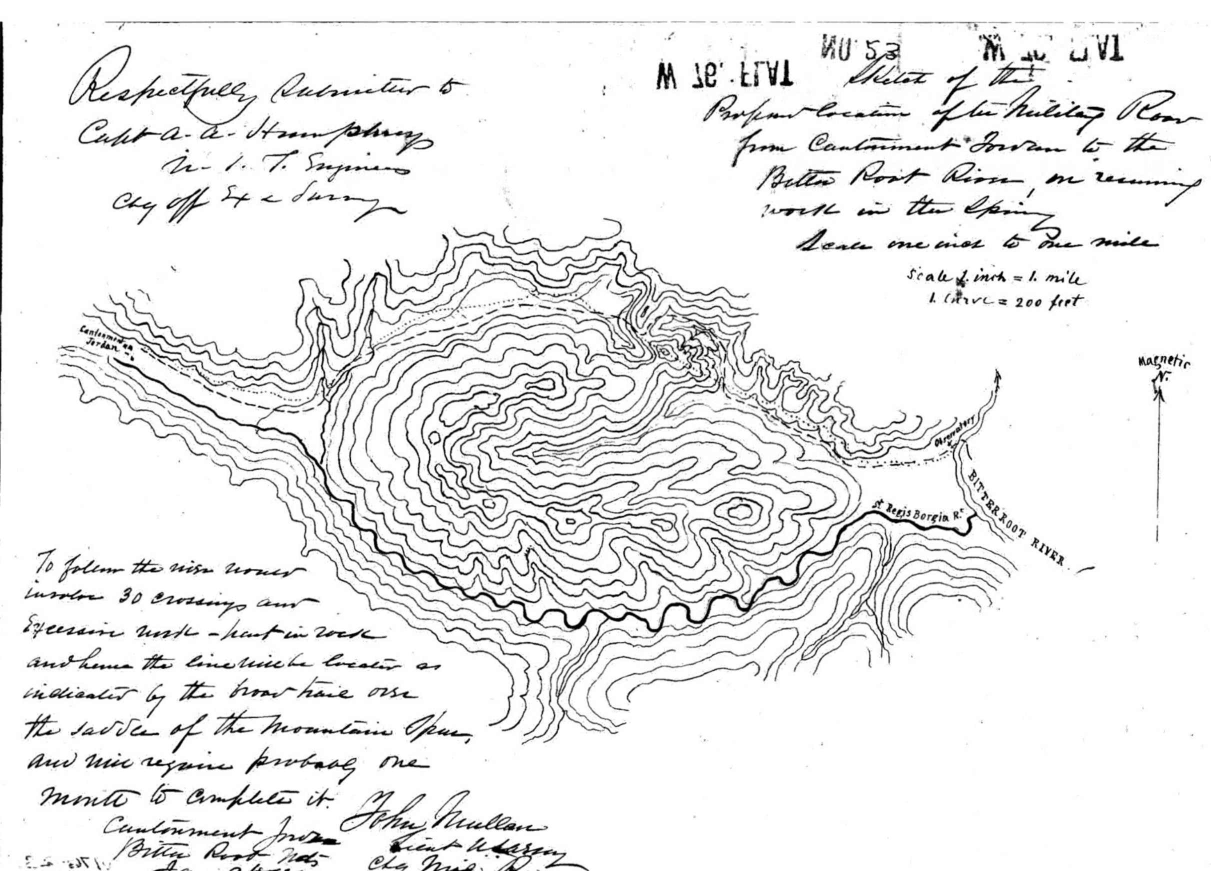

Figure 7-1. *This terrain map, probably drawn in the winter of 1859–60, shows a landform locally known as the "Camel's Hump" in the lower St. Regis Valley, northwest of St. Regis, Montana. Mullan deemed this stretch of the St. Regis River too difficult to construct the road through, so he routed it over the Camel's Hump. The map illustrates the path of the road toward Sohon Pass near St. Regis.* –Courtesy National Archives and Records Administration, Cartographic Section, College Park, MD

like that could get an officer disciplined or court-martialed in a different era–indeed, it probably would have gotten an officer back in the States disciplined in 1860. But in the still-wild West of the mid-nineteenth century, there was room enough–geographically, culturally, and bureaucratically–for a junior officer with ambitions as big as the mountains themselves, and as irrepressible as the changing seasons, to bluntly express criticism and defiance.

By April the season was, finally, changing. On his way back to Cantonment Jordan from Fort Owen that month, Mullan stopped at the confluence of the St. Regis River and what was then identified as the Bitter Root River (a stretch of river today considered part of the Clark Fork), where he had sent a work crew. Here he found that his men had already built six ferryboats for crossing the river and had cleared the road for ten miles east of the ferry. But between the cantonment and the ferry, a distance of fifteen miles, there was still no road, because those fifteen miles, like the cantonment itself, were still under snow. Mullan attributed this inconvenience to the area's thick timber, which prevented the sun from reaching the snowy ground. Recalling the labors of clearing these woodlands, Mullan remarked, "Thus has the dense timber proved our most serious difficulty to contend with, both winter and summer, in the Bitter Root mountains."[4]

Mullan also believed that the thaw along the river might be explained by that isochimenal line that had bedeviled him in so many other ways: "In fact, to compare the climates of these two points [the cantonment and the ferry site] was like the difference of spring and winter, for one was situated within this river of heat and the other without it."[5]

Skipping over the snowy section for the time being, Mullan summoned the rest of his work crew to the ferry site.

The animals still being too weak and emaciated to come back from the Bitterroot Valley, the men trudged those fifteen miles on foot, with two months' supplies and all their tentage and personal baggage on their backs.[6]

Through the end of April, the eighty-man construction party cleared twenty-five miles east of the ferry, along the Clark Fork's northern bank. At the ten-mile point, they were forced to cut through a three-quarter-mile spur of rock that was up to fourteen feet deep in places. This job took about two weeks of blasting and digging.[7] Even so, the men must have been glad to be out of winter camp and working their way east once again.

On the first of May, Mullan dispatched Captain Delany with a twenty-five-man crew to commence work on the fifteen-mile stretch from the ferry back to Cantonment Jordan, which was finally sufficiently–though by no means completely–free of snow. The section required "the crossing of a mountain spur 1,200 feet high, where the continuous grading from base to base will be one mile, mountain sloping at an angle of 50°," Mullan noted.[8] With Lieutenant White working in the other direction, from the Cantonment toward the ferry, it took the whole month of May to clear the timber and grade that towering spur. On June 1, Mullan tested the route with the escort wagons and found it passable. "We therefore abandoned our winter home on the 4th of June, where the fetters of winter and our difficult work ahead of us had bound us for six long dreary months."[9]

While the men were working on the road, Mullan made a decision over which he had agonized for weeks. Thirty-six miles west of the ferry, a mountain reached down toward the Clark Fork River in a series of spurs, rocky and steep, measuring six miles from end to end. Sohon had identified this daunting spot in his reconnaissance the previous year, and Mullan had been pondering the right course of action ever since. Indians had reported a bypass, called Brown's Cut-off, which went up Nemote Creek and around the north side of the mountain, returning to the Clark Fork via Skiotaw Creek. On April 20, Mullan set out with a small party to reconnoiter Brown's Cut-off. The first eight miles were easy going, but as the creek rose higher, the gorge got narrower, until rocks and fallen timber choked it off completely. Still, "after clambering over this exceedingly difficult region for three days," Mullan reported, they were surrounded by snow.[10] On the fourth day, two miles shy of the summit between the two creeks, the snow and timber grew so thick and deep as to prevent their passage entirely, on horseback or on foot. And so, "sick at heart and fatigued with our exertions we were compelled to retrace our steps back to the Bitter Root [Clark Fork]," he wrote.[11]

With no bypass, the road builders would have to either cross the river repeatedly to avoid the spurs or blast through them to the other side. The engineers voted to cross. Their samples indicated that the whole passage was solid rock, and they estimated that cutting through would take upwards of five months.[12] On the other hand, the river was deep and swift here, with steep banks on either side; there would be no easy ford of the type so frequently and expeditiously used on the Coeur d'Alene and St. Regis Rivers the previous fall. Initially, Mullan agreed to the crossing anyway: "I fully appreciated the advantages of a continuous stretch along one bank; but the extensive plateaus on the opposite side formed so inviting a contrast to the rocks and bluffs before us, that at first I allowed my judgment to decide in favor of the additional crossing."[13] And so he ordered the construction of flatboats for the first ferry. Before the materials were gathered, though, he had changed his mind. "Truly," he wrote, "the matter was not of easy solution."[14]

In his report to Captain Humphreys of the War Department, Mullan detailed his dilemma:

> On the one hand, the great length of time (apparently) to work the road, our scanty means, the putting in jeopardy the opening of the remainder of the line this season to Fort Benton, the probable necessity of the line being needed for travel this season, the question of supplies, &c.; while on the other, the disadvantages in our position, the difficulties of additional ferries, the greater inconvenience to the travelling public, the increased taxes of tolls, the liability to have the ferries destroyed by Indians or swept away by freshets, and the delays to travel naturally incident to frequent river crossings.[15]

Never one to shy away from a challenge even when, perhaps, he should, Mullan made up his mind. On May 1, sixty-five men began cutting into what they named, simply and reverently, Big Mountain. When the men working on the road between the cantonment and the ferry finished their task, they joined the crew on Big Mountain, and by mid-May the entire force was focused on those formidable spurs.

The engineers were wrong–the spurs were not solid rock, but they were awfully close to being so; in all cases, the flinty limestone was dense enough to prevent any digging without blasting. The team's four sets of drills were continuously employed.[16] One man lost an eye to a premature explosion and another suffered a severe concussion. In the end, Mullan recorded that the cut through Big Mountain cost the labor of 150 men for six weeks.[17]

On June 20, the road around Big Mountain was opened and the party crossed, wagons and all, to the east side of the cut. That work behind them, Mullan's primary focus became

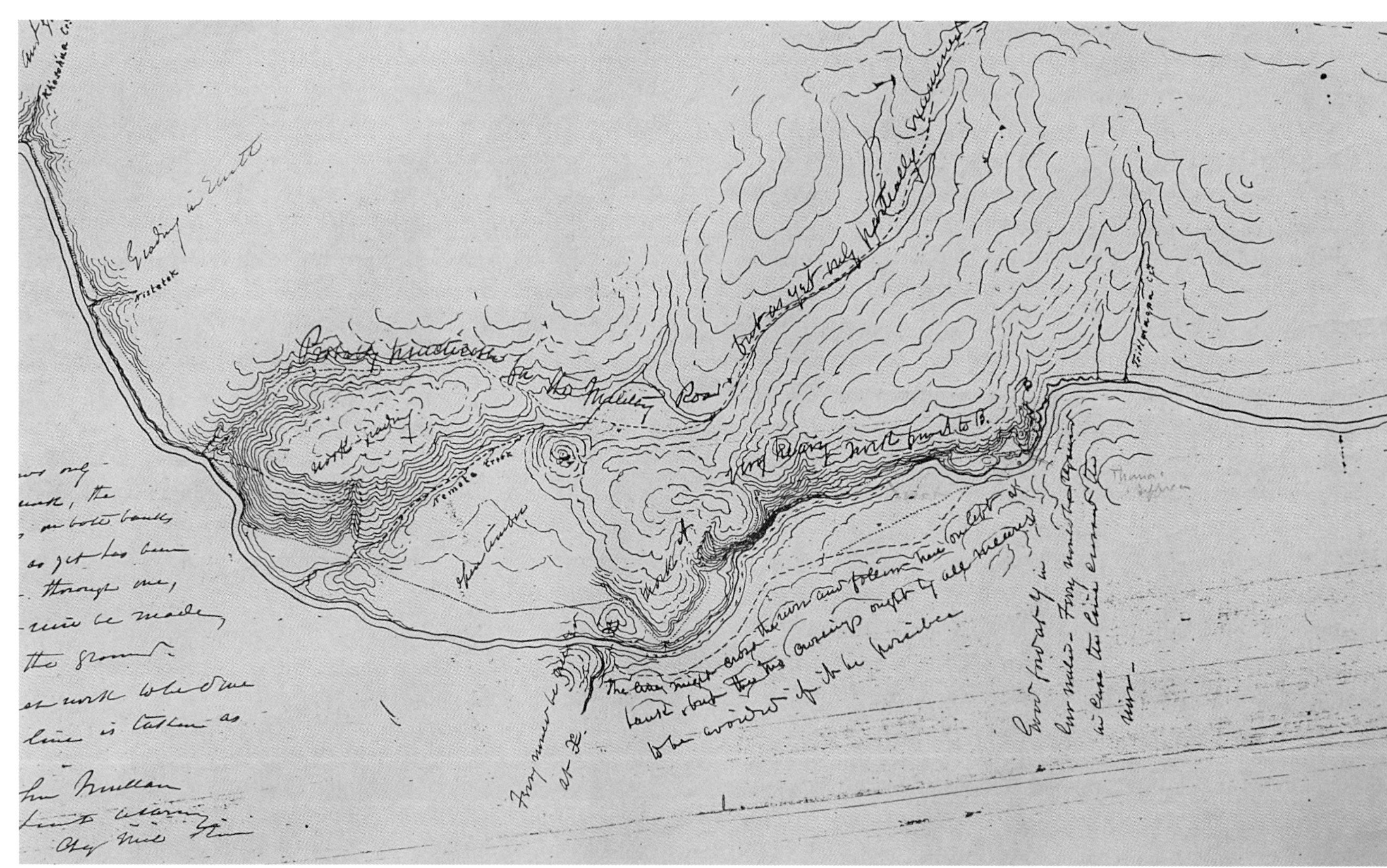

Figure 7-2. *High above the Bitterroot (Clark Fork) River, Mullan routed the road to avoid easily eroded ground. This preliminary map by Gustavus Sohon shows Mullan's original plan; he later changed the route, building the road closer to the river but still high on the mountain (see Figure 7-3).* –Courtesy National Archives and Records Administration, Cartographic Section, College Park, MD

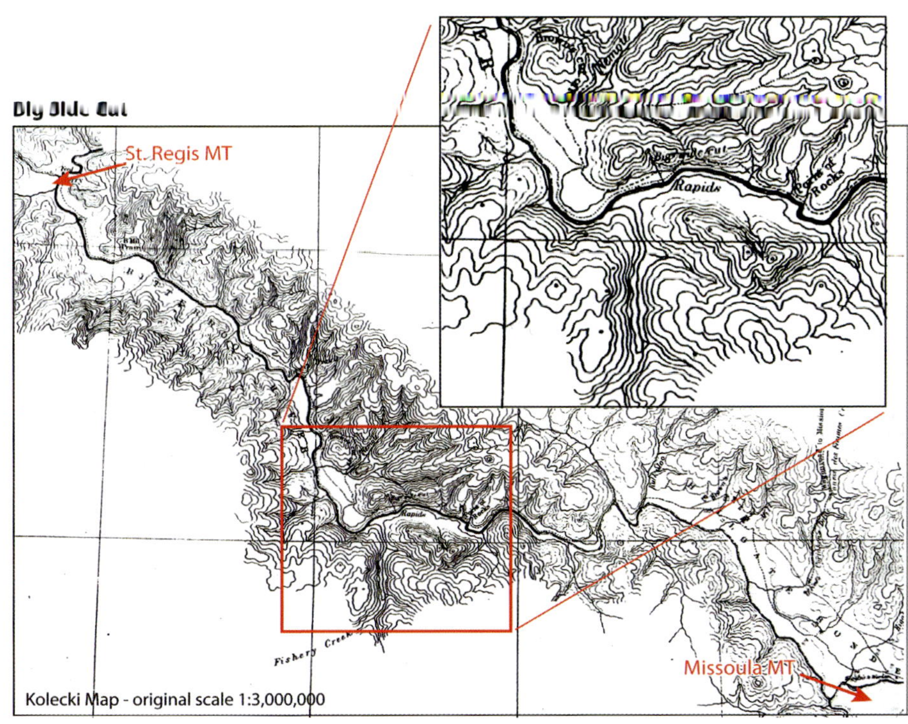

Figure 7-3. *Compare this Theodore Kolecki map of the Big Mountain side cut with Sohon's map (Figure 7-2) to see how Mullan changed the route.* –Courtesy National Archives and Records Administration, Cartographic Section, College Park, MD

Figure 7-4. *Part of the Mullan Road still exists in the Big Mountain area; this modern photograph shows a portion of the road where Mullan's crew blasted.* –Photo by Paul D. McDermott

reaching Fort Benton by the end of the summer. The crew reached the Hell Gate Ronde, fifteen miles east of the end of the cut, by June 23. At this point, Mullan determined to shed some dead weight. He held a public auction to dispose of such salable equipment as he could do without. What he did not sell, he turned over to the Indian Department along with his surplus perishables. Anything he wanted to retain for future work seasons–assuming there would be future work seasons, though he still had no word from Congress or the War Department–he cached near the settlement, where he hoped the local civilians would keep a watchful eye on his gear. He named the cache site Camp Humphreys, in honor of his principal supervisor at the Office of Explorations and Surveys, Captain A. A. Humphreys. Perhaps this bit of flattery was a preemptory bid for Humphreys's approval of Mullan's hasty decision to leave equipment behind.

In his report to Humphreys, Mullan claimed that upon arriving in the Hell Gate Ronde, the expedition had reached "the termination of our heavy work."[18] While it was true that the remaining three hundred miles to Fort Benton were easier going than the one hundred miles the party covered after leaving the mission at Coeur d'Alene, the road ahead, still wilderness, proved to be unpredictable. Mullan's correspondence during the 1860 season is largely a record of work not completed: Three more crossings of the Hell Gate and two of the Little Blackfoot could be avoided with ten days of blasting and grading for thirty men–there was no time to do it now.[19] A stretch over broken ground in the

Deer Lodge Valley ought to be rerouted along the Little Blackfoot to save travel time; this adjustment would take five days' work for fifty men—we'll come back to it.[20] After trending slightly south through the canyons, the road broke to north-northeast, back to the banks of the Little Blackfoot, which was forded two more times; the road probably should be cut a little higher on the bank to avoid flooding next spring, but that would require two days and fifty men—not now.[21] Numerous other such estimates for future work—with some jobs predicted to take months—appeared throughout Mullan's reports.

On July 2, Mullan and his men reached the crossing of the Big Blackfoot River. It should have been bridged, but again, they did not have the time. Working with such haste as to overturn a wagon and lose all its contents in the river, the crossing took two days and left no ferry behind.[22]

While at the Big Blackfoot, an express arrived from Fort Benton with news: Mullan's plan had been approved. Three hundred new recruits, under the command of one Major George Blake, would be waiting at Fort Benton to link up with him and be escorted along the road to Fort Walla Walla.[23] If the Secretary of War approved his bold proposal, Mullan reckoned, he must also, in spite of the long delays in communications, approve of the whole project. Even better, if the War Department sanctioned his plan, Congress must follow suit and eventually appropriate the money to see the road through to completion. If Mullan needed any further incentive to hurry, he now had it.

On the Fourth of July, 1860, the party ferried the first crossing of the Hell Gate River (now identified as another section of today's Clark Fork), then forded it three times in the next twenty-three miles, slogging through waters so deep and swift that they lost an ox on the third ford. "These fords can be avoided by side hill excavation," Mullan contended, "part in rock, which time would not allow us to perform. The amount of work required on this portion of the road to render it practicable at all seasons will be one month's work of fifty men."[24] Moving on, the men crossed the Hell Gate ten more times to reach the 400-mile marker of the road; Mullan observed with pleasure that with every mile, the canyons increasingly gave way to gentle valleys with thinner timber, grassier hills, and more sunlight.

By July 16 the expedition reached the foot of the Continental Divide in the Rocky Mountains, and on the 17th the men crossed over Mullan Pass to the waters of the Missouri River. Mullan's relief at being out of the mountains is palpable in his report:

> Crossing the Rocky Mountain range at Mullan's Pass and descending upon the Missouri slopes, a new climate and new character of country is at once encountered. The spurs of the mountains become lower and less timbered, the country more diversified with hill and dale, and the scope for a wagon-road location enlarged. The climate, too, is warmer, the frosts at night less severe, and the great difference of heat between midday and midnight no longer noticeable.[25]

With the high peaks behind them and knowing Major Blake's command waited ahead, the party made rapid progress down the eastern slope of the mountains, moving along Big and Little Prickly Pear Creeks; here they paused for just half a day to bridge Fir Creek, between the two. Past Fir Creek they proceeded north through rolling prairie, then light timber. The streams they forded were no longer torrents of angry whitewater careening wildly down from the very spine of the continent; rather, they were gentle meadow streams, with gentle names like Willow and Soft Bed, meandering through tracts of rich, tillable soil. Toward the end of July, the party entered a valley of "red sandstone and slate formation"; the soil in this area was less arable than before—Mullan called the stream he found here Hard Bed Creek—but the country held even more promise for development, as it gave "every indication of the presence of gold."[26]

Little Prickly Pear Creek flows into the Missouri River through a narrow, rocky defile with "no berme on its either side."[27] This represented the road builders' last significant obstacle on the path to their destination. The constriction of the passage forced them up and over "a broken section, which we termed Medicine Rock mountain," Mullan wrote.[28] "As thorough an exploration as we could give the country at this time showed that the only practicable line lay over the mountain," he conceded, though he was confident that an alternative route might later be found.[29] Reluctantly, he set his men to building a route over the mountain, which required four days of excavation and grading through "by far the most difficult of any point along the entire line, from Hell's Gate to Fort Benton."[30] On the bright side, in the digging the men found "traces of quartz, and continual indications of gold."[31]

Past the mountain, the men worked their way through ten rocky miles of defile, then nine more of rolling spurs. On July 25, they reached the last crossing of the Little Prickly Pear, and by the 28th they were encamped on the Dearborn River. There, Mullan wrote with a relief that can only be imagined, "We now left both the mountains and their spurs behind us, and emerged upon the broad, swelling prairies of the upper Missouri."[32] While the crew still had better than fifty miles to go before reaching Fort Benton, they were easy miles. This year's battle with the mountains was over.

Adding to Mullan's satisfaction, a messenger from Fort Benton met him at the Dearborn camp with news that the government had approved an additional appropriation of $100,000 and another season of roadwork. Less happily, the dispatches also indicated that Mullan's request to lead the recruits back to Fort Walla Walla had been denied–Major Blake would be in command of the march, and Mullan was to turn his wagons and men over to him.

Mullan arrived at Fort Benton on August 1, exactly as scheduled. Nevertheless, he found Major Blake, who had arrived several weeks earlier, waiting restlessly with his recruits. Blake's three steamers, incidentally, were the first boats to ascend the Missouri River as far as Fort Benton.

Mullan's arrival at Fort Benton was not entirely unhappy. Traveling with Major Blake's command was Mullan's brother James, returning west to serve as physician at the Flathead Agency.[33] Also with Blake was August V. Kautz, Mullan's classmate from West Point. Mullan passed the first evening at Fort Benton talking with Kautz, refusing to come to the main camp to meet with Blake, "much to the Major's disgust."[34] Kautz persuaded Mullan to meet with Blake on the second day, and the meeting was cordial enough, though Kautz commented that "Mullan is quite a monomaniac about his road."[35]

The lieutenant and the major managed to put their differences aside long enough for Mullan to transfer all of his wagons and a number of his crew–including his trusted guide

Figure 7-5. *Steamboats on the Missouri River at Fort Benton. Detail from a lithograph by Gustavus Sohon.* –From John Mullan, *Report on the Construction of a Military Road from Fort Walla-Walla to Fort Benton* (1863)

and interpreter Gustavus Sohon (who was also the expedition's artist) and his wagon master, John Creighton–to Blake for the westward trip over the road. "[T]o their joint good services," Mullan wrote of Sohon and Creighton, "Major Blake was largely indebted for the success of his march."[36]

For the return trip to Fort Walla Walla, it was agreed that Mullan would lead a small detail two days in advance of Blake's troops in order to clear and repair the road. For this Mullan retained just one assistant and a small workforce of twenty-five men, discharging the balance of his crew, whose enlistments had expired. With a fast-moving pack train instead of cumbersome wagons, Mullan was on the move westward by August 5. On the road, he sent runners back to Blake nightly with bivouac instructions, and the runners returned with supplies from the wagons.[37] Speed was of the essence, as the intent of the trip was to demonstrate the value of the road by delivering the recruits to Fort Walla Walla as rapidly as possible. Thus Mullan's roadwork on this trip was limited to clearing fallen trees and rocks that had toppled onto the road since their first pass, repairing bridges that had washed out in the spring freshets, and other minor fixes and improvements, especially in the mountain section, where "repairs became more frequently necessary."[38]

Meanwhile, Blake's men positively enjoyed their excursion across the Rockies. Lieutenant Hardin of Blake's Company B recalled that "the weather was simply perfect the whole time we were in the mountains. Every member of the command except the quartermaster had as pleasant a time as men ever had in crossing the mountains."[39] The recruits and their officers amused themselves hunting and fishing and rhapsodizing about the mountain vistas, as well as second-guessing Mullan's choice of route, judging the quality of his work, and pondering the prospects for a railroad through the region. Another amusement for Blake's command was locating and pillaging Mullan's caches from the previous winter. Despite the reprobations of their officers, the men could not resist helping themselves to these unexpected troves of "free" supplies–particularly those that contained whiskey. On one occasion, an entire company was delayed due to hangover. Major Blake was livid, expressing himself in "vigorous English," and he ordered the whole company to march through the ice-cold Coeur d'Alene River to restore their senses. It was a singular experience for Hardin to see "the total defeat of a command by 'John Barleycorn.'"[40]

For Mullan, however, the journey was hardly a mere walk in the woods. Just a few days out of Fort Benton, he contracted dysentery, and by the time the men began regrading Medicine Rock Mountain, he was reduced to riding in one of the wagons, temporarily turning over his authority to one of his civilian engineers.[41] Although the crew managed to put up a few small bridges and reroute the approach to Deer Lodge Valley, they mostly just cleared the path as best they could while keeping ahead of the rear party. During his wagon-bound convalescence, Mullan took notes on the man-hours and materials that would be required for the next season. Among the improvements he hoped to make were many new bridges, some sidehill cutting to avoid river crossings, and in some places, extensive rerouting. He estimated that a bridge over the main crossing of the Bitterroot River would cost a prohibitive $20,000, a full one-fifth of his present appropriation–the ferry, he conceded, would have to suffice.[42]

While Mullan did not record specifically how many rocks his men had to roll off the road, how many fallen trees they had to cut through, or how many mountain springs, frozen when they built the road, had since thawed and washed out their work, it was clearly a lot. During the winter, one man had counted 165 downed trees just along the mountainous stretch between the Coeur d'Alene Mission and Cantonment Jordan–and that was before melting snow revealed more and caused more still.[43] After the trip, Mullan wrote that "a general overhauling of the whole line should be made before the line can be considered as completed for travel."[44]

By the time the party reached the mission at Coeur d'Alene on September 1, Mullan's health was improving, and he was soon back to his cantankerous, monomaniacal self. His strength revived, he sent Lieutenant Lyon ahead with the work party toward Walla Walla, knowing that they could cross the western prairie with little difficulty, while he remained in the mountains with a smaller crew to do advance work for the next season. Although he had determined that the deepest crossing of the Coeur d'Alene River would require a bridge, he had no clear picture of the extent of its spring flooding, its rise, or its current. Having no block-and-tackle equipment to build a bridge anyway, he settled upon an experimental half-measure. After selecting a crossing site, he built what he intended to be the center pier of a future bridge, a triangular log crib, twenty-one feet long, rising seven feet above the present water level and filled with rock. In the same way, he built abutments on either side of the center crib pier and left them there, hoping they would still be standing and dry on top when he returned the next summer.[45]

Before proceeding westward, Mullan and his party cached some materials at the mission for the next work season, then followed the road around the southern end of Lake Coeur d'Alene. At the crossing of the St. Joseph's (St. Joe) River, Mullan saw the high-water marks from the spring floods and realized that "the entire lower portion of the valley becomes a lake at high water."[46] Apparently anticipating this, he had already dispatched one of his men from the mission with an Indian guide to reconnoiter a route around the north side of the lake. The new route would not be easy to build—Mullan estimated two months' work for fifty men to make this twenty-eight-mile detour—but it would be drier.

By October 5, Mullan and his small crew had caught up with Lieutenant Lyon's advance party at the Touchet River; together the men bridged one last water crossing before resuming the trip to Fort Walla Walla, where they arrived on the 8th.[47] Blake and his troops had reached the fort a few days earlier. En route, about half of the major's recruits were sent to reinforce Fort Colville; the remainder would either stay on at Fort Walla Walla or continue west to The Dalles or Vancouver.[48] Mullan was gratified to report that the Blake expedition, the first major military troop movement over his new wagon road, had made the trip nearly without incident in an efficient fifty-seven days. "Thus ended this military experiment via the upper Missouri and Columbia rivers," Mullan wrote, "and the success that attended it, the good effects that it induced, the economy resulting, and the eulogistic manner in which each officer of the command referred to the trip, all constitute a sufficient commentary upon its feasibility for future military movements towards the north Pacific."[49]

Mullan's pride was justified—the time and money spent moving Blake's detachment overland were indisputably substantially less than the time and expense of taking the ocean route around the continent. And the troops' safe delivery from Fort Benton indicated that Mullan had successfully rendered the previously impossible passage across the northern Rockies safer even than the steamboat trip up the Missouri River from St. Louis. Mullan could not have known that Blake's expedition would be not just the first but also the last major military troop movement over the road.

Mullan regarded his arrival at Walla Walla as the conclusion of not only the "long and tedious work constituting my second expedition," but also of his great military experiment, even though he planned to make another trip over the road to do more repairs and improvements.[50] He might, on this occasion, have reflected on the outlay of labor and capital that would be required, en perpetuity, to keep the line open and unobstructed. He also might have felt a pang of regret at the recognition that on his first expedition, he had made not one but two critical and costly errors of route selection, choosing the south side of the lake rather than the north, and the Coeur d'Alene route rather than the Clark's Fork. He even might have felt an inkling of dread at the prospect of yet another season battling those mountains, correcting those errors, and finishing the work that was so difficult as to be left undone after two seasons. If he experienced any of these feelings, however, he left no indication of it in his correspondence, which brims with optimism and pride.

After disbanding his crew, Mullan set to work developing his strategy for round three. He reported from Fort Walla Walla that he planned to conduct preparations there through the winter and head east on another road-building expedition by April 1, 1861. But he changed his mind when he realized that he had inadvertently sent some critical field notes back East with Kolecki, rendering it futile for him to stay in Walla Walla—he could hardly compile his reports and make plans for the next trip without having this essential information at hand. Furthermore, after making additional careful

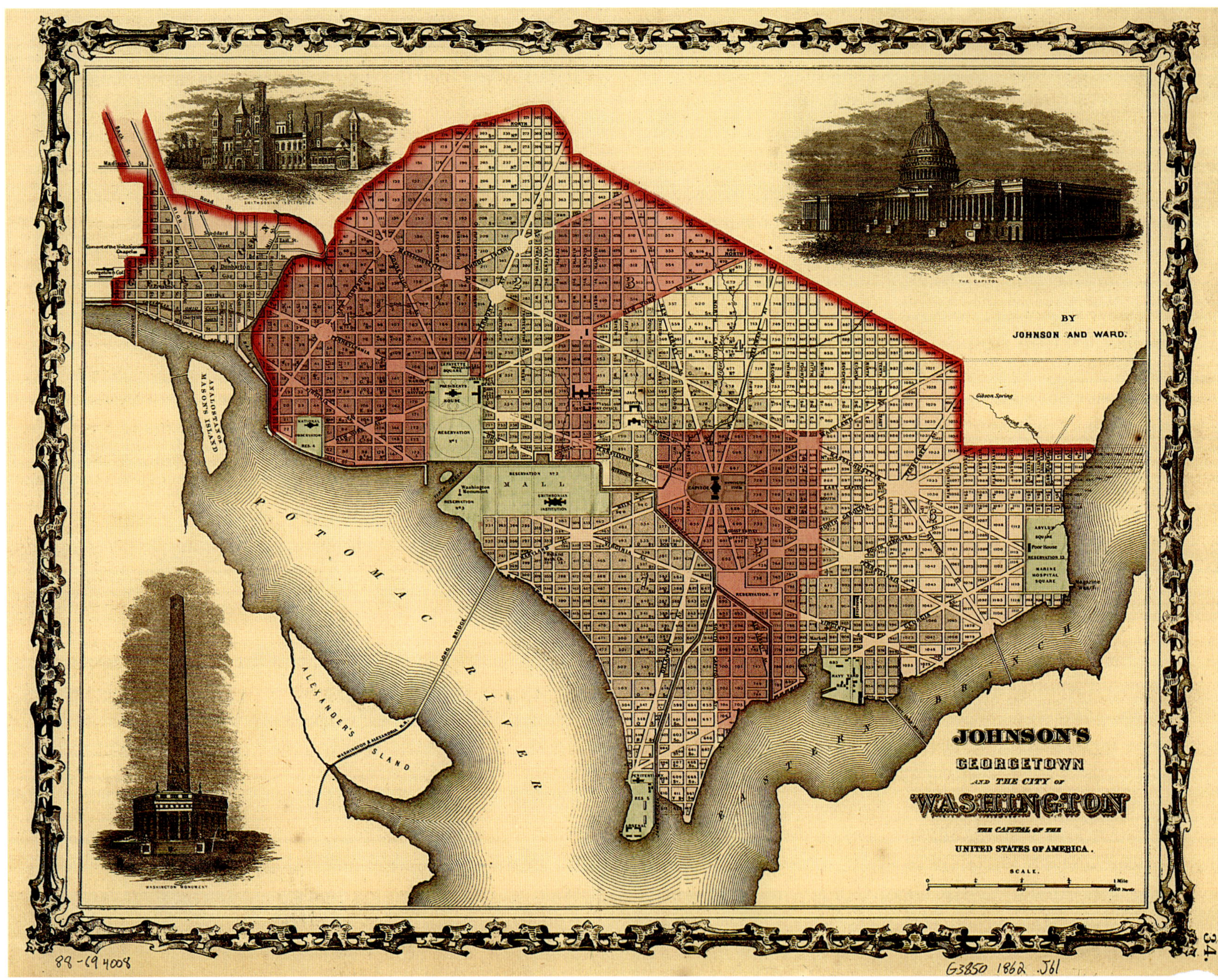

Figure 7-6. *Washington, D.C., in 1862, the same time that Mullan worked on his report for Congress.* –Courtesy Library of Congress, Geography and Map Division

calculations, he concluded that he could save money by starting from Fort Benton in the spring and working westward. Leaving Walla Walla, he made his way to Washington, D.C., where he would instead spend the winter.[51]

Mullan may have had an ulterior motive for making his unexpected and unauthorized trip to the national capital. The convention to nominate Washington Territory's delegate in Congress was scheduled for the spring of 1861. In January 1861, Mullan sent a letter to Indian agent and ethnographer James G. Swan, soliciting his opinion on Mullan's own potential as a candidate for the seat.[52] It is therefore likely that on his visit to D.C., Mullan tested support for projects more personal–and more political–than another season of roadwork.

In his new road plan, rather than return to Fort Benton by steamboat up the Missouri from St. Louis and moving west–the logical route to Washington Territory, the practicability of which justified his whole road project in the first place–Mullan proposed to outfit at Fort Leavenworth, Kansas, and take the Overland Trail to Fort Laramie (in future Wyoming), then go north from there to the Deer Lodge Valley, where he would begin his work westward. Yet it is highly doubtful that this route would have been a less expensive (his original justification) or more effective approach for improving his road than starting at Fort Benton or, as he had initially proposed, at Fort Walla Walla. It would, however, "enable me to test the value of the Laramie and Deer Lodge route, which had always been with me a favorite measure," he explained.[53] This route through the Powder River country would have connected Mullan's military road to the Oregon Trail and opened up a convenient and well-watered land route from the main emigrant trail to Montana.

Mullan was not wrong about this route, but he would not be the one to blaze it. Two years later, at the start of the gold rush in 1863, John Bozeman reconnoitered the same route, and in 1864 he led more than 2,000 settlers over what became known as the Bozeman Trail. In 1860, though, all Mullan knew was that the route ought to be a road, and that he ought to be the guy to build it.

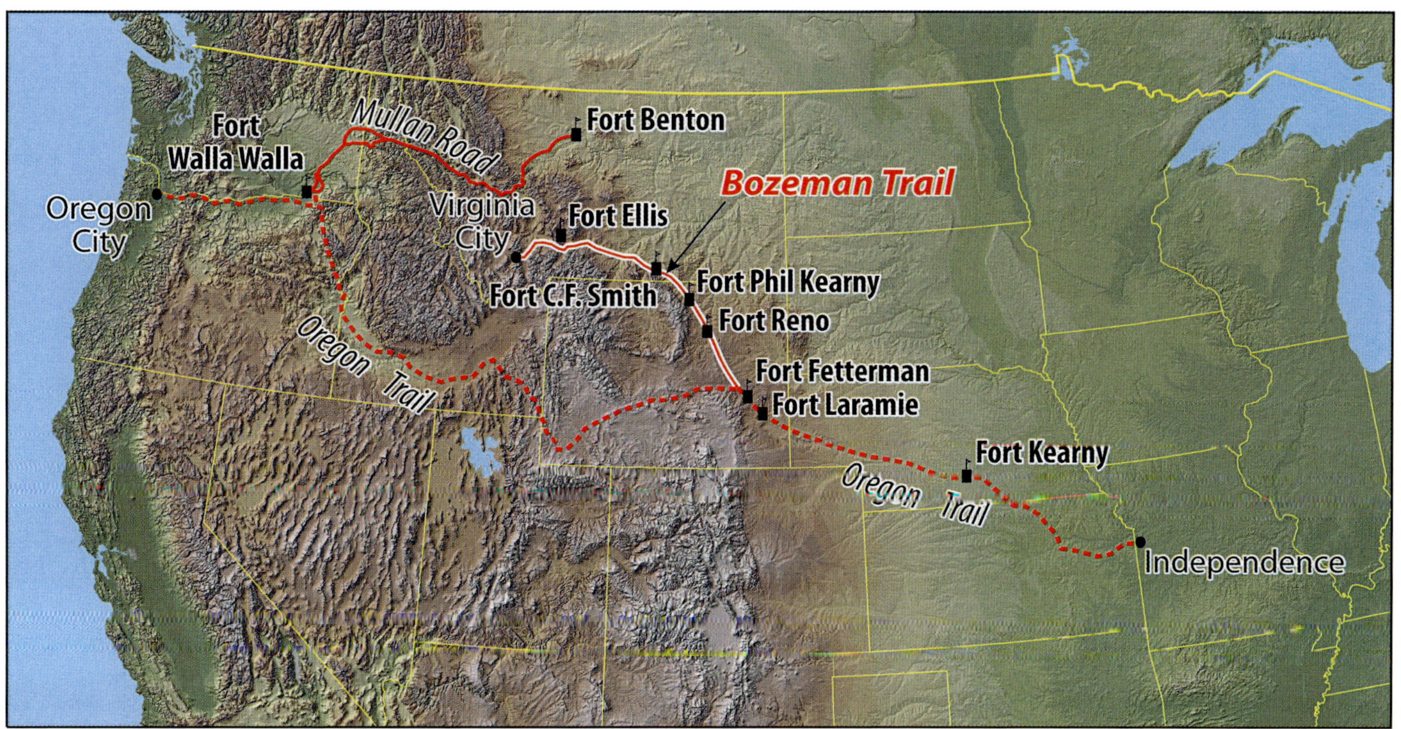

Figure 7-7. *The Bozeman Trail, a route pioneered by John Bozeman and John Jacobs* –Map by Philip Mobley

As it turned out, Mullan's trip to Washington, D.C., in early 1861 proved to be futile–he made no progress in furthering either his political efforts or his new route proposal. Abraham Lincoln's election in 1860 had the city spinning. Beyond the usual tumult surrounding a change in administration, the secession crisis and the seemingly unstoppable descent toward civil war engaged the full attention of all branches of the government. "On my arrival in Washington," Mullan wrote, "I found affairs in a somewhat chaotic state, a change of administration at hand, my first project already indorsed and returned to me, and seeing so many difficulties interpose, I was forced to relinquish my Laramie scheme and proceed back to Walla Walla."[54] He made the cross-country trip from Washington City to Washington Territory in April, for the third time in four years, and by May he was ready to commence work once again, moving east from Fort Walla Walla.

Mullan's aim this season was to clear and refine the road, removing fallen timber and repairing the numerous bridges that had washed out from yet another spring's flooding. Progress was slower than he planned, but he nevertheless called a halt to the work to celebrate the Fourth of July. That Independence Day, he carved the inscription *M.R. July 4, 1861* (*M.R.* for "military road") on a great white pine that stood for more than a century at the head of what is now known as Fourth of July Canyon. Legend tells how, on that day, the crew raised such a racket with their powder and rifles that the Indians stopped harassing them, thinking they had lost their minds.[55] Mullan's journal relates a more reserved, if equally patriotic celebration. The lieutenant issued extra whiskey rations to the men and passed a "pleasant" day dining with Sohon, Captain Marsh, and Dr. Taylor on Wolf's Lodge Prairie. "[O]n returning to camp, my men singing Star

Figure 7-8. *John Mullan and his crew crossed what now known as Fourth of July Pass on Independence Day of 1861. This old photo near the site shows a memorial obelisk honoring the expedition; it reads, "4th of July Pass near old Mullan Tree, erected by the Historical Society of Idaho and the Society of Montana Pioneers. Gift of William A. Clark, Jr., of Butte, Montana, to the city of Coeur d'Alene, 1918." The monument is now located at the Mullan Road Historical Site off Interstate 90, about thirteen miles east of Coeur d'Alene, Idaho.*
–Courtesy Spokane Public Library

Figure 7-9. *Closeup of the inscription that Mullan's men carved into a large pine tree alongside the newly constructed road. The pine was later dubbed the Mullan Tree. A portion of the carving is preserved at the Museum of North Idaho.* –Courtesy Museum of North Idaho, Coeur d'Alene

Figure 7-10. *Old interpretive sign about the Mullan Tree* –Courtesy Museum of North Idaho, Coeur d'Alene

Spangled Banner and Red White and Blue. Having made a few remarks to them and given them a drink they retired quietly and in good spirits." However much they did or did not carouse on the night of the 4th, they were at work again by 5:30 in the morning on the 5th.

As planned the season before, the crew cut a new road around the north side of Lake Coeur d'Alene to avoid the heavy flooding on the original southern route. This task required thirty additional miles of grueling timber-cutting and grading. The workers reached the Coeur d'Alene Mission on August 1, a month later than Mullan had hoped, but he nonetheless pronounced himself satisfied with "a creditable piece of mountain work."[56]

By September 15, working in multiple small teams, the crew had built twenty new bridges over the Coeur d'Alene River and avoided a few more crossings by cutting into sidehills. On the other side of Sohon Pass, they implemented the same system to improve the road along the St. Regis River. But by November, snows were closing in, and Mullan was not about to get trapped in that canyon again over another hard winter. He ordered a halt to the bridging once the framework was up–the decking would wait until next year.[57]

The party proceeded to the site of their new winter camp, at the junction of the Big Blackfoot and Hell Gate Rivers, near present-day Missoula. It seems it was the idleness of the prior winter in the mountains that worried Mullan more than the cold–this time, he was determined to be productive. Dividing his workforce, he set one team to work building the main winter camp at the river junction, while the others would erect four smaller work camps along the Hell Gate (Clark Fork) River. The new main camp–as before, a collection of simple log huts–was dubbed Cantonment Wright in honor of George Wright, the Commander of the Department of the Pacific. The four small work camps were established at locations where Mullan wanted new side cuts dug on the banks of the Hell Gate, in order to avoid the dangerous fords they'd been forced to make on their first pass.[58]

Mullan's main task for this trip was constructing a large bridge over the Big Blackfoot River. By the time his crew had the cabins put up at Cantonment Wright, the ground

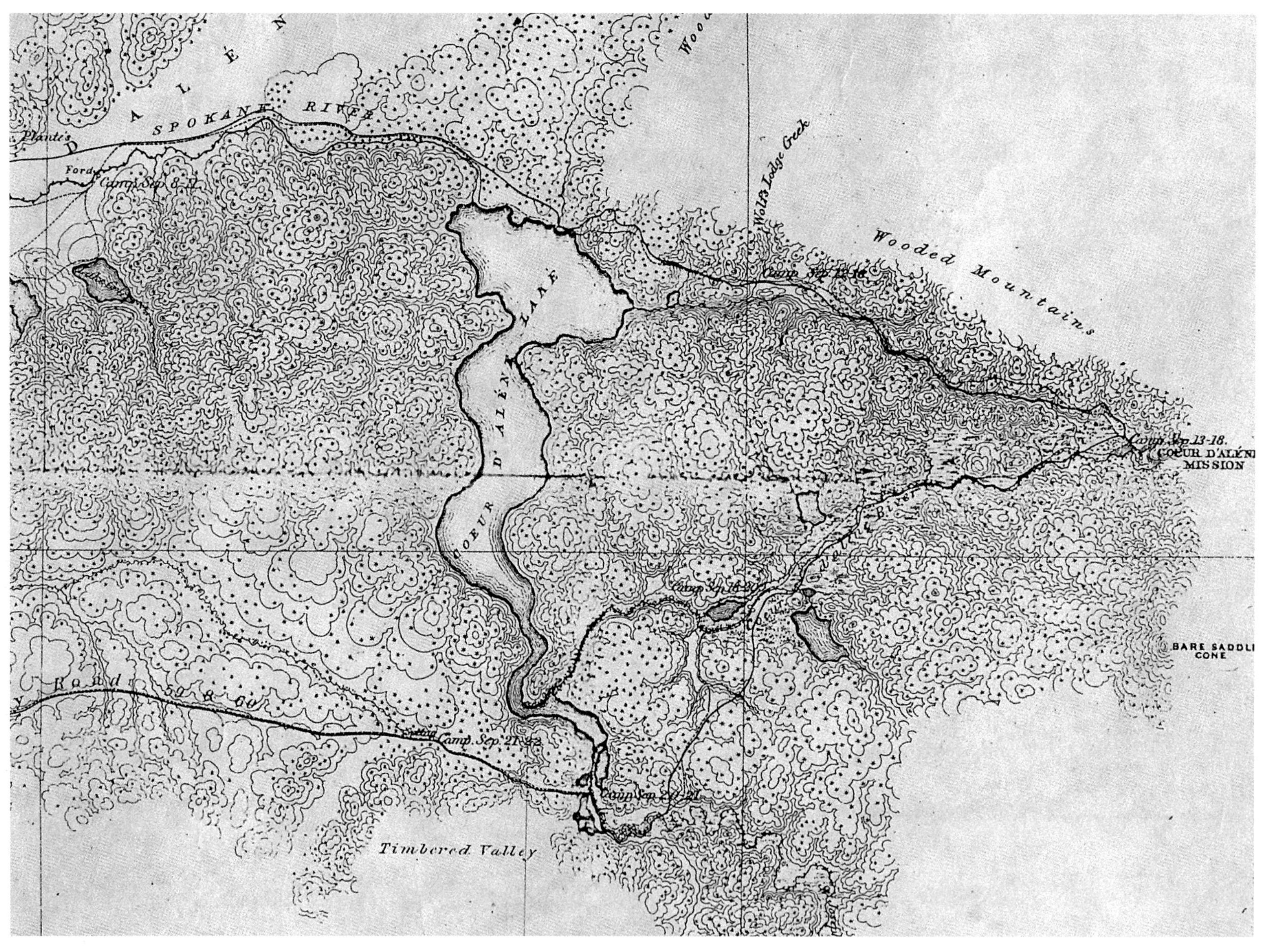

Figure 7-11. *This map by Theodore Kolecki shows the two routes Mullan built around Lake Coeur d'Alene. The first, constructed in 1859, moved eastward around the southern end of Lake Coeur d'Alene, while the second one, built two years later, crossed the Spokane River Plain and followed the north end of the lake.* –Courtesy National Archives and Records Administration, Cartographic Section, College Park, MD

Figure 7-12. *Lithograph of Cantonment Wright by Gustavus Sohon. The cantonment, at the junction of the Big Blackfoot and Hell Gate (Clark Fork) Rivers, was Mullan's base camp during the frigid winter of 1861–62. Note that the buildings were situated in two locations—one above the river terrace and another at the base of the terrace.* –From John Mullan, *Report on the Construction of a Military Road from Fort Walla-Walla to Fort Benton* (1863)

Figure 7-13. *Detail of the bridge at Cantonment Wright, from a Sohon watercolor housed in the National Archives. This bridge over the Big Blackfoot River was one of the longest Mullan built.* –Enhancement by Philip Mobley

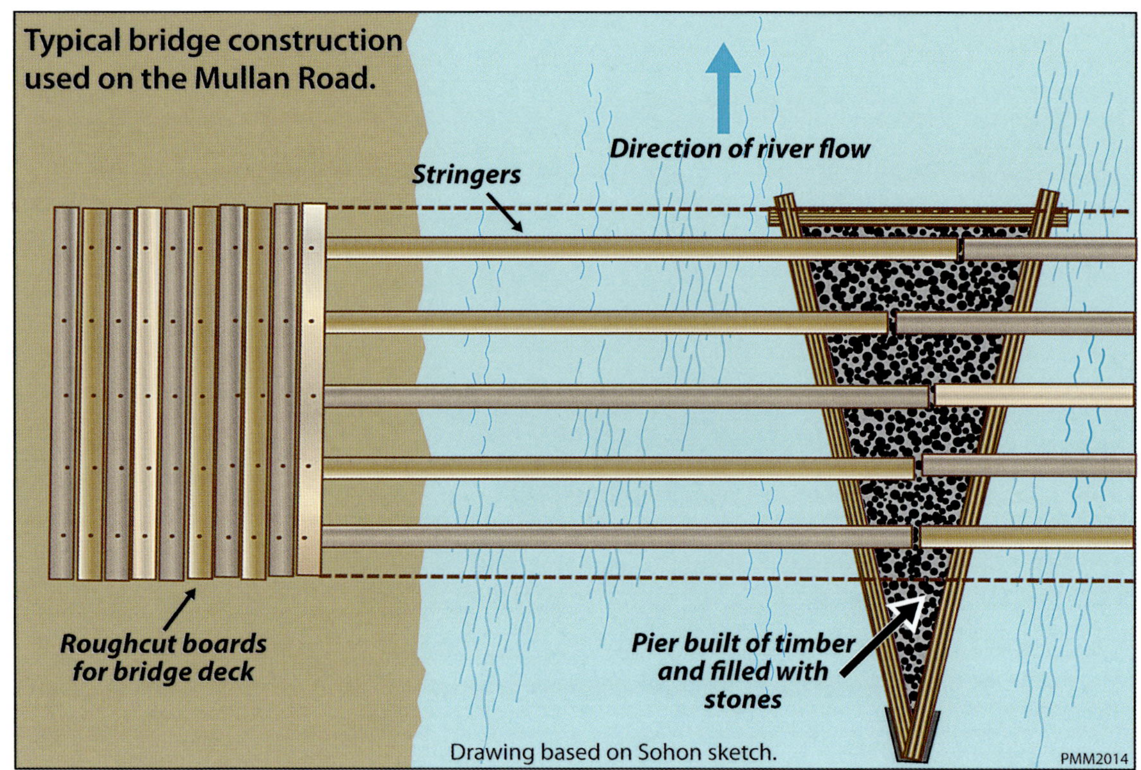

Figure 7-14. *This drawing of a bridge cross-section was based on an illustration by Sohon. In the winter of 1861–62, Mullan built a bridge across the Big Blackfoot using triangular piers filled with rocks, then cutting the ice to sink the piers into the river.* –Illustration by Philip Mobley

was covered in snow and the river was already freezing at its edges. Undeterred, Mullan made a plan: "[W]hen all the timbers [for the bridge] were cut, hauled, hewn and ready to be put together, we threw a boom across the river, which here was two hundred and thirty-five feet broad, six feet deep, and had a current of four miles an hour. By means of this boom we dammed up the floating ice, which in a single night became sufficiently frozen to allow horses to cross."[59] The workers then cut holes in the ice, framed the piers in the holes, and filled the piers with rocks brought in by sled from the surrounding mountains. In this way, they built the entire framework of the bridge over the winter, and in February, the sawyers cut and lay the decking while the ice broke up around them. By the first of March, the four-span bridge was complete, and the bank work done at the other camps had circumvented ten crossings of the Hell Gate.[60]

This victory was all the more remarkable considering that the winter of 1861–62 turned out to be worse than Mullan's previous two in the mountains. The weather was severe, with solid snow from November through April, and temperatures reportedly colder than any in living memory. Cantonment Wright, though conveniently situated for the work, was "exposed to the bleak winds that at times came down the valley of the Blackfoot, and it was found an abode of not over much comfort."[61] Many of the party's stock had wandered off in the fall due to lack of forage on the route. Of those that made it to the winter pasture, most starved or froze to death, and the rest were stampeded by Indians.[62] The bitter weather also made communication between the camps a challenge, as it was often too cold even for traveling by sled. In an eerie repetition of the 1859 incident at Fort Owen, another man lost his legs to frostbite. As Mullan described it:

> I here mention with regret a sad accident that occurred to a citizen passing from one to another of our camps, and which will tend to show the degree of cold we experienced during January. He had left one of the camps with the intention of

> going to the Deer Lodge valley. Night and severe cold overtaking him before he could reach another camp, he halted to build a fire, and being wet endeavored to slip off his moccasins, when he found them frozen to his feet. He became alarmed, and retracing his steps he reached the point he had started from, late at night, but with both feet frozen, and on their being thawed in a tub of water all the flesh fell off. The poor fellow suffered intensely, and his life was only saved by his suffering the amputation of both legs above the knees. . . . A purse of several hundred dollars was raised for him, and he was left to the kind charity of the fathers at the Pend d'Oreille mission, where he remained up to the date of our leaving the mountains.[63]

Contrast this descriptive and sympathetic account with Mullan's muted reaction to the first casualty of the expedition in 1859, an episode that warranted only the briefest mention: "[We crossed] the Snake river without accident, save the drowning of one of my men. . . ."[64] The later passage clearly shows, in addition to "the degree of cold," the degree to which the human cost of his road was beginning to wear on Mullan's psyche.

In the spring, the party returned to bridge building and road repair, but by then the men had become demoralized, and many demanded their discharge. Apart from Captain Marsh and a crew of two dozen men, whom Mullan sent back to finish the bridges on the St. Regis, work ceased for the expedition on May 23, leaving some of the worst bridge repairs undone; Mullan was obliged to pay some local settlers to finish those jobs. He then led the majority of his men back to Fort Benton to be discharged. En route, they performed only the work required for their passage. The party had to travel much of the way by canoe because the swollen rivers had washed over their road. The group reached the fort on June 8. Construction of the Mullan Road was officially complete.

Figure 7-15. *A Mackinac was a lightweight sailboat with two mainsails and a jib. This type of boat was popular on the Missouri River during the fur trade era.* –Illustration by Philip Mobley

At the fort, Mullan had a Mackinac built to transport to St. Louis those men whose enlistments were expired. He then led the remainder back to Walla Walla, again performing "no work beyond what our immediate wants demanded," though he took notes for future improvements.[65]

This last trip on the road constituted, in essence, a victory lap for John Mullan, though it was a limited victory. On the one hand, he found that the road was still in need of significant repairs. Much to his dismay, he discovered that Captain Marsh had misunderstood his instructions and had failed to fix the bridges on his way back through the mountains. Six of the bridges on the St. Regis River were completely washed out, as were another two on the Coeur d'Alene. Even the new bridge over the Big Blackfoot was already "to a certain extent, thrown out of shape."[66] One of the new side cuts on the Hell Gate had been washed away, as well. It must have been disheartening for Mullan to see his road in such disrepair after three years of work. Yet, with his characteristic stoicism, he conceded only that "The freshet in these mountains is a difficulty to be sternly handled, and too much care cannot be taken in selecting suitable abutment sites for bridging."[67]

On the other hand, the project was hardly a total loss—far from it. The road was already serving its primary purpose: opening the Pacific Northwest to American settlement. While camped in the Deer Lodge Valley, Mullan met several wagon trains headed to Fort Benton on the new road. At the same camp, he received the heartening news that four steamers had just arrived at Fort Benton's new trading house from St. Louis, bringing 364 emigrants bound for Walla Walla and points en route. "This cheering intelligence was very gratifying to myself," he related, "seeing in it the commencement of a long line of emigrant travel, that in years to come must course along these rivers in search of new homes toward the Pacific slope."[68]

By the time Mullan left the mountains, the emigrants were overtaking him, and he could not have been more proud that these pioneers were "journeying safely and pleasantly towards the setting sun." He greeted them warmly, augmented their supplies with his own, provided advice on travel and campsite selection, and sent them on their way with a smile. "The safe passage of these emigrants during this season proves the value of this line for emigrant purposes, and will yet cause it to stand in competition with other lines across the continent," he wrote with satisfaction.[69]

Mullan could not have known then that in only a few years, the legislature of Washington Territory would petition the federal government for another $100,000 to repair bridges and clear timber from the road, as it had long become impassable by wagons.[70] He could not have known that Blake's expedition would be the last major troop movement over the road, as the Civil War turned the military's attention back East, and by the time of the Nez Perce campaign of 1877, the road would have deteriorated to the point of uselessness. In the summer of 1877, Generals Sherman and Sheridan, on their tour of Northwest posts, would find the road "blocked up with fallen timber, and its hundred and sixty bridges all gone."[71] He could not have known any of this, but perhaps he should not have been surprised.

At the same time, though, Mullan could see and appreciate the positive effects of his road while it lasted, if not the far-reaching impact it would ultimately have on the western United States. Very soon, the Mullan Road would service the Montana gold rush of 1863 and, some years later, provide access to the most lucrative silver-mining region in America, the Coeur d'Alene district in northern Idaho. In the 1880s, thanks partly to Mullan's pioneering survey work, the transcontinental Northern Pacific Railway would be completed.

Of course Mullan would never know that in the twentieth century, many sections of his road would be paved over to accommodate cars with internal-combustion engines, and eventually many of these sections would form portions of Interstate 90, the nation's longest interstate highway, which after 1991 would provide unimpeded automobile passage between Boston and Seattle. He likewise would never know that Missoula and Coeur d'Alene would become thriving cities with a combined population of more than 115,000 and annual tourist traffic of several times that.

In 1862, as Mullan prepared to make his way back east from Walla Walla to file his reports, he knew only that he had done what he set out to do. The reluctance of the War Department and the stinginess of Congress had not deterred him. The intrigues of the Indians had not derailed him. And while he may not have fully conquered those beguiling, terrifying, wonderful, awful mountains, neither had they conquered him.

". . . as I was sent out to build and construct the road, it *shall* be constructed."

–John Mullan, 1860[72]

---

1 Mullan to Humphreys, January 3, 1860, in John Mullan, P. M. Engle, et al, *Military Road from Fort Benton to Fort Walla Walla: Letter from the Secretary of War . . .* (1861; reprint, Ann Arbor: Michigan Historical Reprint Series, Scholarly Publishing Office, University of Michigan Library, 2005), p. 33; this source is hereafter cited as Mullan, *Report,* 1861.

2 John Mullan, *Report on the Construction of a Military Road from Fort Walla-Walla to Fort Benton* (1863; reprint, Fairfield, WA: Ye Galleon Press, 1994), p. 21; this source is hereafter cited as Mullan, *Report,* 1863.

3 Mullan to Humphreys, April 3, 1860, Mullan, *Report,* 1861, p. 44.

4 Ibid., p. 41.

5 Mullan, *Report,* 1863, p. 20.

6 Ibid.

7 Mullan to Humphreys, June 6, 1860, in Mullan, *Report,* 1861, p. 44.

8 Mullan to Humphreys, April 3, 1860, in ibid., p. 41.

9 Mullan to Humphreys, June 6, 1860, in ibid., p. 46.

10 Ibid., p. 45.

11 Ibid.

12 Ibid., p. 46.

13 Mullan, *Report*, 1863, p. 22.

14 Mullan to Humphreys, June 6, 1860, in Mullan, *Report*, 1861, p. 45.

15 Ibid.

16 Ibid., p. 50.

17 Mullan, *Report,* 1863, p. 23.

18 Mullan to Humphreys, August 2, 1860, in Mullan, *Report*, 1861, p. 50.

19 Ibid., p. 51.

20 Ibid.

21 Mullan to Humphreys, October 12, 1860, in ibid., p. 63.

22 Ibid., p. 51.

23 Mullan, *Report,* 1863, p. 24.

24 Mullan to Humphreys, August 2, 1860, in Mullan, *Report,* 1861, p. 51.

25 Mullan, *Report*, 1863, pp. 24–25.

26 Ibid., p. 25.

27 Ibid.

28 Ibid.

29 Mullan to Humphreys, August 2, 1860, in Mullan, *Report*, 1861, p. 52.

30 Mullan, *Report*, 1863, p. 26.

31 Ibid.

32 Ibid.

33 August Kautz, "From Missouri to Oregon in 1860: The Diary of August V. Kautz," *Pacific Northwest Quarterly* 37, no. 3 (July 1946), p. 218.

34 Ibid.

35 Ibid., p. 219.

36 Mullan, *Report*, 1863, p. 27.

37 George Blake, "Journal of the march of a detachment of U.S. recruits for Oregon commanded by Major George A. H. Blake, 1st Dragoons, from Fort Benton to Fort Vancouver, W.T., 1860 Aug-Oct," n.d., WA MSS S-1721, Beinecke Rare Book and Manuscript Library, Yale University, New Haven, CT.

38 Mullan, *Report*, 1863, p. 27.

39 Martin D. Hardin, *Across the New Northwest in 1860* (Philadelphia: L. R. Hamersly & Co., 1882), pp. 189–90.

40 Ibid., p. 195.

41 Mullan to Humphreys, October 12, 1860, in Mullan, *Report*, 1861, p. 54.

42 Ibid., p. 60.

43 Toohill to Mullan, February 20, 1860, in ibid., p. 137.

44 Mullan to Humphreys, October 12, 1860, in ibid., p. 63.

45 Ibid.; also Mullan, *Report,* 1863, p. 28.

46 Ibid., p. 67

47 Ibid., p. 69.

48 Mullan, *Report*, 1863, p. 28.

49 Ibid.

50 Ibid.

51 Ibid., pp. 28–29.

52 Mullan to Swan, January 12, 186[1] (private collection); a copy of this letter was graciously provided by Thomas Minckler of Thomas Minckler Fine Art. Though dated by Mullan's hand January 12, 1860, that is certainly a mistake, as Mullan was at Cantonment Jordan in January 1860; he undoubtedly made the common January error of dating papers with the previous year. We know that Mullan launched a failed bid for the governorship of Idaho Territory in 1863; this very interesting letter indicates that he harbored political ambitions at an earlier date. In addition to inquiring about the congressional seat, he notes that "I intend to . . . cast my fortunes with the elopment of the territory east of the cascades where yet great wealth awaits more development." Whether this refers to his political fortunes, in anticipation of the creation of Idaho Territory, or his financial fortunes, reflecting the many indications of gold he'd found in the mountains, cannot be known for certain. In the end, though, Mullan did seek political office (unsuccessfully), and he never took up mining.

53 Mullan, *Report*, 1863, p. 28.

54 Ibid., p. 29.

55 Helen Addison Howard, *Northwest Trail Blazers* (Caldwell, ID: Caxton Printers, 1963), p. 159.

56 W. Turrentine Jackson, *Wagon Roads West* (Oakland: University of California Press, 1952), p. 270.

57 Mullan, *Report,* 1863, p. 32.

58 Ibid.

59 Ibid.

60 Ibid., p. 33.

61 Ibid.

62 Jackson, *Wagon Roads West*, p. 271.

63 Mullan, *Report,* 1863, pp. 32–33.

64 Ibid., p. 13.

65 Ibid. p. 34.

66 Ibid., p. 35.

67 Ibid., p. 36.

68 Ibid., p. 35.

69 Ibid., p. 36.

70 Jackson, *Wagon Roads West*, p. 277.

71 Philip Henry Sheridan and William Tecumseh Sherman, *Reports of Inspection Made in the Summer of 1877 by Generals P. H. Sheridan and W. T. Sherman of Country North of the Union Pacific Railroad* (Washington, DC: Government Printing Office, 1878), p. 97.

72 Mullan to Humphreys, January 3, 1860, in Mullan, *Report*, 1861, p. 30.

Completing the Mullan Road from Mullan Pass to Fort Benton

A Harbinger of Change

BY KEN ROBISON

The summer of 1860 was a transformational moment in the history of the Upper Missouri, deep in the heart of native Blackfeet country in present-day Montana. Until that time, the fur-trading posts of Fort Benton and its nearby rival, Fort Campbell, both located at the head of navigation on the Missouri River, had been the focus of white American commercial activity in the region. The Blackfeet Indians, long dominant in the Upper Missouri area, were warily peaceful in their relative isolation from white encroachment.[1]

But during that summer of 1860, three groups converged on Fort Benton and set the stage for change. On July 2, the first steamboats reached the Fort Benton levee, with the first U.S. military unit onboard; on the 14th, the Raynolds Expedition arrived after their extensive explorations of the Yellowstone and Missouri Rivers; and on August 1, First Lieutenant John Mullan arrived with his road-building expedition, having just completed the Fort Walla Walla to Fort Benton Military Wagon Road (a.k.a. Mullan Road). Together, the steamboats from St. Louis and the wagon road to the Pacific Ocean completed the dream of a trade corridor across the northern United States, a "Northwest Passage" from the Mississippi River to the Pacific Ocean and presaged dramatic change in the region.

In St. Louis on May 3, Major George A. H. Blake's brigade of thirteen officers, over two hundred enlisted recruits, and about eighty laundresses and servants from the U.S. Army First Dragoons had embarked three steamboats, the *Chippewa*, the *Key West*, and the *Spread Eagle* to proceed up the Missouri River.[2] Their mission was to arrive at Fort Benton in time to test the feasibility of the new Mullan Road to Fort Walla Walla.

On July 2, one month before Mullan's arrival, these first steamboats ever to fight their way up the Missouri River from St. Louis moored at the Fort Benton levee, completing their two-month voyage. Flying a special flag, "For Fort Benton or Bust," the boats were greeted with great fanfare. When the steamboat *Chippewa*, commanded by Captain William H. Humphreys, and the *Key West*, under Captain John LaBarge, moored at the levee, the steamboat era commenced from St. Louis to the head of navigation on the Missouri River.

Aboard the steamers, which belonged to Pierre Chouteau, Jr. & Company, was the company's owner, Charles P. Chouteau; its Fort Benton agent, or factor, Andrew Dawson; Blackfeet Indian agent, Colonel Alfred J. Vaughan; and a number of company employees, along with military supplies, Indian trade goods, and government annuities.[3] Also onboard the crowded steamboats were Major Blake and his First Dragoons, who were poised to become the first military users of the Mullan Road when they traveled west from Fort Benton to Washington Territory.

After his arrival in the Sun River Valley, Agent Vaughan wrote his impression of the path-breaking trip up the Missouri River:

> We arrived at Fort Union on June 15, and after discharging the Assinaboine annuities, went on our way rejoicing. In due time we made Milk River. . . . the steamer *Spread Eagle*

The steamboat Chippewa*, shown here, together with the* Key West*, were the first steamboats from St. Louis ever to moor at the Fort Benton levee. The steamers brought Major George A. H. Blake and his First Dragoons, who became the first military users of the Mullan Road.* —Courtesy Overholser Historical Research Center, Fort Benton, MT

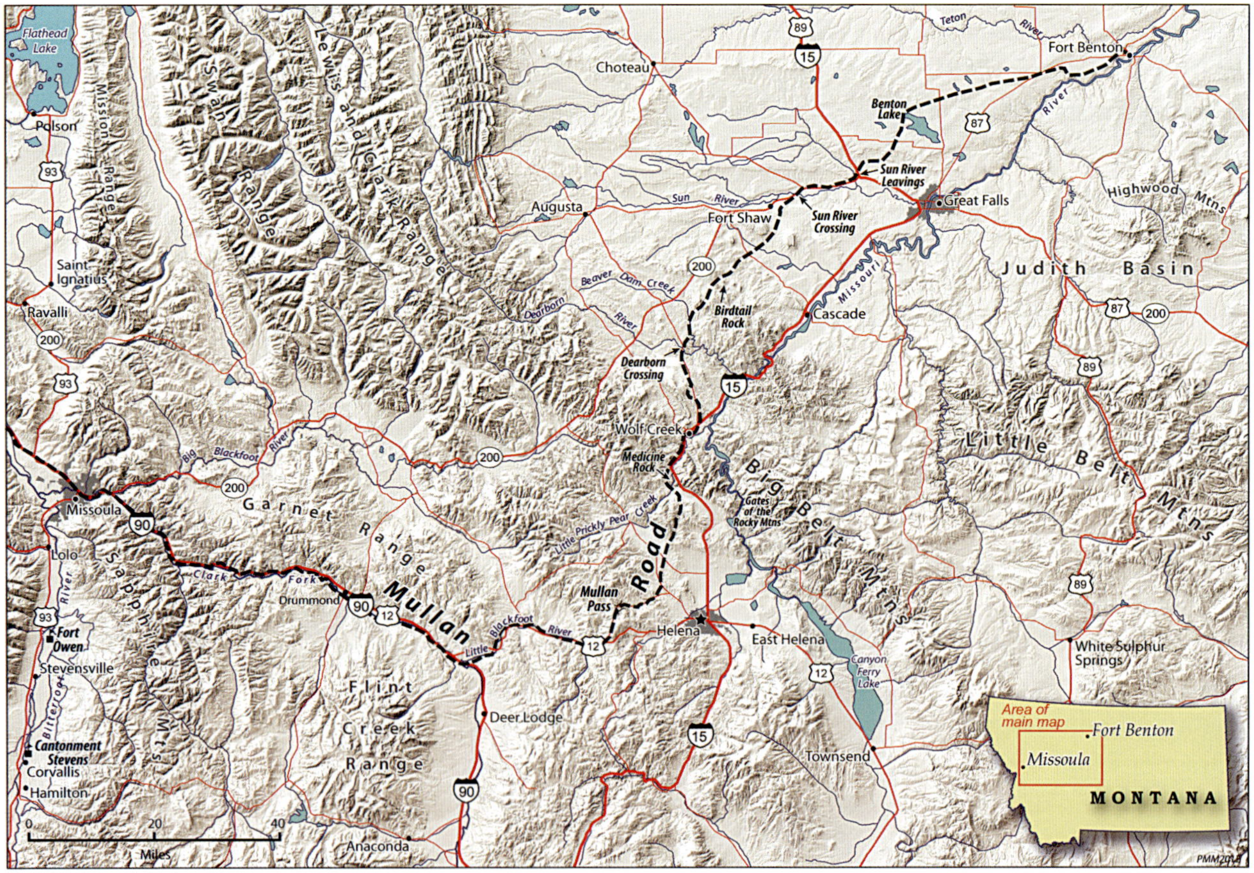

Eastern section of the Mullan Road –Map by Philip Mobley

This is the earliest known photograph of Fort Benton, taken in 1860 by Lt. James D. Hutton, member of the Raynolds military reconnaissance expedition. –Courtesy Overholser Historical Research Center, Fort Benton, MT

Watercolor image of Fort Benton by John Mix Stanley, 1853. –Courtesy Yale University Art Gallery

accompanied us some ten miles further and then returned on her homeward way, having been ten miles further up than any side-wheel boat was before.

Our little fleet, now reduced to two, the *Key West,* commanded by Captain [John] Labarge, in the van, boldly and fearlessly steered their way up what would seem to the uninitiated an interminable trip. At length the long expected goal is made, and on the evening of July 2 the two gallant crafts, amidst the booming of cannon and the acclamations of the people, were landed at Fort Benton with one single accident, and that was a man falling overboard, who unfortunately was drowned.[4]

Meanwhile, a major exploration group, the military reconnaissance expedition of Captain William F. Raynolds, arrived at Fort Benton from the Yellowstone River area on July 14, twelve days after the steamboats and two weeks before Mullan, to await passage down the Missouri. The Raynolds Expedition had spent the previous year exploring the Yellowstone Basin. With the expedition were topographers Lieutenant Henry E. Maynedier and Lieutenant James Dempsey Hutton; naturalist and surgeon Dr. Ferdinand V. Hayden; meteorologist and artist Anton Schonborn; and famed scout and mountain man Jim Bridger. The party was escorted by a detachment of thirty men from the U.S. Second Dragoons, under First Lieutenant John Mullins.[5] Hutton, who carried photographic equipment, took the first known photograph of the Fort Benton trading post, taken from a vantage point across the Missouri River. In the summer of 1860, there was no town of Fort Benton, only the trading post; a short distance upriver stood a rival trading post, Fort Campbell, which had recently been bought by Pierre Chouteau, Jr. & Company.

On July 17, Mullan's expedition ascended the western slope of the Rocky Mountains and crossed the Continental Divide at Mullan Pass.[6] After an arduous thirteen months of road-building in what are now Idaho and western Montana, including his six-month stay at the expedition's winter encampment at Cantonment Jordan, on the St. Regis River, Mullan was confident that the rest of the route to Fort Benton would present no great challenge. His confidence was based on personal knowledge gained during 1853–54, when he had accompanied Washington territorial governor Isaac I. Stevens on his Pacific Railroad Survey Expedition. During this earlier survey expedition, Mullan had traveled some 1,000 miles, crossing the Continental Divide six times while marking a future wagon and railroad route from Fort Benton to Walla Walla. This previous expedition had equipped Mullan with knowledge of the topography, important Indian trails, and Native tribes throughout the Upper Missouri region.[7]

This essay presents the Mullan expedition's activities as they completed their wagon road from Mullan Pass to Fort Benton in the spring and summer of 1860. Throughout the expedition, Mullan made daily entries in his field journal, detailing key events, topography, work, and activities of the journey; these notations were later transcribed for his official reports to the government. In addition, civilian topographer P. M. Engle and civilian expedition member John Strachan recorded insightful observations of the expedition. These reports form the basis of this essay.[8]

In his report written on January 17, 1860, Mullan described his means of acquiring intelligence and communicating with the outside world, emphasizing a persistent theme–justifying expenses:

> Tuesday, January 17. Our two expressmen, [P. E.] Toohill and [George] Young, are each receiving $100 per month for winter service on snow-shoes. . . . Thus do contingencies arise that it is difficult to foresee, and these have to be met, or to be cut off from mails all winter; and I regard the interest of our position too great to let either risk, danger, or expense be considered, when it is necessary to send intelligence to the settlements, the nearest being two hundred and eighty-eight miles.[9]

In this same report, Mullan noted that twenty inches of snow remained on the ground, and he described his plan to resume road building in the spring of 1860:

> I shall have the men on the road as soon as the ground will enable us to work it; at present it is frozen. As I informed the department, I shall go to the Bitter Root valley, between the 15th February and 1st March, to prepare a pack-train to go to Fort Benton for my supplies and to prepare for the location of the road for spring's work.[10]

The long winter at Cantonment Jordan had been hard on Mullan, his men, and their animals, as he noted in this report on March 10, 1860:

The winter was most marked in its character; it began early, precipitately and severe, and, though long, ended mildly and moderately, continuing from 2d November to 20th February. It has proved most disastrous to our animals, which, being thin and reduced when winter set in, were not in a condition to reach the Bitter Root valley in safety through the snow and cold, when grass and browsing were covered up.[11]

In the meantime that winter, to build on the knowledge of the area he had acquired during the Pacific Railway Survey, Mullan sent his topographical engineer, P. M. Engle (sometimes spelled Engel), and his interpreter and expedition artist, Gustavus Sohon, separately on advance trips to Fort Benton. These important trips would give Mullan updated knowledge of the planned route as well as acquire necessary supplies for the expedition. Sohon's trip in particular proved crucial in building relationships with the powerful Blackfeet Indians.[12]

On November 8, 1859, Engle, accompanied by E. Irvine and two laboring men with four pack animals and six mounts, had left the Bitter Root Valley for Fort Benton. Engle's mission was "to ascertain in detail the character of the line for the proper location of the military wagon road, and in order to arrange the working parties for spring's operations, and at the same time collect data regarding the climatology of the main range of the Rocky mountains in midwinter."[13]

Although heavy snows and cold weather slowed the progress of the Engle party, they reached the Continental Divide in twenty days. Engle followed old Indian trails and, in places, remaining portions of an 1854 wagon road. As Engle crossed the divide, he planned for the spring's road construction and reported in detail on the work required:

Monday. November 28. At the foot of the divide, a quarter of a mile from the top, I struck the headquarters of Big Prickly Pear creek [today's Austin Creek]; one mile and three-quarters further, a little branch of running water, coming from the north, crosses the road. Just before reaching this creek a little side-hill cutting will have to be done, and also on the other side of it are points where the hills come close to the creek; the valley widens considerably and affords good camping places. On the east side of Stormy creek [likely today's Greenhorn Creek] the ground is level for three-eighths of a mile; then a side-hill cut of 20 yards will bring the road again in an almost level plain for three-quarters of a mile; at the end of which the wagon road leaves the Indian trail, which leads down the Big Prickly Pear, and takes to the west over a small side hill covered with pedrigal rock, and in a half mile crosses a dry water-run.[14]

Traveling about thirty miles northeast of Mullan Pass, at times battling gale-force winds and snow, Engle's party approached Medicine Rock, just west of Little Prickly Pear Canyon and today's Interstate 15. The Medicine Rock area would prove to pose the greatest challenge of any portion of the Mullan Pass to Fort Benton segment. Foreseeing this, Engle described his plan:

Thursday. December 1. I started at 11 a. m. It was very cold, and the snow was five inches deep; and the road leading to Medicine rock strikes in three-fourths of a mile a little creek, and following it up for half a mile strikes the foot of the hill range. Here the trail battles a very steep ascent, which the wagon road can make more gradually in keeping to a ravine running to the left of the trail. Reaching the top of the first hill, a good road with easy grades leads towards Medicine rock, and from there descends to a small creek called Medicine Rock creek. The descent will require some side-hill work in places where the slope is too steep. From the foot of the hill to the creek is two miles. The next three miles of the road, which brought us to Mullan's creek [today's Lyons Creek], will require work in different places, as the trail keeps along the side hills. Two hollows, the first one pretty deep, will have to be passed, and a few trees to be cut, to bring wagons through with four or five yoke of oxen. The ascent on the other side of the creek is steep, but short, and then the character of the ground for two miles further remains the same as described. At the end of that distance the road leads towards Prickly Pear Creek valley, [with an easy descent].[15]

Battling severe weather, with temperatures as low as forty degrees below zero, Engle's party continued, crossing the Dearborn River and passing Bird Tail Rock to arrive in the Sun River Valley on the afternoon of December 4. There at the Blackfoot Indian Government Farm, Indian agent Alfred J. Vaughan greeted them.[16] Engle reported:

Sunday. December 4. After a ride of three and a half hours we reached Sun River farm, which was twenty-three and one-tenth miles distant from our last night's camp. We were all

> more or less frost-bitten, and have suffered extremely. Colonel Vaughan, Indian agent to the Blackfoot nation, received us very kindly, and with his well-known hospitality offered us the accommodations of the agency for any length of time that we might wish.
>
> Monday–Tuesday. December 5–6. We remained at the farm during the 5th and 6th, and left on the 7th of December. The cold being too severe to take notes on our march, I did so on my return.
>
> . . . I learned at Sun river that the distance to Fort Benton by shortest route was estimated at forty-five miles. At nineteen miles from the farm the road strikes a shallow pond [today's Benton Lake], and seven miles further is a small spring [today's 28-Mile Springs]. . . . The interpreter of Mr. Dawson being at the time at the farm, I accepted his offer to guide us to the fort.[17]

Because of the scarcity of timber and water on the shortest route to Fort Benton, Engle tried to find a better route for the wagon road, one with camping places that had wood and good water. His party traveled along the Sun River toward the Missouri River but determined that the Missouri between the mouth of Sun River and Fort Benton was also poorly timbered.[18]

Engle's group arrived at Fort Benton on December 8 to find a warm reception and very little snow. They remained a week to rest their horses. Malcolm Clarke of Fort Campbell gave Engle a small government-issued mountain howitzer, and Engle affixed an odometer to one of its wheels to measure distances. When Engle departed Fort Benton for his return trip to the Sun River on the 14th, he was accompanied by Colonel Vaughan, who had followed him to Fort Benton, and another party, sent by Fort Benton factor Andrew Dawson.[19]

The route Engle chose for the return trip followed the Teton River rather than the shorter route via the lake and the spring. The Teton route provided wood and water, but it required many river crossings. Striking southwest along Muddy Creek to the Sun River, Engle's party arrived at the government farm on December 19. Along the way, Engle reported, "some Indians," no doubt Blackfeet, had plundered several wagons that he had left unattended for several days because of a heavy snowstorm. Engle sent an unidentified man over to "the Teton mission" (St. Peter's Mission near today's Choteau, Montana) to obtain a cart.[20]

Engle's party continued to battle severe weather, reporting snow increasing in depth every mile, reaching twelve to fifteen inches at Little Prickly Pear Creek. With drifts making passage almost impossible for a vehicle, Engle cached the cart there. As he neared the Continental Divide, Engle reported conditions there:

> Friday. December 30. Started at 10.15 a.m.; wind storm with snow; the snow towards the divide is 15 inches deep; on the divide a drift of 80 yards in length, and five or six feet deep; the crust will bear a man, but our horses sunk through, and have hard work. On the west side of the divide I found only six inches of snow, with occasional drifts of the same depth down the canon of the north fork of the [Little] Blackfoot river.[21]

Engle and his men returned to Fort Owen (at present-day Stevensville, Montana) on January 6, 1860, one week after reaching the summit of the Rocky Mountains. Their experience traveling through harsh winter conditions provided valuable information for planning the spring work on the road. In addition, it allowed for better coordination with the white Americans at the government farm, the Jesuits' Blackfeet mission on the Teton River, and the Fort Benton trading post.

Throughout the long winter at Cantonment Jordan, near today's De Borgia, Montana, in the Bitterroot Mountains, Mullan planned the roadwork, coordinated the crews, and communicated with his supervisors. In an important report dated January 3, 1860, to Captain A. A. Humphreys, the officer in charge of explorations and surveys, Mullan advised Humphreys that he planned to send civil engineer Walter W. Johnson as a special messenger to Washington, D.C., via Fort Walla Walla, to deliver Mullan's master plan for the new military route from St. Louis to Fort Walla Walla, i.e., the wagon road. Johnson would also provide details on the condition of the expedition and acquire the necessary additional instruments before rejoining the expedition at Fort Benton in the summer. Mullan also reported his plan to proceed, with a small party, to Fort Benton in mid-February or early March to scout road locations and pick up his preordered supplies at

Fort Benton. Mullan later changed his mind; unable to find the time to make the trip, he sent Gustavus Sohon to Fort Benton in his place.[22]

In addition to this report, on January 3 Mullan laid out his plan of operations for the coming year, including the movement and disposition of his men and materiel. He argued his case for military use of the new road, proposing that some three hundred army recruits who were needed that year in Oregon be sent up the Missouri River by steamboat to Fort Benton, where they would use Mullan's supply train and proceed over the new road to Fort Walla Walla. Mullan explained the advantages of his new route to the Northwest in purely financial terms—his plan would save some 50 percent in transportation costs:

> The men . . . are generally healthy, and, though our prospects day by day are snow-clad and dreary mountains, are all cheerful, and long for the opening of spring. . . . I shall not be enabled to send another express to Walla-Walla before the middle of March, and this cannot reach Washington before the 1st of June, and I could not get a reply before the 1st of August, at which time I shall be at Fort Benton. Therefore, I will submit to the department my plan of operation in the opening of the spring. . . . I shall push the work with vigor till the 1st of June, when I may send forward a number of men whose services I can best spare, to Fort Benton, to take passage on the steamer. I shall push on the road with the remainder, reaching there by the 1st, or some time in August. I have already at Fort Benton a ninety-foot keel-boat, which I shall use in sending my party down the Missouri.
>
> When reaching Fort Benton I shall send down the remainder of the men, together with a number of the escort, whose times of enlistments will have expired, or near expiration, as an escort to those men as far as Fort Leavenworth. . . . This will not only be necessary, but in the end more economical than to discharge them on the Pacific. . . .
>
> We shall reach Fort Benton with our road, say in August, with a large and empty train, which has cost much. We shall then be compelled either to take this train back empty to Walla-Walla, hiring teamsters for the purpose, or abandon it at Fort Benton. Strict economy, and a desire to see the route practically opened by a new line for military operations, therefore, prompt me to recommend to the department as follows: That a detachment of three hundred recruits, or the number that may be needed to fill up the regiments now in the department of Oregon, be sent, under the charge of a judicious officer, from St. Louis to Fort Benton, in the American Fur Company's steamer, in April or May, to meet our train at the same point, in or about the time of our arrival. Let them start with four months' supplies from St. Louis—sixty days to be used up the Missouri to Fort Benton and sixty days on the march to Walla Walla. Making use of our empty transportations at Fort Benton, we can immediately return to Walla-Walla in safety and security. I would therefore respectfully ask that, if these recruits are sent, they be turned over to my command at Fort Benton by the officer in command, and I will guarantee to guide and take them to Walla-Walla in security and with success, from which point they can be distributed by the department commander as the requirements of the service may at the time demand. On the score of economy alone I will submit to the department the following figures for its consideration and comparison:

300 recruits from New York city to the Columbia river, at the prices now paid, as obtained from books of Quartermaster General's office, Washington, $200 each	$60,000
Passage of 300 recruits from St. Louis to Fort Benton, at $30 each, (price agreed to by Mr. C.P. Chouteau, of St. Louis)	9,000
Cost of transporting 36,000 rations, weighing, say, 108,000 pounds, at the maximum cost prices per pound, 10 cents	10,800
Purchases of horses, &c.,	say 1,200
Hire for forty teamsters for two months, at $50 per month	4,000
	25,000
Contingencies and rates of teamsters	5,000
	30,000

> Saving in cost alone of $30,000, to say nothing of the additional efficiency of the men, the effect upon the Indian mind, and the want of the experiment of testing the value of a new line via the Columbia and Missouri rivers, with a land portage of only 600 miles. During the coming summer this can

be more successfully carried out than at any other time, for we have at hand the land transportation ready at Fort Benton, and thus have the advantage of comparison of a line within the limits of our territory by water, mostly with one made with all the discomforts to troops, especially via the Panama route, through foreign territory.

This will not be to initiate a new movement, but simply a new method or route by which to transport recruits that must be sent out by some route to fill up the regiments now in the department of Oregon.

I am sanguine to believe that if these men start under a discreet and efficient officer, and well clad, and with good shoes, and the number of rations as recommended, they can make the trip in security and safety; and I trust the department will pardon my asking that the command be turned over to me at Fort Benton. Being only guided in this from the fact that my long connexion [*sic*] with the route, and having successfully opened it, and my present knowledge of the details along the line are of such a nature to constitute a sufficient argument for the request, and feeling a special interest in the success of my recommendation. I can carry out the movement without being trammeled by one who is not familiar with the route and the general line of operations between these two points. I trust, therefore, the department will give the matter their special consideration; and whatever the decision may be, I trust to be informed of it at the earliest day.[23]

Mullan concluded his plan for the new transportation route with assurance that "The Indians on this route are still friendly, and at present we have no special fears for the future."[24] Despite these assurances to the War Department, Mullan knew that he could not take the Blackfeet's amicability for granted, and he began planning to cultivate their "friendship."

Near the end of winter, on March 16, Mullan's German-born guide, interpreter, and artist Gustavus Sohon departed Hell Gate Ronde (east of today's Missoula) with a pack train and seventeen Flathead Indians, heading to Fort Benton to bring back 11,000 rations, which had been prepositioned at the fort the previous summer. On March 11 Mullan described Sohon's mission and preparations, noting that the Blackfeet would perhaps be less friendly than originally expected:

I have here secured the services of the friendly Flathead Indians, who, leave here on the 15th instant with Mr. Sohon for Fort Benton, with a pack-train of eighty-five horses, for supplies now at Fort Benton. They will rejoin me by the last of April. In addition to the economy and expedition of this arrangement, much good is anticipated from it, in arranging the unsettled condition of affairs with the Blackfeet Indians, with whom disturbances have again commenced. I write to the Indian agent, Colonel Vaughan, at Fort Benton, to send the chief "Little Dog" and some of the principal men to meet me on a visit, in order that I may assure their tribe that we come as friends, and in order to guarantee our safe conduct while working the road through their country. Unfortunately for the country, there are numerous ill-disposed whites and half-breeds among them, who, by a studied course of evil design, are daily instilling into their minds things well calculated to produce the most pernicious results. The Indians hence are much exercised, and being in open war with the Pend d'Oreille Indians, are disposed to be troublesome. I shall, however, by a judicious course, assure them of our good intentions, and am sanguine to believe that the many conflicting elements may be so governed that much good may yet accrue from our mission.

The Indian affairs in this quarter are *not* in the condition that I would well wish them, and I fear, unless measures, speedy and remedial, are taken, that we shall have much trouble ere long in this region. The Blackfeet are, I believe, if not meddled with, a good people, and might be made better; but the evil elements are so directed that it is evident that it is only a question of time that they shall be included with other miserable and treacherous tribes. But I hope for the best.[25]

Thus Sohon's trip to Fort Benton assumed crucial importance. The expedition needed the supplies to continue the spring work, and Mullan needed to establish direct contact with the Blackfeet. Sohon rejoined Mullan in mid-April, accompanied by a delegation of Blackfeet led by their principal chief, Little Dog, who sought assurance of "the true objects" of Mullan's mission. Mullan summarized the Blackfeet visit in just a few words:

They [the Blackfeet] were well received, treated kindly, and, receiving a few presents, returned much pleased to their people. Certain malicious white men had, by false reports, spread quite a panic among them, and they had augured anything but favorable objects in our approach to their country. I am confident that by a prudent and judicious course with them our passage will be fraught with neither delay nor difficulty.[26]

Expedition civilian John Strachan described more completely and colorfully this critical first encounter of the expedition with the Blackfeet under Chief Little Dog, including a rich depiction of the Indians' attire:

> We are now in the Pikani, or Black Feet Indian country. We were informed by some of the Flat Head Indians that the Black Feet were opposed to our passing through their country; the Chief was at once summoned and arrived several weeks after with eight of his principal men. The Chief, whose name is Little Dog, was well dressed in black cloth and Beaver hat. This suit he received as a present from the Indian Agent. The rest of the party were decked out in all their finery, feathers and flowing ribbons, ornamented leggings and sashes. Their feet were covered with moccasins richly ornamented with beads and red cloth, and the usual worked tobacco and pipe pouch. The latter is shaped like a lady's reticule, and is generally prettily worked with beads. The simple bag does not, however, give sufficient scope for ornament, and usually it has several long tails to it, which are worked with silk of gaudy colors. The party were all mounted on fine horses, and the Chief at their head had a fine American flag floating in the breeze as an emblem of peace. Lewis and Clark, when in that country, presented an American flag calling it a flag of peace.
>
> After having a hearty shake of the hands, a custom which all Indians through this country strictly adhere to, they dismounted, had supper, and in the evening a council took place. Most of the Indians spoke, and the Chief made an able speech, which through three interpreters of different languages at last came in English. He was willing that we should pass through their country, but was opposed to the whites killing their game. He was unwilling to go to war with the whites and kill them off, as he was depending on them for guns, powder, paint, beads &c. Lieut. Mullan told them his object in their country, and in conclusion asked them if they would promise not to go to war with any of the different tribes around them. He [Little Dog] was willing not to fight with any but the Snake; as it was a powerful tribe and always committing depredations upon their tribe, they had to fight. This Chief was the most intelligent looking Indian I had seen. He presided over more Indians than any other Chief in the Territories, by some estimated at twenty thousand. After following us three days they left.[27]

After six long months in camp, the weather moderating and the grass growing, on June 4 the expedition set out from their winter encampment to begin construction on the road. Working hard, by July 2 they reached the Hell Gate Ronde. Here Mullan received a welcomed letter from Major George Blake dated June 15, which provided Mullan's first confirmation that his plan for the Oregon recruits was under way.

By mid-July the expedition was well up the North Fork of the Little Blackfoot River (today's Dog Creek) and poised to cross the Continental Divide. Mullan reported this landmark crossing:

> On the morning of the 17th we crossed the range at Mullan's Pass [whose distance from Walla-Walla was 470 miles] without difficulty, and encamped upon the waters of the Missouri. You approach and descend this pass by a gradual slope; its summit is not timbered; the mountains on both sides are much higher and densely timbered with fir and pine. It is six thousand feet above the level of the sea, and is evidently one of the lowest depressions in the whole range. . . .
>
> Crossing the Rocky Mountain range at Mullan's pass and descending upon the Missouri slopes, a new climate and new character of country is at once encountered. The spurs of the mountains become lower and less timbered, the country more diversified with hill and dale, and the scope for a wagon-road location enlarged. The climate too, is warmer, the frosts at night less severe, and the great difference of heat between midday and midnight no longer noticeable.
>
> The first stream touched upon the eastern slope is the Big Prickly Pear creek [today's Austin Creek; Mullan and Engle used the name Prickly Pear Creek for the overall drainage as well as for tributary creeks that today have individual names]. It rises at the foot of the pass, and, draining a region twenty miles in width, empties into the Missouri. Its valley for two miles partakes of a cañon like character, where, however, with moderate work, we secured a good location; at the end of this the valley widens, until, near its junction with the Missouri, it has become a prairie ten by fifteen miles in extent [today's Helena Valley]. Many fine, though small tracts of tillable land are found within it, and, at present, game abounds. Once in the valley of the Big Prickly Pear we had made our extreme point of southing, and were enabled to turn all the eastern spurs and take a direct line for Fort Benton.[28]

July 17, 1860, was a frustrating day for Mullan, as he had expected, with much anticipation, to observe a solar eclipse with his astronomical team, but it did not work out that way—that team had been sent downriver from Fort Benton, without Mullan's knowledge, by Major Blake. The dispassionate comments Mullan made in his report regarding this turn of events did not reveal his rage:

> The eclipse of the sun occurred at an early hour on the morning of the 17th of July, but as our astronomical party had now descended the Missouri for St. Louis we were deprived of the opportunity of observing it from the summit of the Rocky Mountains, as originally contemplated.[29]

As the expedition proceeded along Big Prickly Pear Creek, only light grading and clearing were required. The next morning, July 18, the men crossed the Big Prickly Pear (today's Austin Creek and Greenhorn Creek) three times, then left its valley, moving northeasterly for four miles over rolling hills to camp on Fir Creek (now called Skelly Creek). The next day the expedition made a sixty-foot side cut at Fir Creek and built a thirty-foot bridge over that stream. Then, moving over a rolling prairie, they reached Silver Creek (now Three Mile Creek), five miles north of Fir Creek, where they camped. There Mullan noted, "We had now left the more difficult sections of the mountains, and skirted along their eastern bases over the long lateral spurs making out from the main range; these spurs were untimbered, and here became reduced to easy rolling hills; only light work was needed in this last section . . ."[30]

Moving forward on July 20 and 21, after crossing Silver (now Three Mile) and Willow (now Silver) Creeks, tributaries of Big Prickly Pear Creek, the expedition reached Soft Bed Creek (now Willow Creek). None of these streams posed obstacles to travel except during spring periods of flooding. Soft Bed Creek was a minor stream flowing into Little Prickly Pear Creek, one of the larger tributaries of the Missouri River. Little Prickly Pear passed through a prairie interval (today's Sieben Ranch) and entered a deep rocky canyon, today called Lower Little Prickly Pear Canyon.[31]

After moving along Soft Bed Creek, which required only light work, the expedition crossed the hills in the Baldy Mountain drainage on July 21. Completing moderate work there, they descended to Hard Bed Creek (today's Clark Creek), then approached Medicine Rock, the last major obstacle before reaching Fort Benton. Strachan described this formation:

The route of the Mullan Road as it descends along Clark Creek before crossing Little Prickly Pear Creek. —Photo by Ken Robison

> Here is a curious site—what the Indians call a medicine rock. It is of a conical shape, and about twenty feet high. Here the Indians, in case of death or sickness, deposit pieces of cloth, bows, arrows, bears' claws, and buffalo horns, which they believe will prevent any disease or trouble.[32]

Mullan reported on the hard work required in the Medicine Rock area:

> Sunday. July 22–25. We remained three and a half days in the Small Prickly Pear region, being stopped by severe work on the Medicine Rock mountain. As thorough an exploration as we could give the country at this time showed that the only practicable line lay over the mountain, and accordingly we were obliged to do considerable work, principally in earth excavation. After crossing this mountain, the country in advance was still broken and knobby, the river flowing through an impracticable cañon. We followed the trail for ten and a half miles, when the valley becoming wider it was

This tough grade lies along the north side of Medicine Rock Mountain along the Mullan Road. –Photo by Ken Robison

The "Medicine Rock" was described by John Mullan and John Strachan as conical shaped and twenty feet high; the landform in question has never been located. The rock shown here is along the path of the Mullan Road. –Courtesy Leland J. Hanchett, Jr.

> followed down for four and eight-tenths miles, making twenty crossings–the trail being very rocky, and requiring four to six days' work to make it passable by the high water trail.
>
> On the 25th of July we reached the last crossing of the [Little] Prickly Pear, and also the last point where work of any account will be required. In reference to the region of the Medicine Rock mountain, I would state that it is truly the most difficult between Hell Gate and Fort Benton, and compares unfavorably with any section of the same length on the entire road.[33]

On July 26 the expedition left today's Wolf Creek area and moved over prairie hills for nineteen miles to the Dearborn River.[34] While camped on the banks of the Dearborn, about twelve miles above its mouth, Mullan was overtaken by W. W. Johnson, who was returning directly from Washington with dispatches, including the happy news that the expedition would receive an additional appropriation of $100,000 for another season of roadwork. The dispatches also brought the less welcome news that upon reaching Fort Benton, Mullan would not assume command of Major Blake's Dragoons, thus dashing the lieutenant's dream of leading the first military train over his road.

Mullan's disappointment over being passed over for the command added to his resentment toward Blake, who had already angered him by dismissing his astronomical team before the eclipse. But in fact, Blake had discharged the astronomers under the mistaken impression that Mullan's appropriation had not been renewed and that the road project was about to end. Incensed at this action, which he viewed as a total betrayal by the War Department, Mullan immediately sent Blake an express letter full of "reproaches" and "prolific of suggestions."[35] Thus, as historian Ryan Shaw notes, "Mullan and Blake, both of whom had a reputation of being difficult to get along with under normal circumstances, had developed an acrimonious relationship before they even met."[36]

The party of civilian topographer Theodore Kolecki, who had been exploring the Big Blackfoot Valley, also rejoined the main expedition at the Dearborn camp.[37] Here Mullan observed, "We now left both the mountains and their spurs behind us, and emerged upon the broad, swelling prairies of the Upper Missouri."[38] He recorded the width of the Dearborn River as being 120 to 200 feet, "always fordable, with the exception of about two weeks in the spring."[39]

At the Dearborn camp, John Strachan reported that the banks of the river were covered with timber, mostly cottonwood, and that game was plentiful. He also noted the presence of grizzly bears:

> Grizzly bears are more numerous, one of which I saw lassoed by our Indian guide, a boy of about fourteen years of age. These ferocious animals are often killed in this way. The lasso is thrown over the head, the rider at once puts his horse to his speed and by this means the bear becomes choked, when another Indian rides up and throws a lasso over a leg; they then ride in opposite directions, and in this way he is subdued.[40]

Leaving the Dearborn River on the morning of July 27, the expedition traveled over the prairie, crossed Beaver Creek (today's Flat Creek) with little work, and after an easy climb camped at Bird Tail Rock, having covered 16.5 miles. At this prominent landmark, Mullan noted that the site had good grass and water, but no wood.[41]

Following the gradual descent of Bird Tail Creek on the morning of July 28, the expedition proceeded sixteen miles to

Bird Tail Rock, a prominent landmark on the Mullan Road between Dearborn Crossing and Sun River Crossing, as it looked in 1868. This large igneous intrusion reaches a height of 4,934 feet. –Courtesy Overholser Historical Research Center, Fort Benton, MT

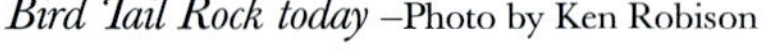

Bird Tail Rock today –Photo by Ken Robison

Sun River Crossing, the first bridge spanning the Sun River on the Mullan Road. The Blackfeet Government Farm stood on the north bank of the river, near the crossing. –Courtesy Overholser Historical Research Center, Fort Benton, MT

the ford at Sun River Crossing. On the north bank of the Sun River, near the ford, was the Blackfoot Indian Government Farm and Indian agency, under agent Alfred J. Vaughan. Mullan described the place as follows:

> The Sun river rises in the main chain of the Rocky Mountains, in latitude 48°, and empties into the Missouri about nine miles above the falls. Its border is fringed with cottonwood, and its valley is from one to three miles broad, possessing many tracts of arable land. The Blackfoot agency has large fields under cultivation, where wheat, oats, and every character of vegetable is [*sic*] raised. There are many beautiful agricultural tracts to the south of the Bird Tail Rock, along the many little creeks that rise in the broken section lying between it and the Missouri. Fuel is abundant, though timber for building purposes is scarce.[42]

Strachan was less impressed with the Sun River farm, suggesting that this large holding, under cultivation at government expense, "has proved almost a complete failure; and I have no doubt they will have to change the location to some far-distant place, or abandon it altogether."[43]

As the expedition prepared to move on to its final leg, Mullan observed that from this point on "our work proper ceased, for the remaining distance of fifty-five miles to Fort Benton was over an easy and almost level prairie road, with no running streams." He continued with a detailed description of the landscape:

> The prairie is thirteen hundred feet above the Missouri at the fort, and broken at its southern edge by deep coulées and ravines making into the river, and on its northern edge by similar formations making into the Teton river. A knowledge of the topographical face of this country would therefore show that a feasible line would lie over the high table land and between the heads of the coulées, on its either side, provided water was supplied. This is afforded by a lake [today's Benton Lake] sixteen miles to the eastward of the Sun river and by springs [today's 28-Mile Springs] seven miles to the

Benton Lake, on the Mullan Road northeast of the Sun River Leavings at Muddy Creek, provided a source of water along the road. –Photo by Leland J. Hanchett, Jr.

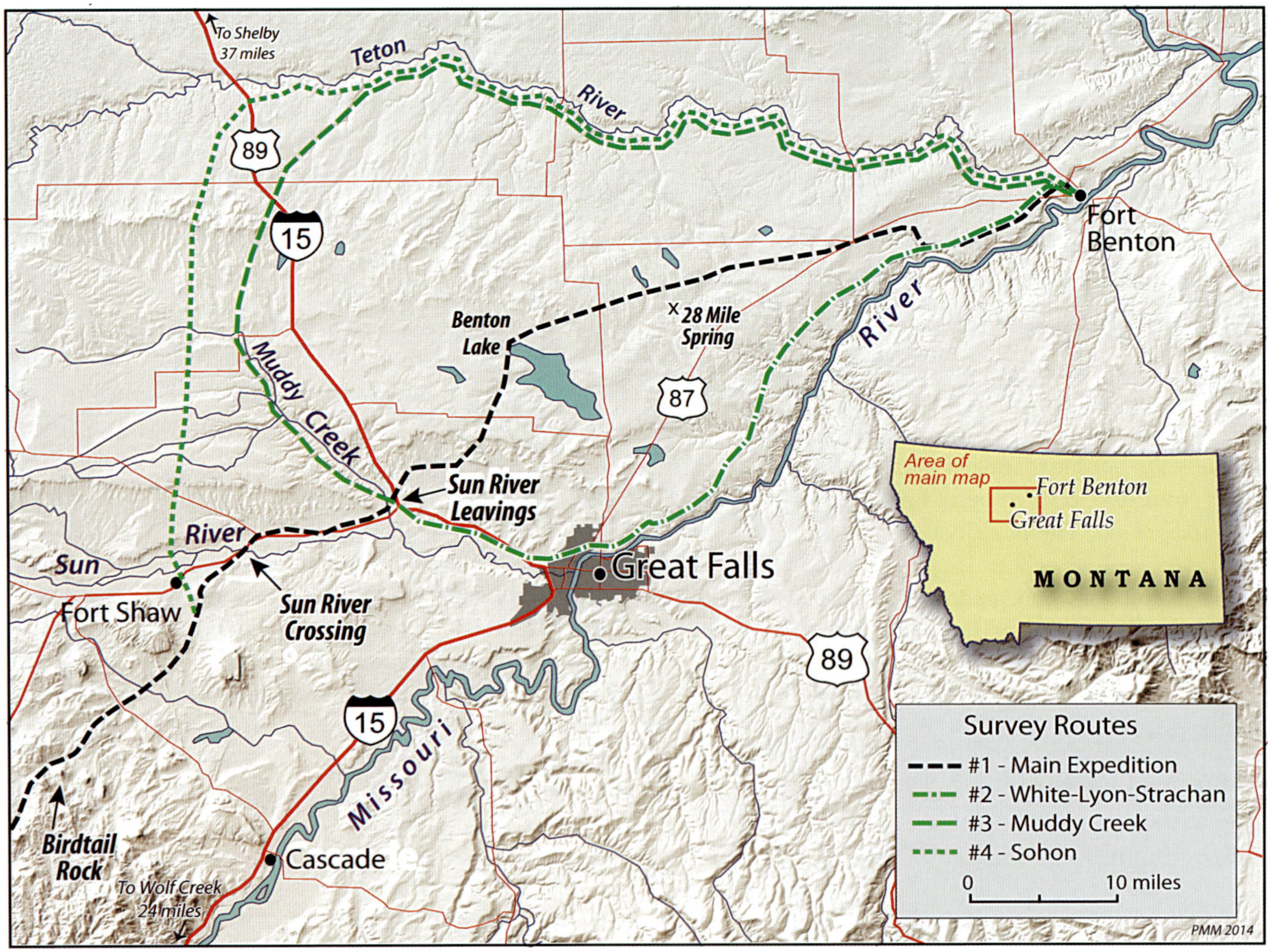

The four alternative survey routes explored by the Mullan expedition from the Sun River to Fort Benton –Map by Philip Mobley

east of the lake. These points subdivided the distance of fifty-five miles from the Sun river to Fort Benton into convenient day's marches. It will be readily seen that the character of the region immediately bordering the Missouri and Teton rivers precludes the possibility of carrying a road by either of these sections.[44]

On the morning of Sunday, July 29, 1860, Mullan left a portion of his party in camp on the Sun River and divided the remainder of his men into four groups; each was to proceed to Fort Benton by a different route, observing the country along the way. The first route, by which the main expedition train would travel in two parties, was the shortest and most direct route; it led down the Sun River, then northeast to Benton Lake and 28-Mile Springs, then east to Fort Benton. The second route, which a party led by lieutenants James L. White and Hylan B. Lyon and including John Strachan would follow, proceeded southeast to the Missouri River near today's Tower Rock, then followed the river, passing the five falls of the Missouri and Giant Springs ("the spring" in Mullan's words), to Fort Benton.[45] The third route, the one Mullan himself took, followed the Sun River east for about twenty miles, then turned north up Muddy Creek to the Teton River and along that river to Fort Benton. The fourth route, on which a small party led by Gustavus Sohon would proceed, led directly from Bird Tail Rock to the Sun River at the "old Blackfoot mission" (a short-lived mission near today's Simms, Montana), then struck north to the Teton River and along that river to the fort.[46]

On the morning of July 31, White and Lyon's Missouri River route party arrived at Fort Benton, followed in the afternoon by the main train. The next day, August 1, 1860, thirteen months after departing Fort Walla Walla, Lieutenant John Mullan at last led his expedition down the bluffs into Fort Benton exactly on schedule. Mullan did not record his feelings at that moment, but he must have been thrilled that his arduous, 624-mile, trailblazing expedition had succeeded. He no doubt said to himself, "I have done it!" Yet in his official report he simply wrote: "Wednesday, August 1. Our different parties that left the Sun river on the 29th of July all reached Fort Benton by the 1st of August. Here we found Major Blake encamped, with three hundred recruits awaiting our arrival."

While Mullan, upon his arrival at Fort Benton, failed to record his impression of the remarkable scene before him, two weeks earlier Captain William F. Raynolds had described his own:

> Fort Benton was not visible until we ascended the summit of the bluff opposite, when it burst upon us as the central point of an inspiring picture. It is located in a beautiful valley amid an amphitheatre of lofty hills. The substantial trading-houses, the shining tents of troops, and several hundred Indian lodges, filled the small plain before us, the signs of life and business contrasting forcibly with the vast solitudes through which we had for weeks been journeying. After enjoying the beauty of the prospect we descended from the bluff and encamped opposite the fort.[47]

In a letter written to his brother, civilian John Strachan described Fort Benton in the summer of 1860 even more colorfully:

> Fort Benton at last appeared in sight, and the prospects for home now began to brighten. Although yet distant over three thousand miles from St. Louis, the river from here is navigable, and we could now see the prospects of an outlet from this desolate region. Fort Benton belongs to the American Fur Company, is upon an extensive scale, and is worthy of the vast interest of which it is the center. Everything may be had within the Fort. They have a bakery, blacksmiths', carpenters' and coopers' shops; trade offices for buying, others for selling, for keeping accounts, and for transacting business; and also shops for retail. Goods are sold at enormous prices, the stock consisting of cotton and woolen goods, ready-made clothing, ship chandlery, tin and iron ware, fancy articles, and, in short, everything of every kind and description, including all sorts of groceries. Sugar is sold at one dollar and upward per pound, and everything else in proportion. The business here amounts to about $160,000 a year; buffalo robes the staple of the trade. All is arranged in the best order, and I should think, with great economy.[48]

Writer James Willard Schultz, who lived among the Blackfeet for many years, provided the most detailed description of the Fort Benton trading post as it appeared in the late 1850s:

> Entering the big gate in the wall facing the river, one found on the right the carpenter shop and blacksmith shop. On the left, first a long warehouse and then the trade room for the

Indians, where, behind breast-high counters, were tiers and tiers of shelves upon which the various trade goods were displayed. Along the west side of the great inner court were three houses, the lower stories of which were: another warehouse; the store for company employees; and the kitchen of the bourgeois or superintendent of the fort, who in my time was Andrew Dawson, a Scotchman, and a partner in the great company. He lived and had his office in the upper stories of the westernmost of the houses on the north side of the court, and the upper of the other three houses in the row were quarters of our clerks, Matthew Carroll and George Steell, and our Father [Thomas Jackson] who was the tailor. Practically all the lower stories in this row were reserved for the use of the Indians who were continually coming to the fort to trade. All the houses on the east side of the court were occupied by the *engagés*, or laborers and their families, and the gunsmith, the post hunter, and the general foreman and their families.[49]

When the four parties rendezvoused at Fort Benton, Mullan used the observations reported by each to evaluate the different routes based on distance, terrain, availability of wood and water, and various other factors. Although he was concerned about the lack of wood and water on the route the main party had taken, in the end he concluded that it was the best route for the permanent wagon road.[50] He had originally thought that the permanent road would go along the Teton River route, but in the end, the direct route, via 28-Mile Springs and Benton Lake, became the main Mullan Road.

Mullan arrived at Fort Benton to find Major Blake waiting impatiently with his detachment of new recruits. When Blake landed at the fort, he was already out of sorts from the long steamboat ride from Jefferson Barracks in St. Louis. Moreover, the trip had been particularly taxing–along the way, twenty-nine men had deserted his command, two others had drowned, and one of his company commanders had been shot in an altercation with one of the steamboat pilots, obliging Blake to conduct an en route board of inquiry.[51]

Before convening with Blake, Mullan met with his former West Point classmate, Lieutenant August V. Kautz of Blake's brigade. In his own account, Kautz captured Mullan's obsession with the road and his troubles in completing it:

Wednesday. August 1. Lt. Mullan arrived in the afternoon by the Teton route. A heavy rainstorm prevailed this afternoon. It rained very hard and the wind blew with great violence. I had a long talk with Mullan about his road and his troubles. He is decidedly more monomaniacal in his demonstrations than I ever knew him. He imagines everybody who is not in favor of his road to be against it. He has no doubt been very badly treated by Genl. Harney and by the Secretary of War. White and Lyon came up, and spent the evening in camp, but Mullan did not come up, much to the Major's [Blake's] disgust.[52]

Despite his antipathy, Mullan met with Blake the next day to organize the movement of the Oregon recruits over the Mullan Road to Fort Walla Walla. The conference went more smoothly than Kautz, who attended the meeting, had expected:

Thursday. August 2. A very cool day. I went down at Mullan's request, and came up with him to see the Maj. [Blake]. Their interview proved more amicable than I anticipated. . . . Lt. M. has turned over all the wagons he has in possession, and takes our pack train in exchange. The Major intimidated Mullan into letting him have all the wagons by telling him that he would not move without them. Mullan is quite a monomaniac about his road.[53]

While at Fort Benton, Mullan began preparing his report to the U.S. War Department, summarizing his thirteen months of road-building from Fort Walla Walla to Fort Benton and noting the various types of terrain and conditions he and his men had encountered:

Thursday. August 2. Our road involved one hundred and twenty miles of difficult timber-cutting, twenty-five feet broad, and thirty measured miles of excavation, fifteen to twenty feet wide. The remainder was either through an open, timbered country, or over open, rolling prairie. From Walla-Walla eastward the country might be described in succinct terms as follows: First one hundred and eighty miles, open, level, or rolling prairie; next one hundred and twenty miles, densely timbered mountain bottoms; next two hundred and twenty-four miles, open timbered plateaus, with long stretches of prairie; and next one hundred miles, level or rolling prairie. Thus it is seen that the Rocky and Bitter Root mountains rise midway in our route, with long prairie slopes on either side; that the latter are intersected in every direction by streams flowing from both water-sheds, and rising in the heart of the mountain system; that these prairie stretches interpose but slight obstructions to the location of a road, and it is only in the more elevated central sections where our sterner engineering problems are to be met.[54]

On August 3, Colonel Vaughan began distributing government annuities to the Piegan Blackfeet assembled at Fort Benton. John Strachan described the colorful scene as thousands of Indians gathered to receive the annuities:

On our arrival at this point [Fort Benton] the Indian Agent Col. Vaughn [*sic*], had just got ready to distribute the annual presents [which had come up on the steamboats] to the Indians. They soon appeared in sight. They came from the south side of the river and passed through between two high bluffs. A continued train of Indians and horses kept moving through this pass for nearly five hours, and I should judge the number of men, women and children could not be less than ten thousand with an average of two horses to each Indian. They have encamped opposite the Fort, and their lodges were mostly of a conical shape and of buffalo skins. They next day crossed the river [at the ford], which is about six hundred yards wide, every rider swimming his own horse, but on many occasions there were from two to three on each horse. The pack-horses were loaded with tent-poles and buffalo robes, together with their provisions, consisting mostly of buffalo meat. With their load they swam the river without much difficulty.

After all had crossed, the Agent at once proceeded to distribute the different articles. The chief, Little Dog, takes command, with from eight to ten of his principal men. The whole party is ordered to form a circle, which was at once done. Inside are the men, on their knees, and all arranged according to their respective positions; behind them the women and children are placed in like manner. The Agent has his goods placed in the center, and hands them to the Chief, who distributes them to his subordinates, and they pass around the circle giving to those whom they believe best entitled the most, some receiving more than others, while many received nothing. The Chief's and subordinates' shares seemed almost enormous. The articles distributed consisted of navy bread, pork, blankets, fancy dry goods, guns, steel for arrow-points, powder and lead. This proceeding lasted a whole day, and good order was kept throughout. During the evening they made a most tremendous noise, which might be heard a long distance, singing and beating with sticks, and splinters of rails, together with a few old camp kettles we had thrown away, which served them as a substitute for musical instruments.

Next day pork was to be seen in all directions which they had cast away as of no value. Gambling and horse-racing now commenced and was kept on constantly. After two horses were selected to run, all who felt disposed might bet on the race, some betting everything they had, even to their clothing—all was put up before the horses started. Boys, mostly from six to ten years of age, were selected to ride; the distance was about one mile. The horses are never put to full speed until the last half of the distance. After the race is decided, the winner takes his goods, and there are no sympathizers with the losers.

These Indians are remarkable for their fine buck-skin dresses, which are richly ornamented with ribbons and beads. They had with them a large number of dogs, which are kept about their lodges. These animals have but little to recommend them, being ill-shaped, of a dingy color and of no use in hunting. The Indians seldom eat them, and never kill them off, however numerous they may become; their keeping costs nothing, for their food is the offal of fish and game, which accounts for the wretchedness of their appearance.[55]

During Mullan's brief stay at Fort Benton, most of his men were discharged and boarded steamboats headed back to St. Louis. The expedition's artist, guide, and interpreter, Gustavus Sohon, and the wagonmaster, John Creighton, were transferred to Major Blake's command.[56] Mullan himself, leaving ahead of Blake, would lead a small civilian crew to Walla Walla, making repairs and improvements to the road along the way. In his report Mullan gave details regarding the departure of the majority of his expedition:

We remained here [Fort Benton] until the 5th of August. A Mackinac boat was built for a party to descend the Missouri to St. Louis, composed of discharged civilians and such soldiers as were near the expiration of their term of service, all under charge of Lieut. J. L. White. This party made the trip without accident, and for a portion of the distance in company with one of the boats of Captain Raynolds, of the topographical engineers, who for three years had been exploring the country from Fort Laramie to the head waters of the Yellowstone and Missouri, and who had reached Fort Benton only two weeks in advance of our own party. The arrival of Captain Raynolds was of advantage to ourselves, as he was

A historic photograph of John Creighton's wagons at The Dalles
–Courtesy Overholser Historical Research Center, Fort Benton, MT

Wagonmaster John Creighton, who led the civilian wagon train that supported the expedition.
–Courtesy Thomas Minckler Collection, Overholser Historical Research Center, Fort Benton, MT

enabled to add to our transportation a large number of pack animals.[57]

In his journal entries, Mullan also discussed his plans and preparations for the next year:

> On reaching Fort Benton I had sent Mr. Kolecki to Washington with my field note books, giving him instructions to commence the compilation of our maps. My instructions for the winter contemplated my compiling certain maps at Walla-Walla, but, as the note books necessary for the purpose had been sent off, I was left comparatively idle. The instructions from the War Department required that I should lay before them a plan of operations for the next season's work. Accordingly I prepared one that looked toward resuming work from the Pacific slope, and sent it to the department; but, on comparing a table of statistics of cost of labor and material on the Pacific with a similar table for the Missouri, I became convinced that the latter was the more judicious.
>
> My second programme looked toward outfitting at Fort Leavenworth, and proceeding via Fort Laramie to the Deer Lodge valley, where I should spend the winter of 1861, and then press the work vigorously toward Walla-Walla, reaching there by the winter of 1862.[58]

On August 5, 1860, Mullan departed the fort for his return trip to Walla Walla, on which he and a small crew, taking sixty-five pack animals and four light wagons carrying tools and supplies, would repair and improve the wagon road in advance of Blake and the troops: "Sunday. August 5. Every available means of transportation being turned over to Major Blake for the use of his command, and having completed all the arrangements that our mission called for, we left Fort Benton on the morning of the 5th of August, on our return to Fort Walla-Walla."[59]

Moving rapidly past the springs and the lake, Mullan reached the Sun River on August 7, camping at the Indian agency farm. That same day, Blake's brigade had departed from Fort Benton, arriving at the farm the morning of August 10.

Leaving the farm, Mullan's party traveled swiftly over the road, past Bird Tail Rock and the Dearborn River, reaching Medicine Rock the evening of August 12. By then, Mullan had become sick with dysentery and confined to a wagon, so he assigned civil engineer Walter W. Johnson to supervise the repair work. For two days, the men worked to improve the grade at Medicine Rock Mountain. After the steep ascent, the road was good until the sharp descent to Little Prickly Pear Creek. They spent August 13 and 14 grading both the ascent and descent.[60]

Still a day behind Mullan, Major Blake's brigade doubled-teamed their wagons over Medicine Rock Mountain on August 15. That same day, Mullan's party built a new thirty-five-foot bridge over Silver (Three Mile) Creek. The next day, Mullan's men reached the summit of the Continental Divide of the Rocky Mountains without double-teaming. Along the way, plans were made for additional work to be done to further improve the road the next year.[61]

Blake's brigade crossed the divide on August 18. Lieutenant Kautz observed, "The divide is the most practical pass I ever saw in any mountainous country." Kautz also added another happy note: "Lt. Mullan spent the afternoon in camp, and took dinner with me. He is quite well again."[62]

Mullan and his party reached Fort Walla Walla on October 8, 1860, a day behind Blake's brigade, who had passed Mullan's repair crew on the road.[63] In the end, despite their disagreements, the lieutenant and the major joined together for this successful first use of the road. Much credit was due to two of Mullan's men, guide Gustavus Sohon and wagonmaster John Creighton. As Mullan reported, "to their joint good services, Major Blake was largely indebted for the success of his march."[64]

Pending further refinements to be made in 1861–62, Lieutenant John Mullan and his men had completed their wagon road from Fort Walla Walla–the head of navigation on the Columbia River–to Fort Benton–the head of navigation on the Missouri River–on time and only slightly over budget. Supported by Mullan's transportation, Major Blake's dragoons, having traveled Mullan's military wagon road over Mullan Pass to western Washington Territory, were the first military users of the new road.

Later, Mullan was able to complete two more years' work improving the road. Although his plan for continued military use of the road was disrupted by the Civil War and other factors, the Mullan Road did become an important freight and stage road after the discovery of gold in southwest Montana in 1862. Over it passed much of the commerce of the new territory. At the eastern end of the road, Fort Benton boomed as an important river port and staging area. At its peak in 1867, forty-one steamboats landed at the Fort Benton levee with over 80,000 tons of freight and thousands of passengers, most of them headed to the mining camps of the newly organized Montana Territory. By then, the section of the Mullan Road between Fort Benton and Helena had become known as Benton Road. As the Mullan Road evolved, eventually the railroads, foreseen by the Pacific Railway Surveys, arrived, and today, modern highways and railroads follow much of the old Mullan Road. John Mullan would be proud.

---

1 The Blackfoot Nation comprises four tribes: Siksika, or Blackfoot; Kainai, or Blood; Pikanii, or Northern Piegan; and Piikuni, or Southern Piegan (also commonly called Blackfeet). By 1831 fur-trading posts with the Blackfoot had extended to the upper Missouri. In that year the American Fur Company (after 1834 known as the Upper Missouri Outfit of Pierre Chouteau & Company) opened Fort Piegan, followed by Forts MacKenzie (1833), Chardon (1844), Lewis (1845), and Benton (1846). In 1846 an opposition trading post, Fort Campbell, named for St. Louis fur trader Robert Campbell, was established two miles upriver from Fort Benton. Trading posts were more than businesses. They also served as social centers where news, ideas, and customs were exchanged among native Indians and between white traders and Indians. Prior to the arrival of steamboats, trading posts were supplied by keelboats and mackinaws. See William E. Lass, *Navigating the Missouri: Steamboating on Nature's Highway, 1819–1935* (Norman, OK: Arthur H. Clark, 2008), p. 192; Lesley Wischmann, *Frontier Diplomats: The Life and Times of Alexander Culbertson and Natoyist-Siksina* (Spokane, WA: Arthur H. Clark, 2000), p. 291; and John G. Lepley, *Blackfoot Fur Trade on the Upper Missouri* (Missoula, MT: Pictorial Histories Publishing, 2004).

2 The steamer *Spread Eagle* did not proceed above Fort Union. This fort, on the Missouri River at its confluence with the Yellowstone River, near today's Montana–North Dakota border, was a major trading post for Pierre Chouteau & Company.

3 The previous year, 1859, Charles Chouteau had made a concerted effort to reach Fort Benton with the newly constructed light-draft steamboat the *Chippewa,* but he fell short, reaching Brule Bottom, twelve miles below Fort Benton. See William E. Lass, *A History of Steamboating on the Upper Missouri River* (Lincoln: University of Nebraska Press, 1962), pp. 15–18; John E. Sunder, *The Fur Trade on the Upper Missouri, 1840–65* (Norman: University of Oklahoma Press, 1993), pp. 211–13; the "Robison-Wahlberg List" of Upper Missouri River Steamboat Passengers, 1859–89 (Schwinden Library and Archives and Joel F. Overholser Historical Research Center, Fort Benton, Montana); and Lawrence H. Larsen and Barbara J. Cottrell, *Steamboats West: The 1859 American Fur Company Missouri River Expedition* (Norman, OK: Arthur H. Clark, 2010).

4 Vaughn quotation from Albert Watkins, "The Oregon Recruit Expedition," *Mid-West Quarterly* 1, no. 1 (October 1913), pp. 47–48.

5 William F. Raynolds, *Journal of Captain W. F. Raynolds, United States Army Corps of Engineers* (Washington, DC: Government Printing Office, 1868), pp. 78, 109–10.

6 John Mullan, *Report on the Construction of a Military Road from Fort Walla-Walla to Fort Benton* (Washington, DC: Government Printing Office, 1863), p. 24. This source is hereafter cited as Mullan, *Report,* 1863.

7 See Isaac I. Stevens, *Reports of Explorations and Surveys to Ascertain the Most Practicable and Economical Route for a Railroad from the Mississippi River to the Pacific Ocean* (Washington, DC: A.O.P. Nicholson, Printer, 1855–60), hereafter cited as Stevens, *Reports of Explorations and Surveys*; and Martha Edgerton Plassmann, "Lieut. John Mullan's Montana Explorations in 1853–54: Remarkable Work of Lieut. Mullan, Who Blazed Trails Ahead of Railroad" (*Grass Range Review*, February 4, 1924).

8 See John Mullan, *Letter from the Secretary of War, transmitting the report of Lieut. Mullan, in Charge of Construction of the Military Road from Fort Benton to Fort Walla-Walla,* U.S. Congress, 36th Congress, 2d sess., H. Ex. Doc. no. 44 (Washington, DC: Government Printing Office, 1861), hereafter cited as Mullan, *Report,* 1861 (this source also includes reports from P. M. Engle and others); Mullan, *Report,* 1863; and John Strachan, *Blazing the Mullan Trail: Connecting the Headwaters of the Missouri and Columbia Rivers . . .* (New York: Edward Eberstadt & Sons, 1952), p. 53. Strachan's was the first published narrative of the Mullan expedition; his letters to his family were published in the *Rockford* [IL] *Register* from April 1860 to April 1861.

9 Mullan, *Report,* 1861, pp. 36–37. During the winter, Mullan used messengers on snowshoes to coordinate his various groups and activities.

10 Ibid.

11 Ibid., p. 37.

12 Along with Mullan, Sohon had accompanied Isaac I. Stevens on his Pacific Railway Survey of 1853–55, and his landscape art was prominently featured in Stevens's *Report of Explorations and Surveys.*

13 Mullan, *Report,* 1861, pp. 117–18.

14 Ibid. The author would like to offer special thanks to Dr. Richard S. Buswell for his assistance in relating Mullan's geographic terms with modern usage.

15 Ibid., pp. 119–20.

16 A provision of the 1855 Lame Bull Treaty provided for the establishment of a government farm for the Blackfeet. In 1858 Agent Vaughan selected the Sun River Valley, near today's town of Sun River, as the location for this farm. See Joel Overholser, *Fort Benton: World's Innermost Port* (Helena, MT: Falcon Press, 1987), pp. 306–7; and Albert Watkins, "The Oregon Recruit Expedition," *Midwestern Quarterly* 1, no. 1 (October 1913), pp. 47–48.

17 Mullan, *Report*, 1861, pp. 122–23.

18 Ibid., p. 123.

19 Ibid., p. 124.

20 Ibid., p. 126. The "Teton mission" was the original location of St. Peter's Mission, a Jesuit Blackfoot mission founded in 1859 on the banks of the Teton River, near Priest's Butte and today's town of Choteau, Montana. The mission would later be moved to the Sun River. See e.g. Great Falls, Montana, *Daily Leader,* July 28, 1928.

21 Mullan, *Report,* 1861, p. 127.

22 Ibid., pp. 31–34.

23 Ibid.

24 Ibid., p. 34.

25 Ibid., pp. 39–40. On Stevens's 1853–54 Pacific Railway Survey, Sohon had gained valuable knowledge of the Indian trails and tribes, including their languages. In addition, Sohon, along with fellow artist John Mix Stanley, sketched many scenes of the survey route, published in the government reports.

26 Ibid., p. 47.

27 Strachan, *Blazing the Mullan Trail*, p. 50.

28 Mullan, *Report,* 1863, p. 25.

29 Ibid. Mullan had earlier laid out his plans for observation of the eclipse. His astronomical party, headed by John Weisner, were collecting observations at Hell Gate until May 10, when Mullan was ordered to send Weisner and his men to Fort Benton, where they arrived May 20.

30 Ibid.

31 Ibid.

32 Strachan, *Blazing the Mullan Trail*, p. 49.

33 Mullan, *Report,* 1861, p. 52. While Mullan was not satisfied with the Medicine Rock route after three days of work, he could not afford further delay. He described plans for future work in this area to select a better route and perform additional work, estimating it would require fifty men for ten to twelve days. In his later report he added, "On the Medicine Rock mountains we found traces of quartz, and continual indications of gold; one of my men found ten cent prospects in the Big Prickly Pear" (Mullan, *Report,* 1863, p. 26).

34 Strachan, *Blazing the Mullan Trail*, p. 52.

35 August Kautz, "From Missouri to Oregon in 1860: The Diary of August V. Kautz," *Pacific Northwest Quarterly* 37, no. 3 (July 1946), p. 216.

36 My thanks to Maj. Ryan Shaw for his input. Much of the information regarding Mullan and Major Blake was based on the manuscript of Shaw's forthcoming dissertation, "West from West Point." See also "'A Creditable Piece of Mountain Work,'" parts 1 and 2, in this book.

37 Mullan, *Report,* 1863, p. 25.

38 Ibid., p. 26.

39 Mullan, *Report,* 1861, p. 52.

40 Strachan, *Blazing the Mullan Trail*, pp. 49–50.

41 Mullan, *Report,* 1863, p. 26. A description of the prominent Bird Tail Rock was provided by Jesuit Father F. X. Kuppens: "Bird Tail is a very peculiar landmark near the Mullan road about midway between the Dearborn and Sun Rivers. It is a high isolated and very steep hill, and the many large fragments of rock all about its sides give it a formidable aspect. The top appears to be one Solid Mass of Stones, and at its very highest point there juts out bold against the sky perhaps [seven] tall monoliths of colossal size. The Indians, in designating the hill, would raise their open hand above their head and extend the fingers. Very little effort of the imagination was required to find that the name Bird Tail was appropriate." See Genevieve McBride, *The Bird Tail* (New York: Vantage Press, 1974), p. 13.

42 Mullan, *Report,* 1863, pp. 25–26.

43 Strachan, *Blazing the Mullan Trail*, p. 52.

44 Mullan, *Report,* 1863, p. 25. The lake and springs described by Mullan developed into stage stations as the Mullan Road (later called Benton Road) evolved during the 1860s. The lake was later named Benton Lake and the springs became 28-Mile Springs.

45 In *Blazing the Mullan Trail*, p. 52, Strachan colorfully described the Great Falls of the Missouri as "forming the grandest view perhaps in the world, surpassing in beauty of scenery and magnitude the Falls of Niagara."

46 Mullan, *Report,* 1863, p. 26; and Mullan, *Report,* 1861, p. 53. The Jesuit Blackfoot Mission moved from the Teton River to the banks of the Sun River in March 1860.

47 Raynolds, *Journal of Captain W. F. Raynolds*, pp. 109–10.

48 Strachan, *Blazing the Mullan Trail*, p. 53.

49 James Willard Schultz, *William Jackson, Indian Scout* (1926; reprint, Springfield, IL: William K. Cavanaugh, 1976), pp. 1–2.

50 Mullan, *Report,* 1863, p. 26.

51 Kautz, "From Missouri to Oregon," p. 205.

52 Ibid., p. 218.

53 Ibid., pp. 218–19.

54 Mullan, *Report,* 1863, p. 37.

55 Strachan, *Blazing the Mullan Trail*, pp. 54–55.

56 John A. Creighton, born about 1836 in Scotland, served as wagonmaster on the expedition, leading twenty-nine teamsters, cooks, saddlers, and blacksmiths; see *The River Press,* Fort Benton, Montana, "150th Anniversary Mullan Road Conference, 1860–2010," special edition, May 2010.

57 Mullan, *Report,* 1863, p. 26.

58 Ibid., p. 28.

59 Ibid., p. 27.

60 Mullan, *Report,* 1861, p. 55.

61 Mullan, *Report,* 1861, p. 56.

62 Kautz, "From Missouri to Oregon," p. 223.

63 Major Blake's dragoons successfully reached Fort Walla Walla on October 4, 1860, in good health except for one death and some fifteen cases of scurvy and other diseases.

64 Mullan, *Report,* 1863, p. 27.

Part Three

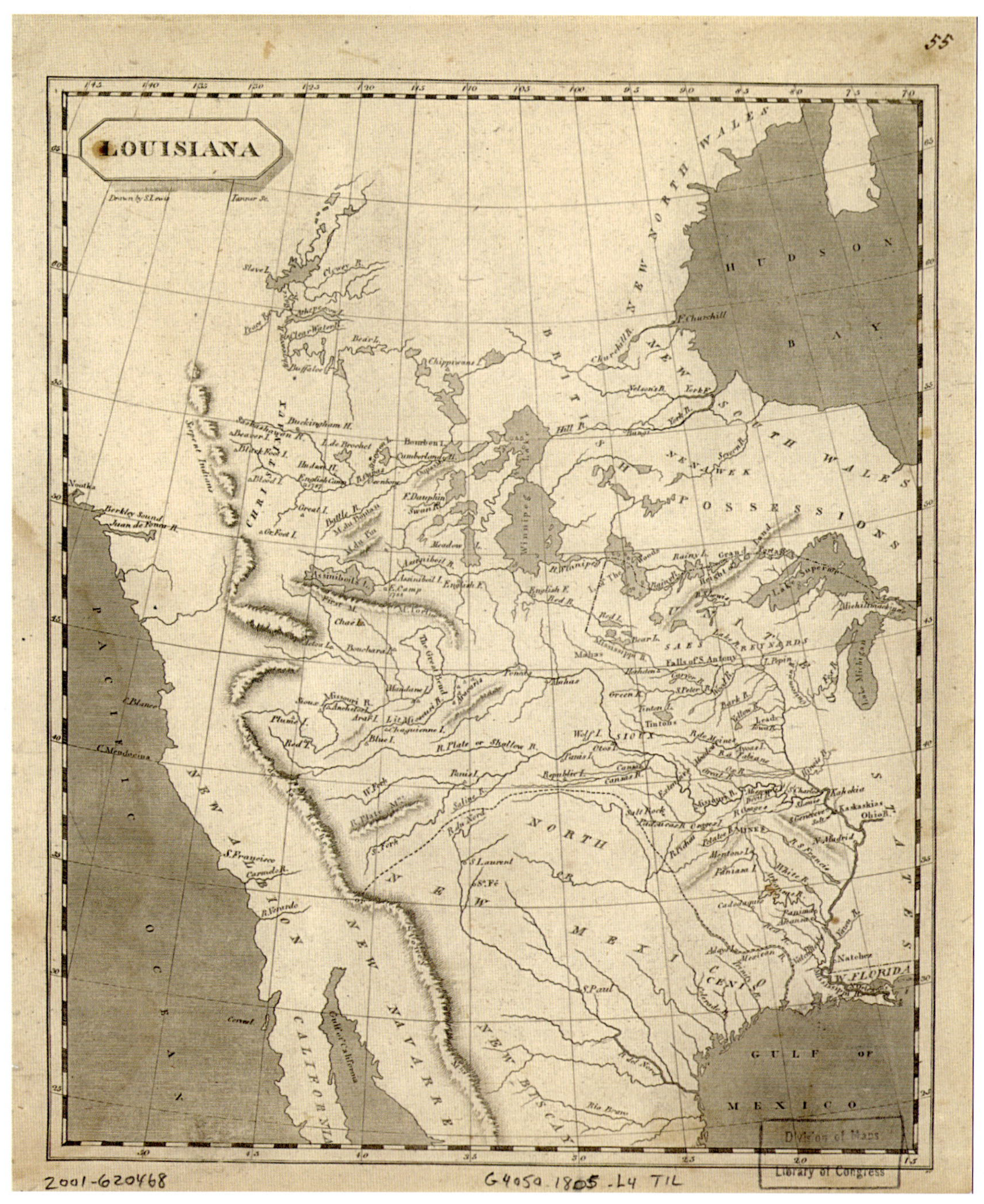

Figure 9-1. *This 1804 map by Samuel Lewis embodied what Thomas Jefferson and his contemporaries envisioned about the West. Scale approximately 1:21,000,000. Published in Arrowsmith & Lewis,* New and Elegant General Atlas*, 1804.* –Courtesy Library of Congress, Geography and Map Division

Surveying the Mullan Road

BY WILLIAM "MONTANA BILL" WEIKEL

WITH PAUL D. MCDERMOTT, RONALD E. GRIM, MAJ. RYAN L. SHAW, AND PHILIP MOBLEY

Editor's note: This essay is derived from a lecture presented at Fort Walla Walla on April 14, 2012, by William "Montana Bill" Weikel. In his presentation, Weikel portrays a surveyor and demonstrates surveying equipment that was used during the nineteenth century. Some of the graphics in this paper came from his PowerPoint presentation.

At the beginning of the nineteenth century, very little was known about the western United States, particularly the Rocky Mountains and its environs. It was assumed that moving between the Missouri River and the Columbia River would require only a few days' travel across the Rockies. No one realized how extensive the Rocky Mountain system was and how difficult it would be to traverse it. What was known and perceived about these mountains was portrayed on an 1804 map by Samuel Lewis (Figure 9-1). Lewis's map, which appeared in a general reference atlas published by Aaron Arrowsmith, reflected what Thomas Jefferson and his contemporaries envisioned about the West. It suggested that river systems flowing from the mountains would provide easy passage over the single ridge that defined the Rockies. This is the perception that guided the famous Lewis and Clark Expedition. As it was a flawed idea, it could easily have led to their failure.

Lewis and Clark's journey represented the first major step in gaining an understanding of the geographical and topographical character of the interior West. These young explorers, investigating the northern region in 1804–6, traveled down the Ohio River and up the Mississippi and Missouri Rivers until they reached the Rocky Mountains. From there they proceeded across the Bitterroot Range through Idaho and eventually to and down the Columbia River to the Pacific Coast. In the process of exploring this vast region, they gained new knowledge about its topographical character, its native inhabitants, and its vegetation and wildlife. Most of this survey was accomplished using simple instrumentation–a compass and a sextant. Distances traveled were estimated.

The cartographic results of the Lewis and Clark Expedition were first published in 1814, nine years after the completion of the journey. The revised map, also drawn by Samuel Lewis, used information derived from Lewis and Clark as well as data provided by others (see Figure 9-2). On this map, Lewis delineates a series of north-south ridges and provides more detail about the Cascade Mountains and other smaller ranges punctuating the Great Plains. Many rivers in the West were given names to demonstrate that Lewis and Clark or others explored them. Most significantly, the cartographer depicted the mountain complexes with a greater areal extent than had been previously known. In both the 1804 map and 1814 map, mountains were represented using symbols that resembled long caterpillars, a technique known as hachuring.

Other new maps also resulted from Lewis and Clark's fine work. Over the next forty five years, additional information about these areas was acquired from traders, fur trappers, and military personnel who traveled through different

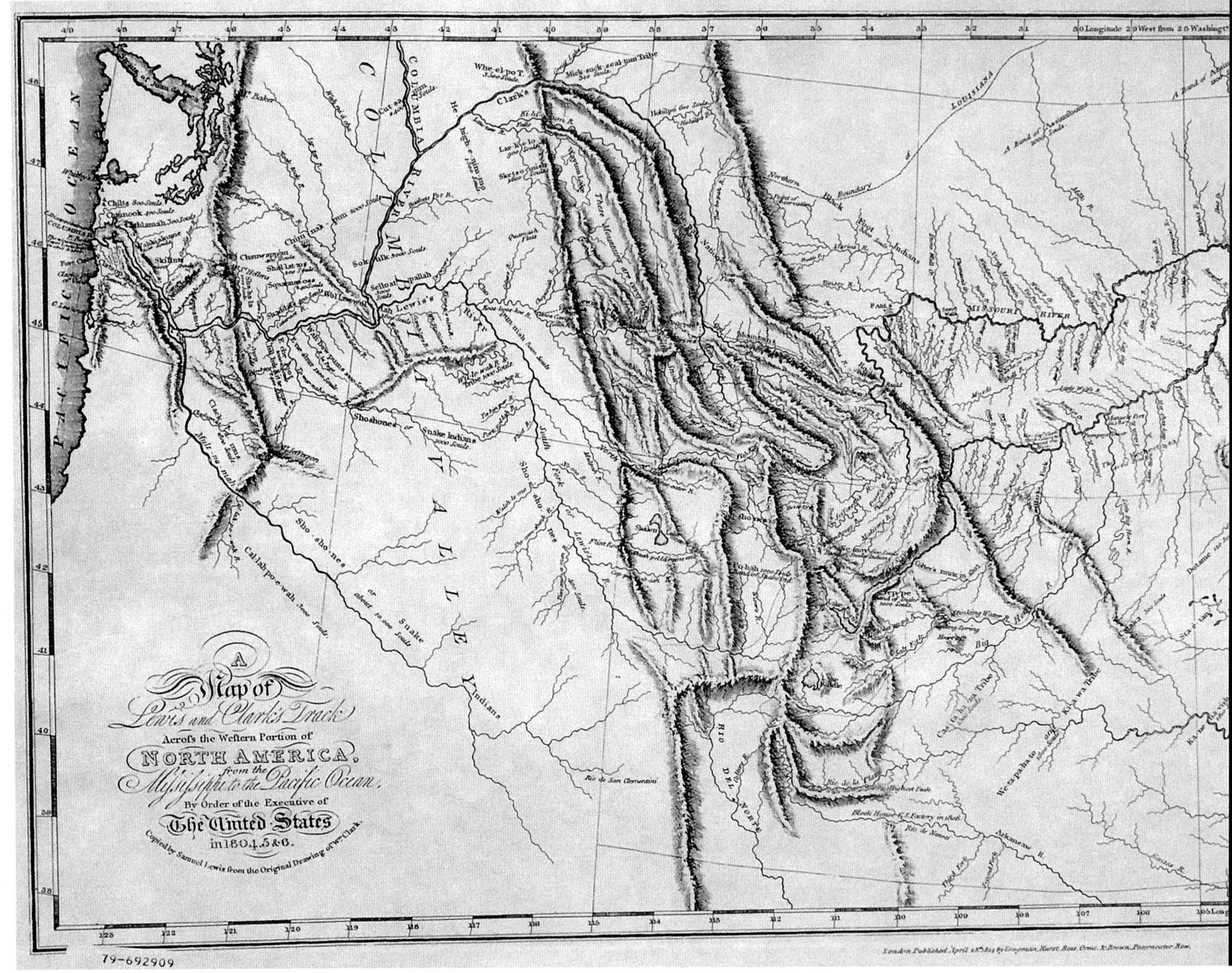

Figure 9-2. *Samuel Lewis's 1814 map of the western United States. This map was based on information supplied by the Lewis and Clark Expedition. Notice the increase in complexity in the depiction of the Rocky Mountains. Scale approximately 1:4,400,000. Published in London by Longman, Hurst, Rees, Orme & Brown in 1814.* –Courtesy Library of Congress, Geography and Map Division

sections of the West. Gradually, maps were filled in with greater detail and more information. An 1843 map by John C. Fremont (Figure 9-3) shows one portion of the West that he had explored.

A greater amount of geographical data was accrued after the U.S. Congress authorized five major western expeditions in 1853. The purpose of these surveys, the government's earliest intensive investigation of the West, was to explore potential routes for the nation's first transcontinental railroad. Four of them explored east-to-west routes, while the fifth followed a north-south route along the West Coast, demonstrating the possibility of connecting California with the Pacific Northwest. The most fruitful of these expeditions was the northern survey led by Isaac I. Stevens, which explored the region between the 47th and 49th parallels. All together, these explorations yielded tremendous amounts of geographic information, including topography, flora and fauna, hydrography, natural resources, and climatological data.

This presentation focuses on the Stevens survey of 1853–54 and its bearing on John Mullan's subsequent road-building expedition in 1859–62 (see Figure 9-4). Stevens and his party began in 1853 near St. Paul and proceeded west traveling through what would become Minnesota, South Dakota, North Dakota, Montana, Idaho, and Washington–to the Pacific Ocean. Significantly, much of Stevens's route overlapped with the one blazed by Lewis and Clark. Like the other expeditions, Stevens's was undertaken to determine whether or not a railroad could feasibly be built across this region, as well as to estimate the cost of building one. The estimates were made by one of the most able railroad engineers of the time–Captain George B. McClellan.[1] In addition to investigating the topographical character of the region and ascertaining the resources required for building and supplying a railroad, Stevens would, as importantly, determine how best to obtain permission for building a railroad through the lands occupied by different groups of Native Americans.

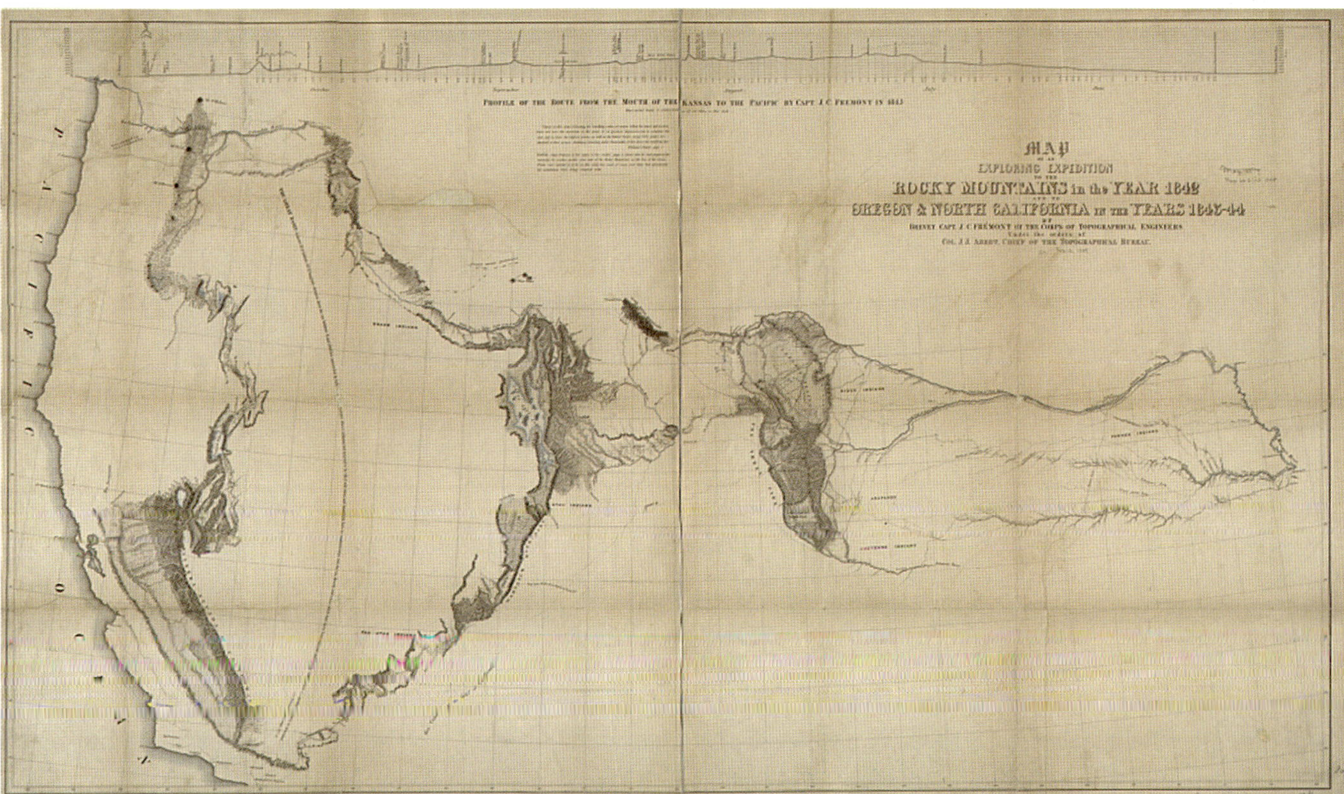

Figure 9-3. *John C. Fremont's version of the West, 1843* –Courtesy Norman B. Leventhal Map Center, Boston Public Library

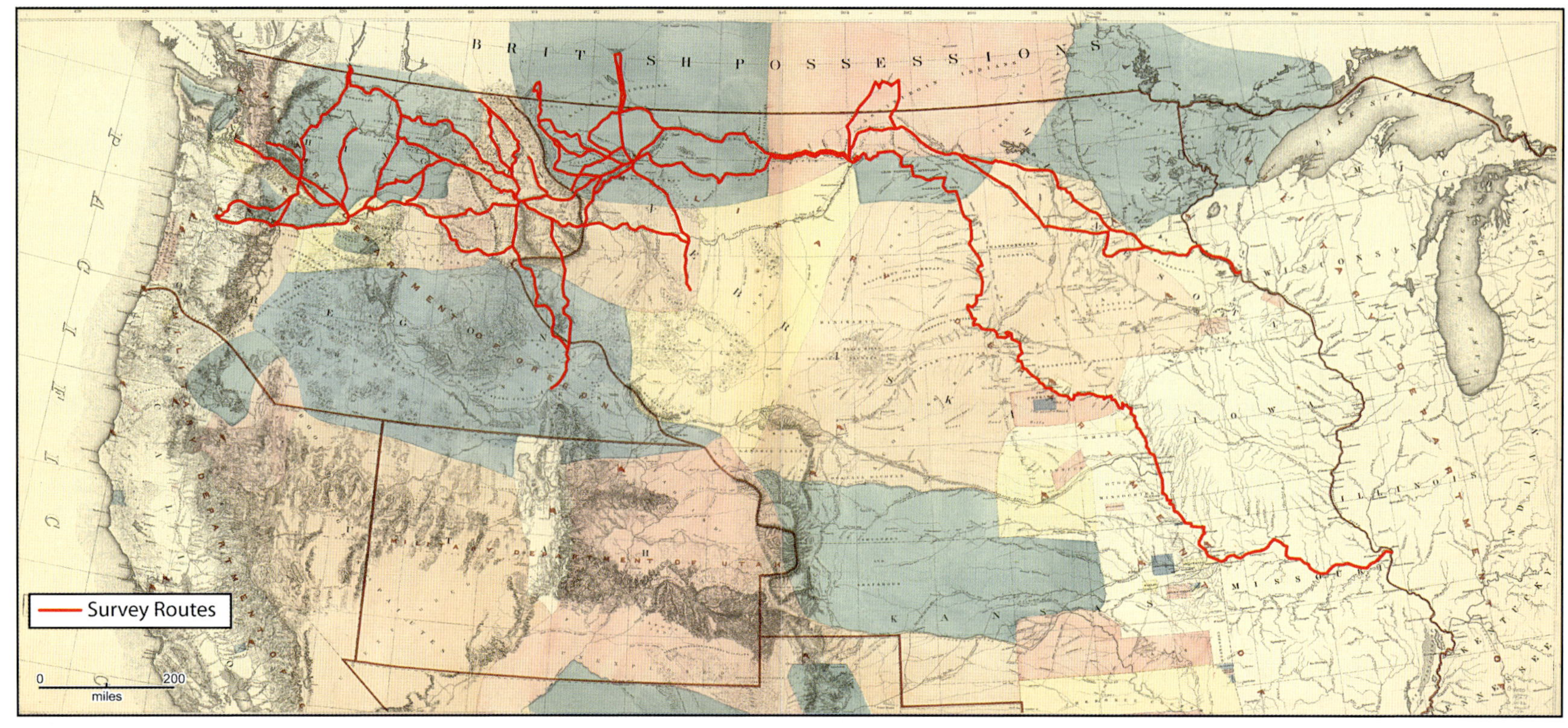

Figure 9-4. *Routes explored by Isaac Stevens's Northern Pacific Railroad Survey of 1853–54. Map by G. K. Warren, 1858. Original scale approximately 1:3,000,000. Notice the intensity of the explorations made by the Stevens party.* –Courtesy Library of Congress, Geography and Map Division

In 1854, additional work was conducted by Stevens's subordinate, Lieutenant John Mullan, who explored the intermontane region west of the Continental Divide and through the Bitterroot Mountains. In the process, Mullan gained considerable knowledge, which was eventually used to construct his military wagon road between Fort Walla Walla and Fort Benton. Further information was derived while Mullan was actually building the road, between 1858 and 1862.

Early Survey Methods

In the early nineteenth century, survey data was gathered with simple instrumentation. Usually only two items were involved–the magnetic compass and the surveyor's chain (see Figure 9-5). The latter was a chain of predetermined length, used to determine distance; alternatively, surveyors used a rod, also known as a perch, typically 16.5 feet long. Usually one surveyor's chain was equal to 4 rods, or 66 feet, but they came in different standardized lengths depending upon need and preference. These early surveys were generally crude, and their accuracy could easily be questioned.

As Stevens's survey party moved along the route, they measured and recorded directions and distances based on a series of stations they set up as orientation points. Compass directions, or bearings, were established from these stations and distances between them were measured. Frequently, directions to nearby landmarks were also noted. Later, exact distance for each bearing was determined using trigonometry. This type of survey, known as an open traverse, measures a strip of land

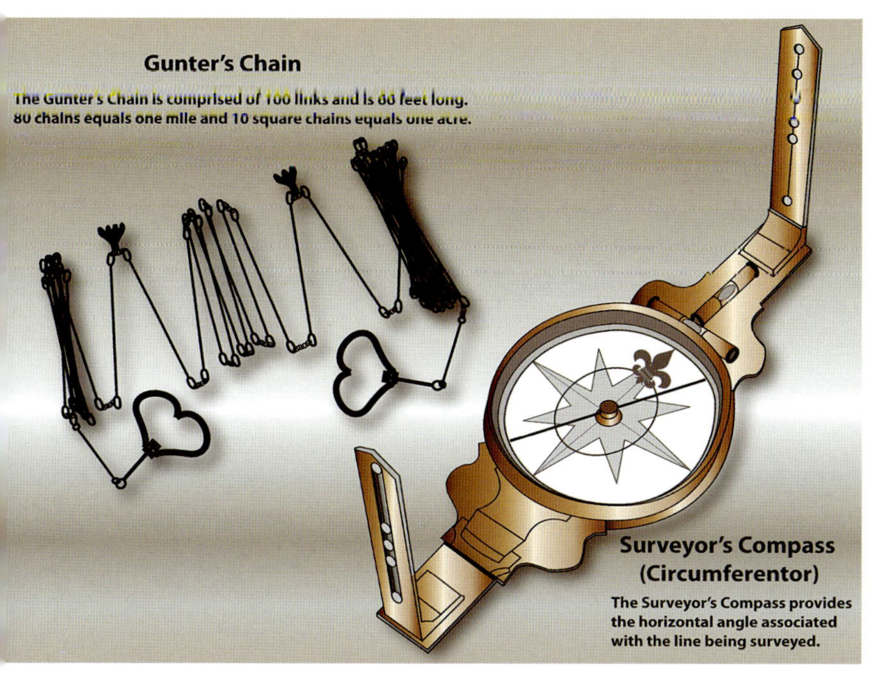

Figure 9-5. *Magnetic compass and surveyor's chain* –Illustration by Phil Mobley

Traverse Table

Way Point	Compass Bearing	Distance
1 start	S 10 E	340 yds
2	S 50 E	510 yds
3	S 05 E	410 yds
4	S 35 E	175 yds
5	S 24 E	425 yds
6	S 70 E	560 yds
7	S 40 E	340 yds
8	S 65 E	470 yds
9	S 23 E	710 yds
10	N 80 E	380 yds
11	S 82 E	360 yds
12	S 20 E	180 yds
13	S 72 E	575 yds
14 end		

Direction of Traverse

PMM 2012

Figure 9-6. *A hypothetical example of an open-traverse survey. Data is collected at various stations, or way points. The compass bearing defines the direction from that station, and the distance between stations is measured.* –Illustration by Philip Mobley

(see Figure 9-6). By contrast, a closed-traverse survey ends at the point where it started. A classic example of a closed survey is the kind used to establish property boundaries.

East of the Appalachian Mountains, much of the country was surveyed using the metes-and-bounds system.[2] In this system, the surveyor defined boundaries using a series of calls, each of which included the compass direction and the distance along a specific bearing. The calls had a specified format, beginning with the compass direction (e.g., N30E) and followed by the distance along the line (e.g., 70 perches), probably measured by a rod or a chain. An example of a land patent with boundaries defined by the metes-and-bounds system is shown in Figure 9-7. This particular patent, named "Little Water," came from Washington County, Maryland. Notice how irregular its shape is. Incidentally, Maryland had one of the most complex landholding systems of any of the original thirteen colonies. Figure 9-8 shows the layout of land patents in eighteenth-century Washington County.[3] The application of the system and its inherent errors resulted in many legal disputes regarding boundaries. Sometimes a patent's contested legal status was reflected in its name; for example, the name of a plat might include the word *strife or conflict.*

As settlement proceeded into the Old Northwest Territory (today's Ohio, Indiana, Illinois, Michigan, Wisconsin, and northeastern Minnesota), and later the lands of the Louisiana Purchase, the survey system changed. The Public Lands Survey System of 1785 was more systematic, and its application created mostly rectangular-shaped landholdings, rather than the puzzlelike configurations of holdings east of the Appalachians. The new system ultimately resulted in fewer disputes about who owned what land where. Typically, surveys in the West preceded significant settlement.

Figure 9-7. *A sample land patent, with measurements taken using the metes-and-bounds survey system.* –Illustration by Philip Mobley

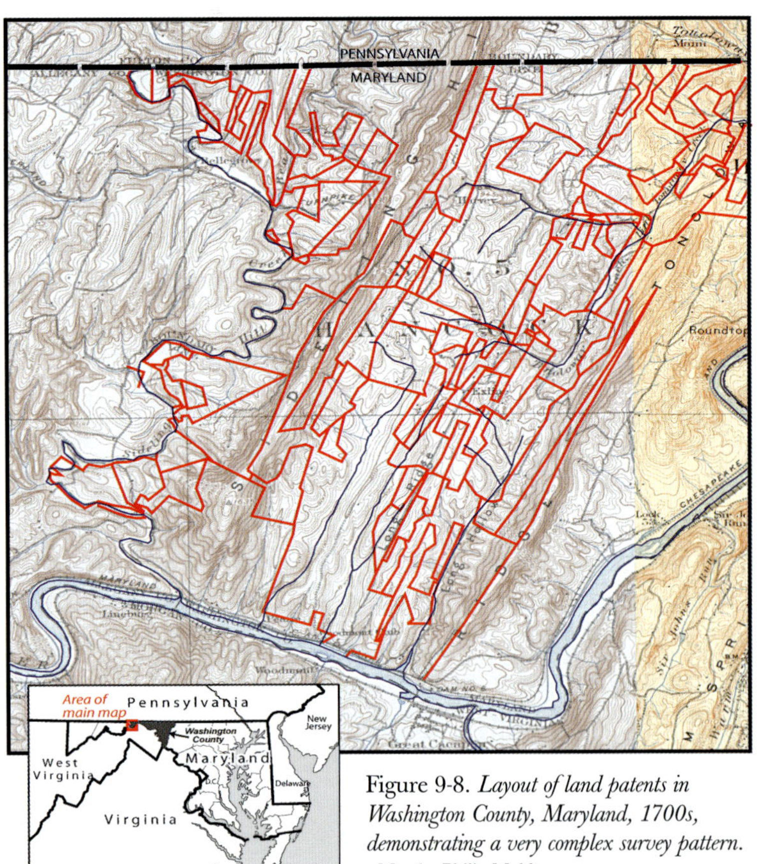

Figure 9-8. *Layout of land patents in Washington County, Maryland, 1700s, demonstrating a very complex survey pattern.* –Map by Philip Mobley

The various localities in the West were defined by a series of large rectangles, each of which represented a specific township and range designation–for example, Township 10 North, Range 5 East. Each rectangle, or block of land, encompassed thirty-six square miles. Within the blocks were smaller rectangles referred to as sections. Each section was one square mile, or 640 acres. Normally, for landownership purposes, sections were then subdivided into quarter sections, each encompassing 160 acres–the size of a typical homestead after 1862. The average farm was further portioned into four units that could be used for crops and/or livestock, each measuring forty acres. The only conditions that made it difficult to apply this system were those in which the terrain was very rugged or mountainous. The diagram in Figure 9-9 shows the application of the public-land survey system near Corvallis, Montana, in 1872.[4]

The Stevens Survey

Maps created from Isaac Stevens's Northern Pacific Railroad Survey data began to appear in 1855, when the first preliminary report was printed. The sample map in Figure 9-10 was created at a small scale–1:200,000, or about twenty miles to the inch. It was compiled by two men–John Lambert and J. K. Duncan. Among other features, the map shows wormlike corridors that represent the paths followed by significant streams and rivers. Stevens's survey route is symbolized by a thin dashed line, which delineates the courses traveled by the main party as well as the routes explored by John Mullan, F. W. Lander, A. W. Tinkham, J. Doty, Lieutenant Arnold, Captain McClelland, G. Gibbs, and others. Most of these supplemental parties comprised a single individual or a very small group. The dates marked along these routes indicate when and where the main party camped. Heavier black lines depict the suggested path for the transcontinental railroad–if and when it was constructed. The map also indicates, with relative precision, the location of the Continental Divide, from which rivers flow in opposite directions either to the Pacific Ocean or to the Gulf of Mexico. Lambert and Duncan's positioning of this divide was used as base data for later maps.

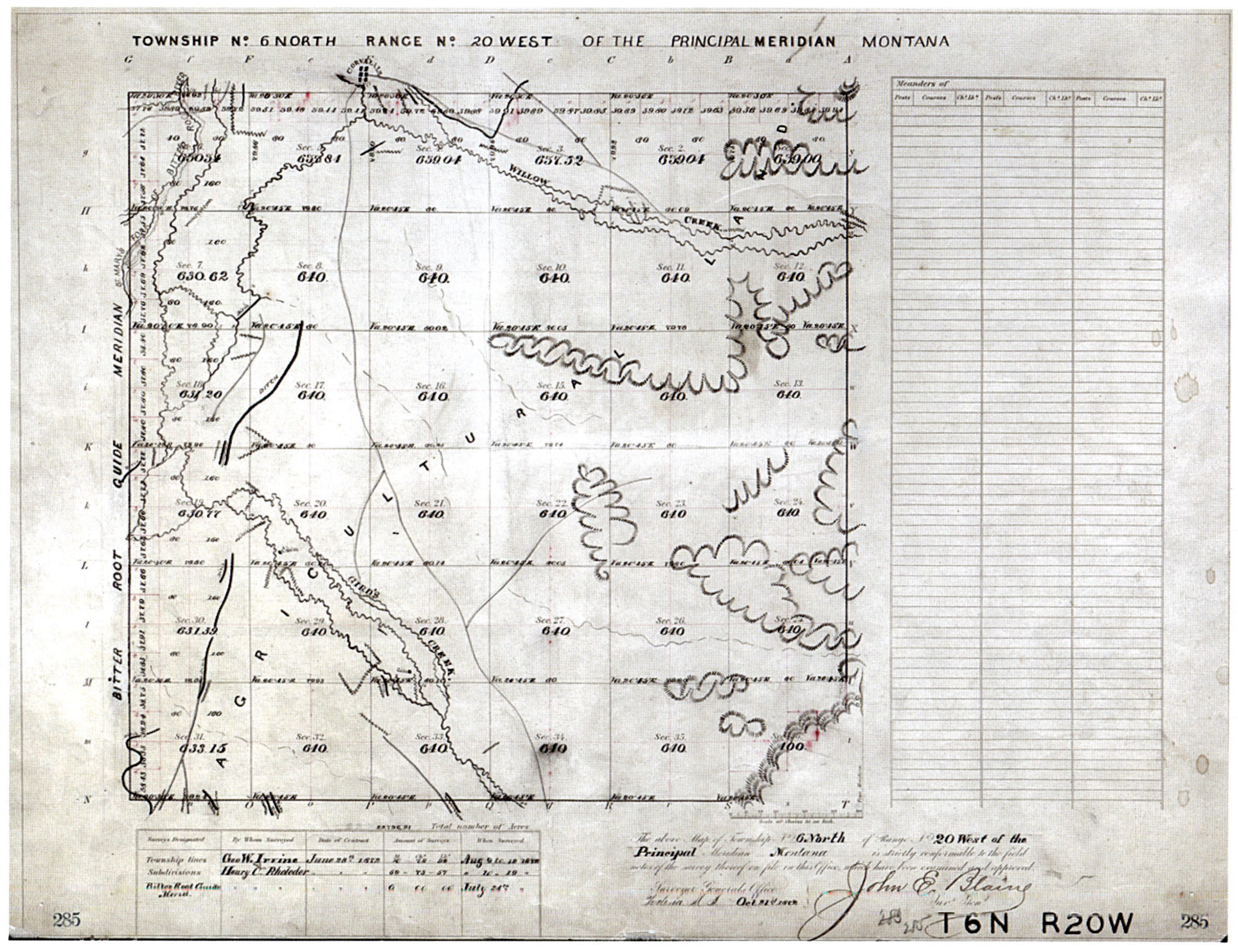

Figure 9-9. *Public Land Survey plat of Township 6, Range 20, Corvallis, Montana, 1872* –Courtesy U.S. Bureau of Land Management

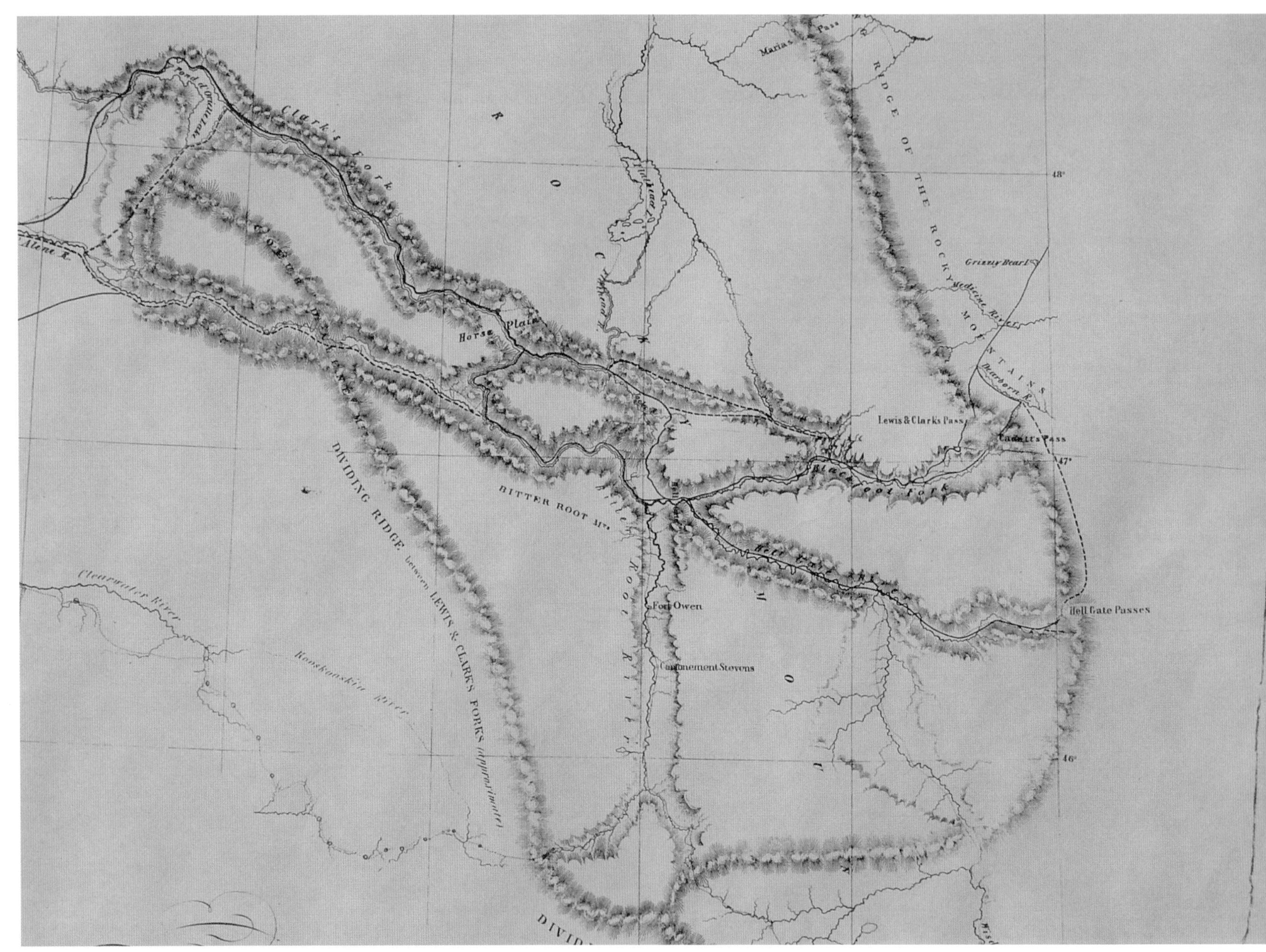

Figure 9-10. *Detail from Isaac Stevens's "Preliminary sketch of the Northern Pacific Railroad Exploration and Survey," Map 3, 1855. Scale approximately 1:200,000.* –Courtesy Library of Congress, Geography and Map Division

Another interesting bit of information provided by the Stevens maps are terrain profiles, or cross-sections, which define the location, the shape, and often the elevation of significant landforms–in this case, the peaks of the Rocky Mountains–along a particular route. The vertical profile of the Rockies shown in Figure 9-11, was taken from the Stevens expedition's map of the region. The inclusion of this profile in Stevens's report demonstrates that the survey party was also collecting elevation data. A key instrument for determining elevation was the barometer, which measures changing air pressure; this measurement is then used to calculate elevation.

The G. K. Warren Maps

One of the ablest topographers of the mid-nineteenth century was Gouverneur Kemble Warren. Born in 1830, the same year as John Mullan, Warren also attended West Point like Mullan and graduated in 1850, two years before Mullan did. Joining the Corps of Topographical Engineers, Warren went to work at the Office of the Pacific Railroad Explorations and Surveys in 1854. He was soon assigned to compile a comprehensive map of the lands west of the Mississippi River, including all known points of geographical significance. To fulfill this assignment, Warren created two maps. The first, titled "Map of the Routes for a Pacific Railroad," was made in 1855 (Figure 9-12). It was, in his own words, "a hurried compilation."[5] This map was not very large, measuring only 12 x 21 inches. Size, however, does not always signify importance. This small map represented a significant stage in the evolution of geographic knowledge about the western United States.

As more information was obtained from the Pacific Railroad surveys, Warren compiled it and incorporated it into his cartography, eventually creating a comprehensive new map of the western United States, titled "Map of the Territory of the United States from the Mississippi to the Pacific Ocean ordered by the Hon. Jeff'n Davis" (see Figure 9-12a). This map was much larger than his 1855 work, measuring 45.5 x 42 inches. The original scale of the newer map was 1:3,000,000, or forty-eight miles to the inch. A synthesis of the most reliable information available at the time, it was Warren's masterpiece.

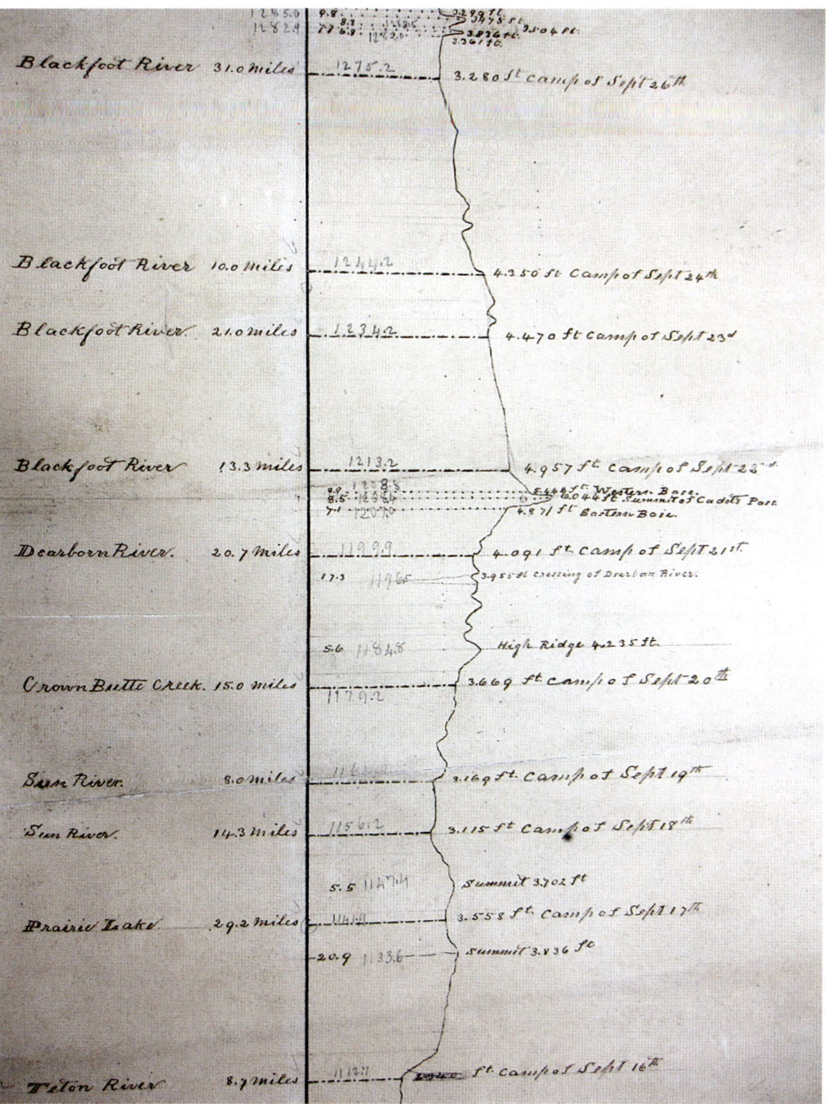

Figure 9-11. *Vertical profile of the Rocky Mountains; illustration created by the Stevens survey*
–From Isaac I. Stevens, *Narrative and Final Report of Explorations for a Route for a Pacific Railroad* (1860)

From a geographic perspective, the Pacific Railroad maps were an important contribution to our understanding of the West. Using Warren's 1858 map as a source, we overlaid the paths taken by the five major transcontinental railroad surveys on a modern terrain base (Figure 9-13).

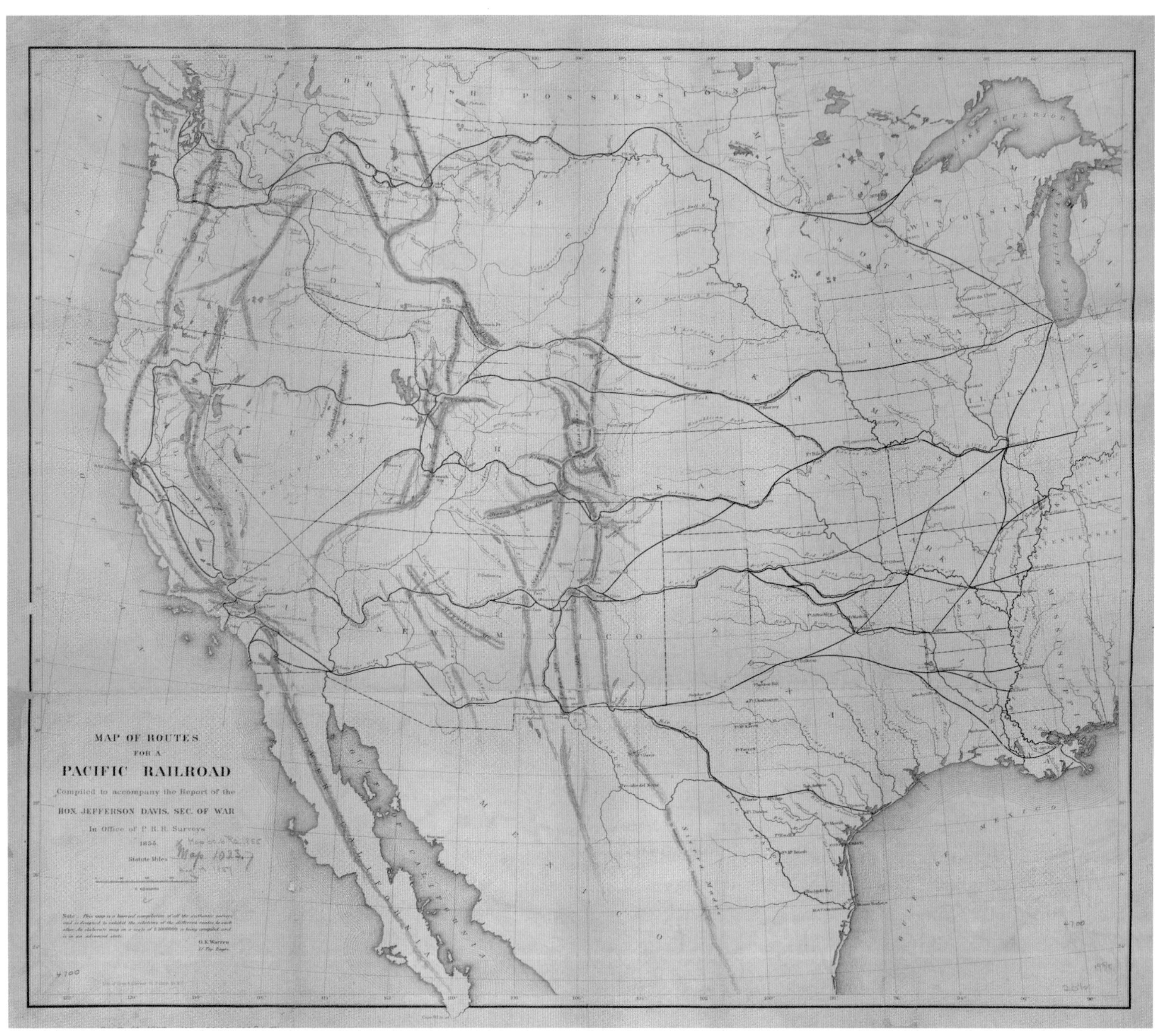

Figure 9-12. *G. K. Warren's 1855 map (scale approximately 1:6,000,000)* –Courtesy Norman B. Leventhal Map Center, Boston Public Library

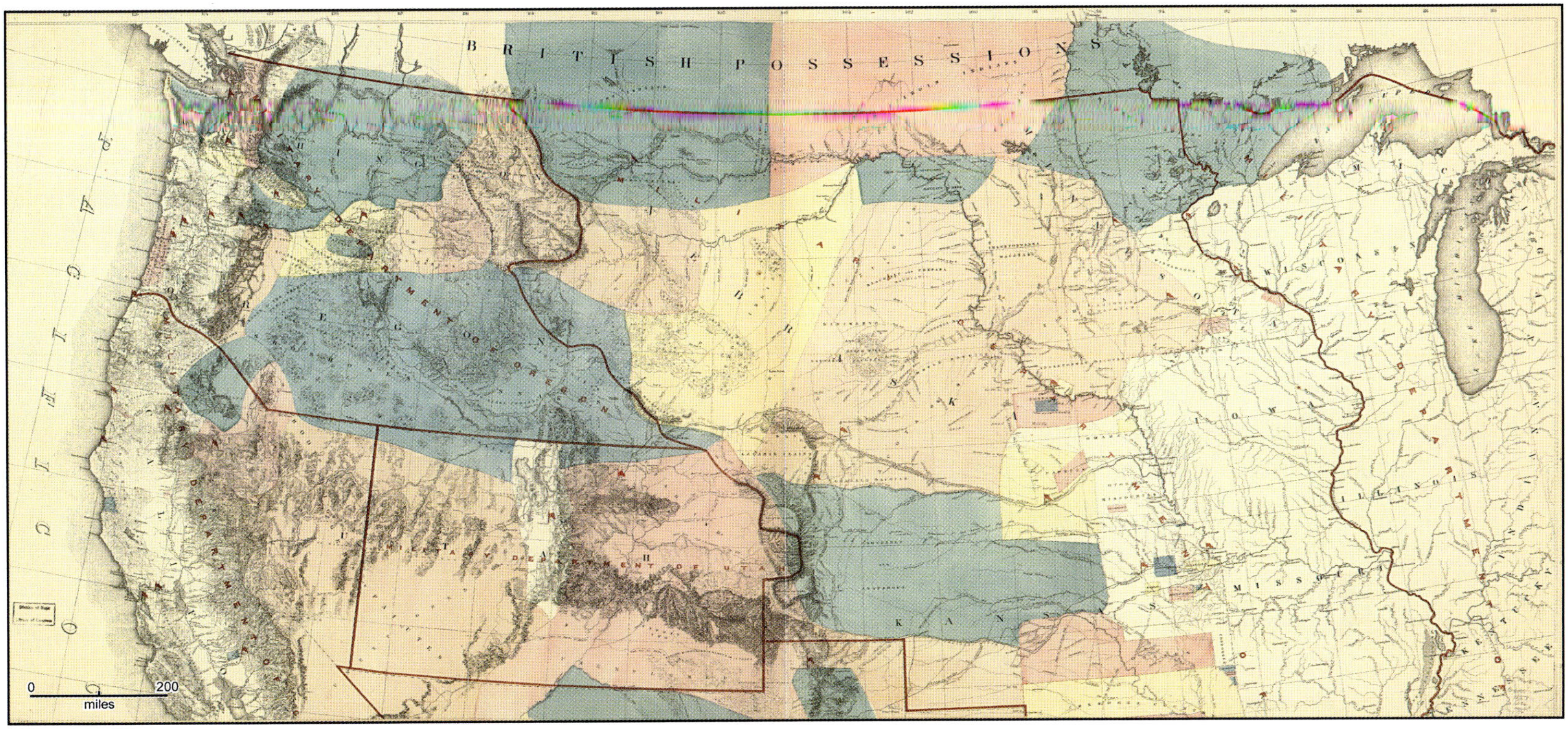

Figure 912a. *"Map of the Territory of the United States from the Mississippi to the Pacific Ocean, ordered by the Hon. Jeff'n Davis" by G. K. Warren, 1858* –Courtesy Library of Congress, Geography and Map Division

Figure 9-13. *The five routes explored for the Pacific Railroad Survey*
–Map by Philip Mobley

The Transcontinental Railroad: Costs and Benefits

The most significant information from the railroad surveys was published in thirteen volumes, which, interestingly, cost considerably more to produce than the surveys themselves had cost.[6] The original surveys were radically underfunded; for example, Stevens had already spent his original provision of $40,000 by the time he reached Fort Benton in September 1853.[7]

When Stevens's northern survey was completed in 1854, Captain George B. McClellan estimated the cost of building the transcontinental railroad on that route—$108 million. Was that figure reasonable? Probably so. At the time, the cost of building one mile of railroad was approximately $40,000 to $50,000. To build 2,500 miles of railroad would cost between $100 million and $125 million. When Secretary of War Jefferson Davis reviewed Stevens's report, however, he revised this estimate to a much higher figure–$147 million![8] Davis's estimate created the impression that the northern route was not a financially viable choice. But was it? That question is difficult, if not impossible, to answer.

Ultimately, the route chosen for the first transcontinental railroad was a compromise position; the line traversed the country's midsection from Omaha, Nebraska, to Oakland, California. We can say that this was not an inexpensive choice, but it is very difficult to calculate what the actual final cost was because so much was involved in the construction and some of the real costs were concealed. In addition to the building costs was the value of the considerable amount of land the government gave away to the railroad companies.

The question of which route would have been the best choice remains. In terms of cost and ease of construction, the southernmost route would win. In fact, this was the route Jefferson Davis favored, but it had the disadvantage of not directly connecting the country's industrial region in the Northeast with the natural resources of the West.

In the end, portions of all five routes explored in the Pacific Railroad Surveys became the basis for railroads, roads, and eventually, interstate highways. The men who blazed these paths, however, would not live to see the ultimate fruition of their labors.

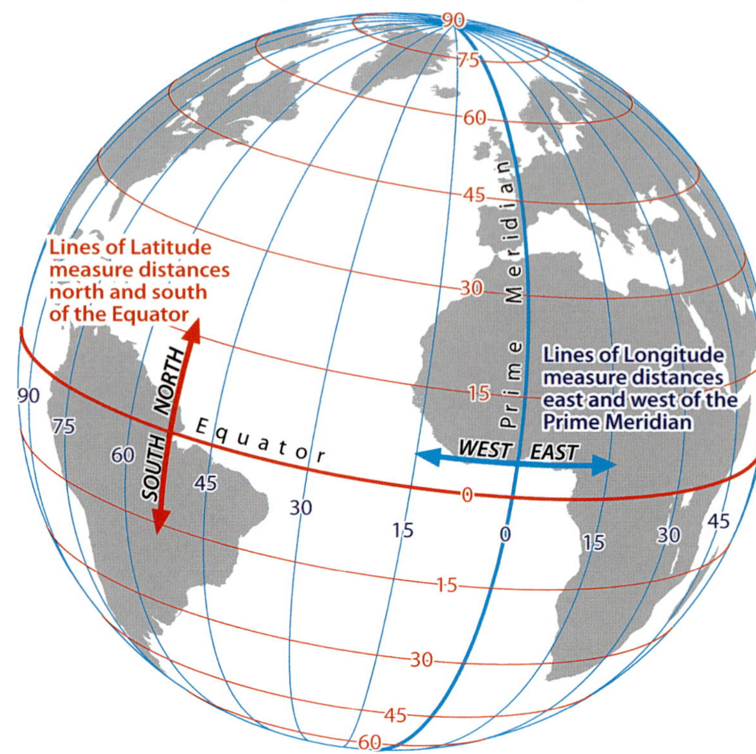

Figure 9-14. *Illustration of longitude and latitude* –Illustration by Philip Mobley

The Accuracy of Nineteenth-century Surveys

One question we particularly want to address in our analysis of the Stevens and Mullan surveys is, how accurate were they? Did the surveyors determine distances and elevations correctly? How precise were the form and contour lines on their maps? Our conclusion is, not very. At best, the locations they recorded were within twenty miles of their true geographical position. While we assume that the surveyors' longitudinal calculations were reasonably accurate, these measurements were the ones most subject to error. In terms of their relationship to modern geographical positions, the longitudes calculated in the surveys were considerably off the mark.

Longitude and *latitude* are the terms used to define a geographic position on the earth's sphere. Positions are calculated based on spherical distances, expressed as degrees, minutes, and seconds from the starting points, either a prime meridian or the earth's equator–the former being 0 degrees east-west, and the latter, 0 degrees north-south. Longitude is measured as the distance either west or east of the prime meridian, and latitude is the distance north or south of the equator. For example, a location might be 34 degrees, 21 minutes, and 50 seconds west longitude and 15 degrees, 10 minutes, and 58 seconds north latitude. Figure 9-14 shows longitude and latitude positions; longitude lines are shown in blue, latitude lines in red.

In assessing the accuracy of the surveys, we must take into account the spherical datum that was employed for determining longitude and latitude. Cartographers in the nineteenth century used a different measurement for the earth's spheroid than is now used in modern mapping. Unless this change of datum is factored in, the work appears to be less accurate than it really is.

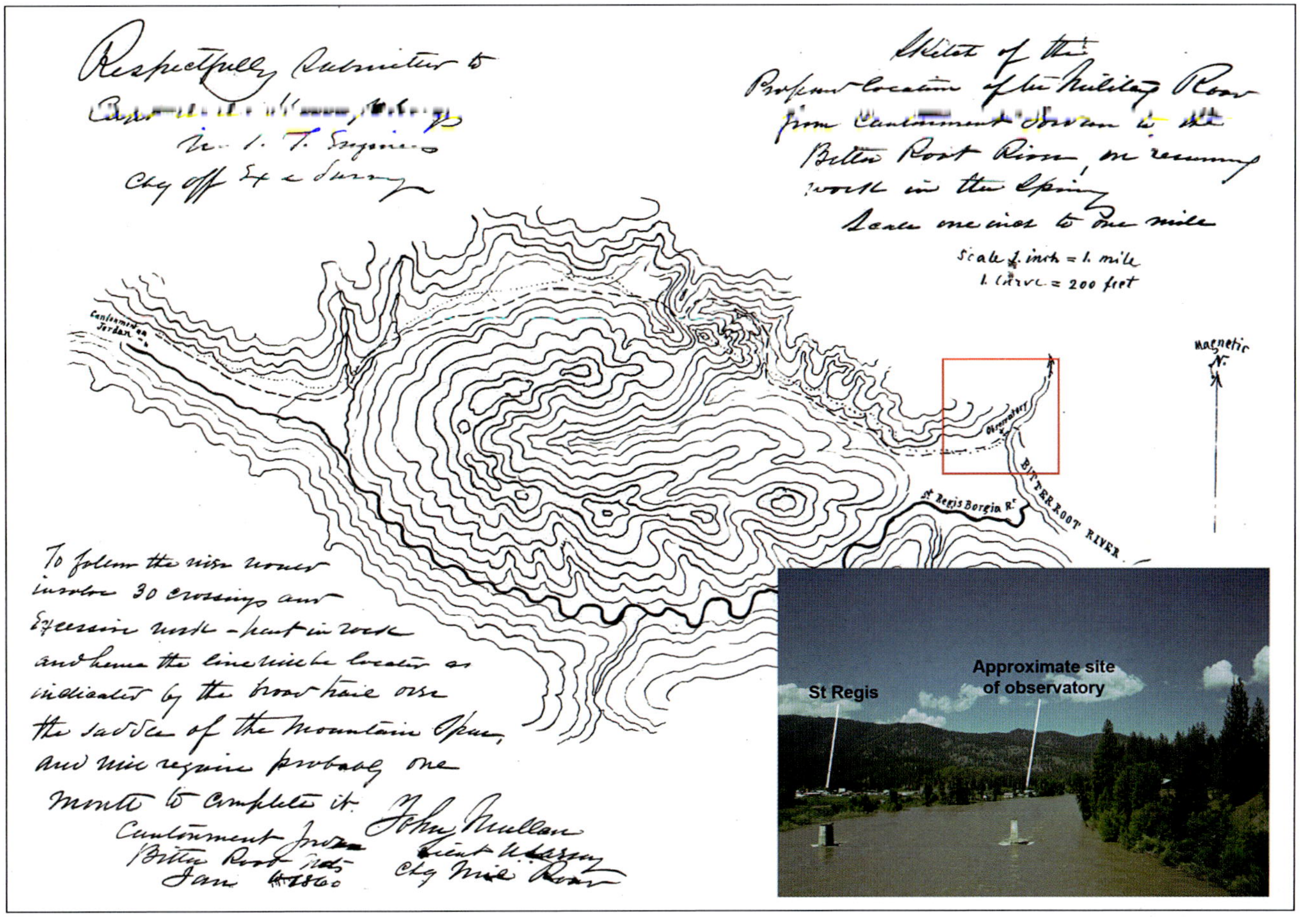

Figure 9-15. *Location of the observatory at St. Regis, Montana. Astronomical readings taken at such observatories helped improve the accuracy of maps. The inset photo shows St. Regis today.* –Map courtesy National Archives and Records Administration, Cartographic Section, College Park, MD; photo by Paul D. McDermott

Another data reference that has changed from Mullan's time is the one used to determine sea level. It was a complex calculation, based on the difference between high and low tide, a figure that varies among the different oceans. We now have a standard baseline that equalizes sea level from one part of the world to another. Thus, one should evaluate differences in elevation figures between nineteenth-century maps and twenty-first-century maps based upon the different calculations of mean sea level. Similarly, the elevation of Sohon Pass was calculated by the Mullan party as 4,932 feet above sea level, while its actual elevation is 4,950 feet above sea level. Of course, for all of this, we must bear in mind that Stevens, Mullan, and their surveyors made their measurements as accurately as they could with the technology available at that time. The survey crews made repeated observations, using a barometer, to ensure correct measurements of elevations. They calculated longitudinal positions with chronometers, using a recognized meridian, such as St. Louis.

One extremely interesting aspect of Mullan's expedition is his effort to take astronomical readings in addition to topographical measurements. Mullan set up observatories at various points along the road from which he recorded repeated astronomical observations (see Figure 9-15). The observatories were generally established in open areas where a greater expanse of night sky was visible. Today, the National Archives (Record Group 77, Records of the Office of Chief of Engineers) houses many of the Mullan team's field notebooks, in which they recorded their celestial observations.

The Surveyor's Instruments

Sextant: For hundreds of years, navigators have been able to determine latitude (a position north or south of the equator) by determining the angle of the sun above the earth's horizon at noon. Over time, various devices were invented to accomplish this task efficiently, and eventually, the sextant evolved (see Figure 9-16). To figure latitude, the viewer makes separate observations of the horizon and the sun or other celestial body. Once those two points are sighted, he turns a knob that optically lowers the image of the celestial object until it is aligned with the horizon (see Figure 9-17). In the process, the sextant's arm moves down along a metal arc, stopping at the point where the objects appear to touch. The user then reads the number that is etched on the arc, or scale; this number represents angular degrees. The process is repeated several times to assure greater accuracy. The results are compared with the figures on a table that gauges latitude.

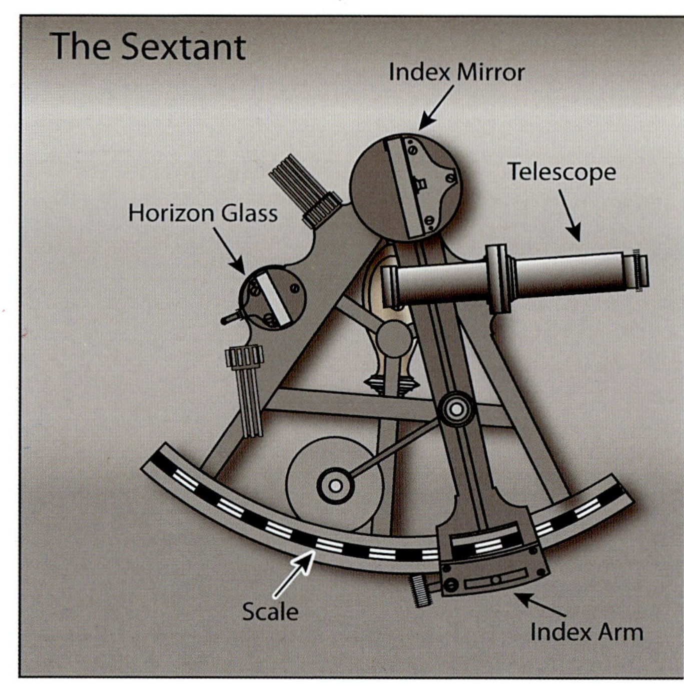

Figure 9-16. *A Sextant measures the sun's angle above the earth's horizon, a calculation that is used to determine latitude.* –Illustration by Philip Mobley

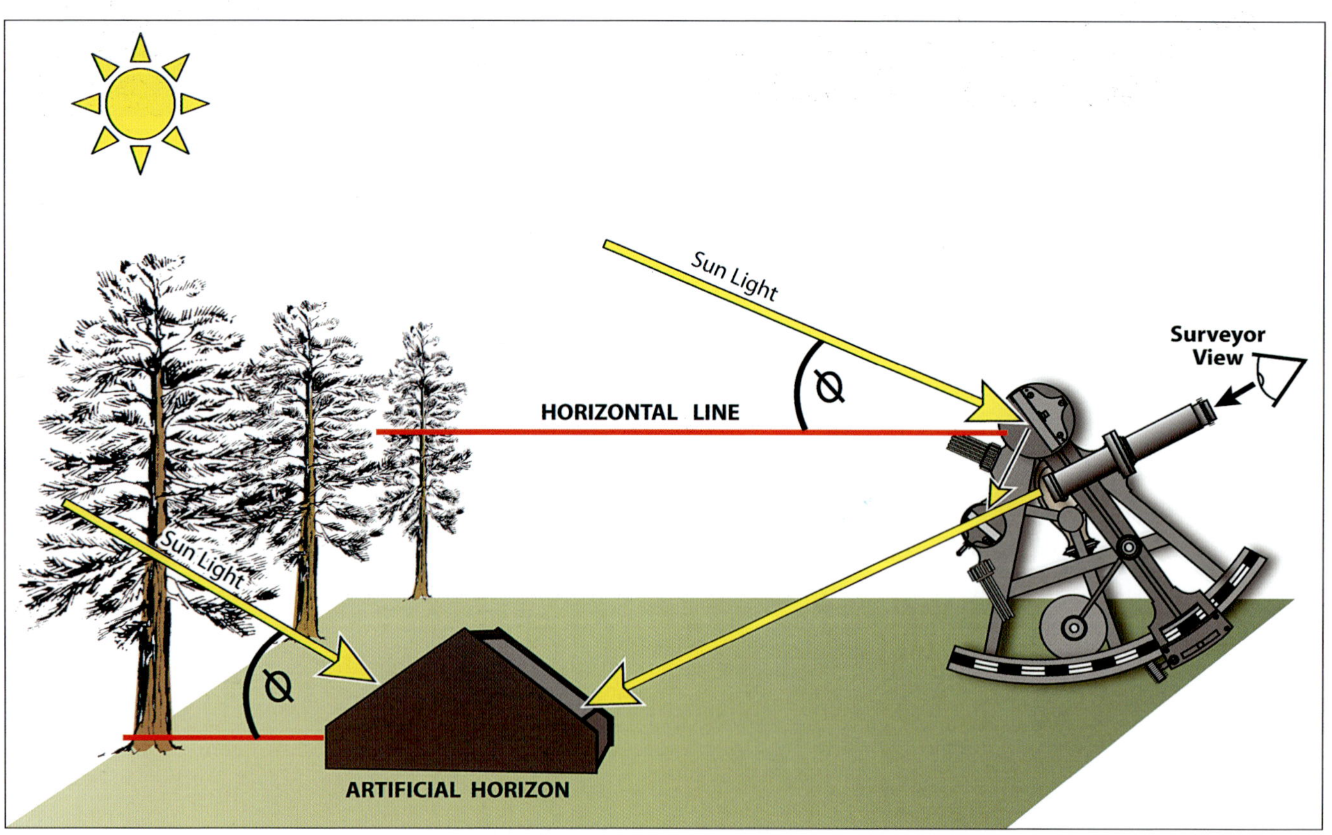

Figure 9-17. *This shows how an artificial horizon is used in sighting a celestial body with a sextant. The user created an artificial horizon when the natural horizon was obscured by vegetation or other obstructions.* –Illustration by Philip Mobley

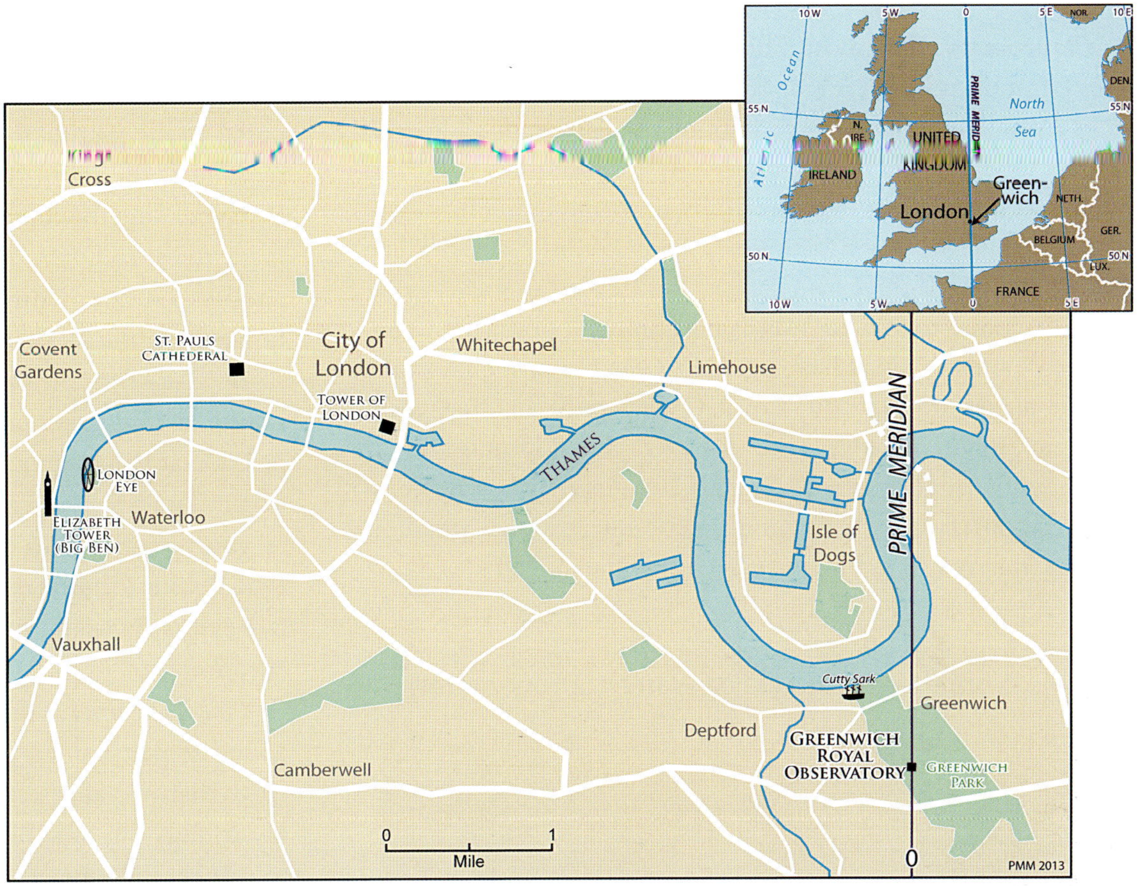

Figure 9-18. *Location of the Royal Observatory in Greenwich, England* –Map by Philip Mobley

Figure 9-19. *Dr. Ronald E. Grim at the Prime Meridian marker in Greenwich; a chrome line embedded in the pavement indicates the meridan.* –Photo courtesy Ronald E. Grim

Chronometer: The most difficult component to determine in geographic position is longitude, i.e., the distance east or west of a prime meridian, or 0° longitude. What is a prime meridian? One can begin at literally any location one wishes to use, as longitude is determined by time, measured from any given position. For a long time, many countries used a line that transected their own capital, but this led to great confusion because each place was using a different prime meridian. Today, most nations have agreed to use the prime meridian that passes through the Royal Observatory at Greenwich, Great Britain, creating a standardized global location for 0° longitude (see Figures 9-18 and 9-19). Another key function of the observatory is to fix a precise international standard time, referred to as Greenwich Mean Time (GMT).

Figure 9-20. *The chronometer, an instrument used for determining longitude, was originally developed by John Harrison in 1761.* –Illustration by Philip Mobley

Longitude is determined based upon the measurement of time, i.e., the position of the sun, in any given location compared to the meridian time. The more accurately the time is measured, the more accurate the longitude position will be. By calculating the difference in the measured time (expressed in degrees, minutes, and seconds) between one place and another, one can figure out how far east or west one's location is from the meridian, or starting position. So how does it work? We know the earth rotates 360 degrees in twenty-four hours, or 15 degrees per hour. Thus, if according to the sun's position at your location, it is two hours later than Greenwich Mean Time, your longitude is 30 degrees east. Likewise, if at your position it is an hour and 15 minutes earlier than GMT, your location is 15 degrees, 15 minutes west. But how did explorers know the time in Greenwich? They set their chronometer!

This instrument, developed in the eighteenth century by a clockmaker named John Harrison, was basically a special clock that was able to accurately keep time on a moving ship in virtually any weather (see Figure 9-20). The clock was set to display the time at the meridian, and the user determined his longitude by comparing the time at his location (expressed in degrees, minutes, and seconds) to the meridian time on the chronometer. So how did explorers know what time it was at their location? Often they used a sextant to determine high noon, the point where the sun's position was highest above the horizon; they could then compare it to the time shown on the chronometer. Before the chronometer became commonplace, significant errors in longitudinal position occurred, which often proved costly.

With the advent of the global positioning navigating system (GPS), longitude and latitude are now determined using satellite positions as the satellites revolve around the earth (see Figure 9-21).

Odometer: Over the centuries, people have invented various devices for measuring distance. In the 1800s, many travelers used a wagon odometer–an instrument that counted how many revolutions a wagon wheel made over the course of

Figure 9-21. *The determination of location using the Global Positioning System (GPS)* –Illustration by Philip Mobley

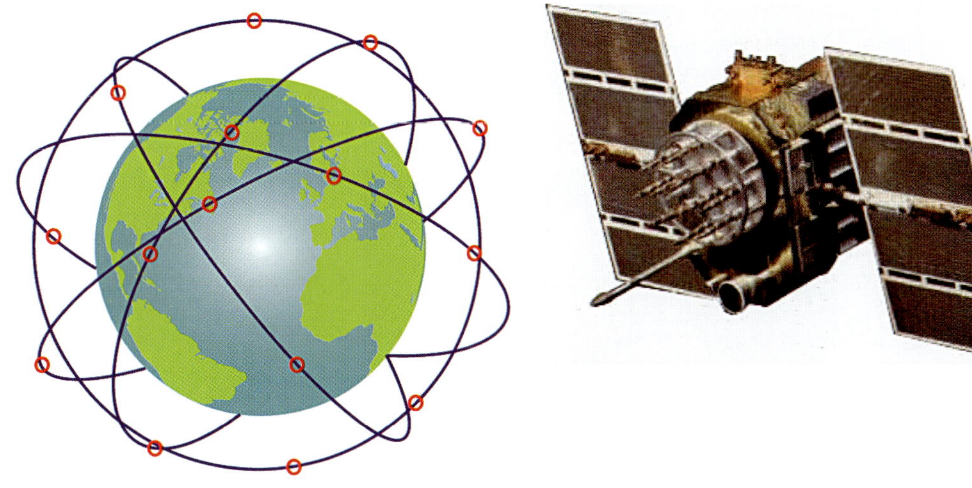

a trip (see Figure 9-22). With this type of odometer, affixed to a wagon wheel, the inner plate of the device flipped over with each rotation of the wheel and advanced the numbered dial. To determine daily mileage, the traveler recorded the dial reading at the start of the journey and recorded it again at the end of the day, then subtracted the first number from the second to get the number of revolutions. Because larger wheels make fewer rotations than smaller ones, the size of the wheel also had to be factored in. Thus, to calculate the distance traveled, he multiplied the number of revolutions by the wheel's circumference. Likewise, the total distance for an entire trip could be calculated by the same method.

John Mullan often used a wagon odometer as he traveled through the West. To make calculations easier, he and his assistants created meticulous tables matching revolutions to distance, allowing for different wheel circumferences. Figure 9-23 shows several tables actually used by Mullan to determine distances traveled.

We still use odometers in our vehicles to measure distance traveled, though readouts have increasingly become digitized. Today, GPS is the usual means of determining how far one point is from another. Rather than wheel rotation, GPS uses a system of triangulation based on satellite signals. Extremely accurate, it is a new tool in the modern surveyor's toolbox.

Figure 9-22.
Wagon odometer
–Illustration by Philip Mobley

Mullan Military Wagon Road Odometer Table

Odom.	Ft. & In.	Mi.	Odom.	Ft. & In.	Mi.	Odom.	Ft. & In.	Mi.	Odom.	Ft. & In.	Mi.	Odom.	Ft. & In.	Mi.
1	10' 11"		26	283' 10"		51	556' 9"		76	829' 8"		145		3/10
2	21' 10"		27	294' 9"		52	567' 8"		77	840' 7"		194		4/10
3	32' 9"		28	305' 8"		53	578' 7"		78	851' 6"		242		5/10
4	43' 8"		29	316' 7"		54	589' 6"		79	862' 5"		290		6/10
5	54' 7"		30	327' 6"		55	600' 5"		80	873' 4"		338		7/10
6	65' 6"		31	338' 5"		56	611' 4"		81	884' 3"		385		8/10
7	76' 5"		32	349' 4"		57	622' 3"		82	895' 2"		436		9/10
8	87' 4"		33	360' 3"		58	633' 2"		83	906' 1"		485		1
9	98' 3"		34	371' 2"		59	644' 1"		84	917' 0"		970		2
10	109' 2"		35	382' 1"		60	655' 0"		85	927' 11"		1455		3
11	120' 1"		36	393' 0"		61	666' 11'		86	938' 10"		1940		4
12	131' 0"		37	403' 11"		62	676' 10"		87	949' 9"		2425		5
13	141' 11"		38	414' 10"		63	687' 9"		88	960' 8"		2910		6
14	152' 10"		39	425' 9"		64	698' 8"		89	971' 7"		3395		7
15	163' 9"		40	436' 8"		65	709' 7"		90	982' 6"		3880		8
16	174' 8"		41	447' 7"		66	720' 6"		91	993' 5"		4365		9
17	185' 7"		42	458' 6"		67	731' 5"		92	1004' 4"		4850		10
18	196' 6"		43	469' 5"		68	742' 4"		93	1015' 3"		5335		11
19	207' 5"		44	480' 4"		69	753' 3"		94	1026' 2"		5820		12
20	218' 4"		45	491' 3"		70	764' 2"		95	1037' 1"		6305		13
21	229' 3"		46	502' 2"		71	775' 1"		96	1046' 0"		6790		14
22	240' 2"		47	513' 1"		72	785' 0"		97	1058' 11"	2/10	7275		15
23	251' 1"		48	524' 0"	1/10	73	796' 11"		98	1069' 10"		7760		16
24	262' 0"		49	534' 11"		74	807' 10"		99	1080' 9"		8245		17
25	272' 11"		50	545' 10"		75	818' 9"		100	1091' 8"		8730		18
												9215		19
												9700		20

Figure 9-23. *Table showing the calculation of distance based on wagon odometer readings* –Table by William Weikel

Barometer: A barometer is an instrument used to measure atmospheric pressure, which decreases as one moves from a lower altitude to a higher one (see Figure 9-24). Thus, by knowing the air pressure, one can determine one's elevation, or altitude. In moving from sea level to mountaintops, air pressure drops because the higher one goes, the thinner and less dense the atmosphere is. Atmospheric pressure is expressed in inches. At sea level, it measures 29.96 inches.

Similar to a mercury thermometer, early barometers used tubes containing mercury to produce readings, which were taken from a scale printed on the side of the tube. At the bottom of the tube was a reservoir of mercury, while the top end was open to allow the air pressure to exert its influence on the mercury at the bottom. Moving upward in altitude, the air pressure decreases, which allows more mercury to flow back into the reservoir, thus draining the mercury in the tube and producing a lower reading.

The aneroid barometer (see Figure 9-25) functions in the same way as the mercurial barometer except that instead of mercury it relies on a vacuum-canister mechanism that reacts to the air pressure. The expansion or contraction of the air in the canister moves the pointer that indicates the pressure change. Not only was the pressure-change meter easier to read than the mercury scale, but the aneroid barometer was also easier to handle, as there was no concern about handling mercury or breaking the glass tube.

Mullan and his topographers, civil engineers, and surveyors needed to measure elevation changes along the route of the road in order to figure out how much grading would be

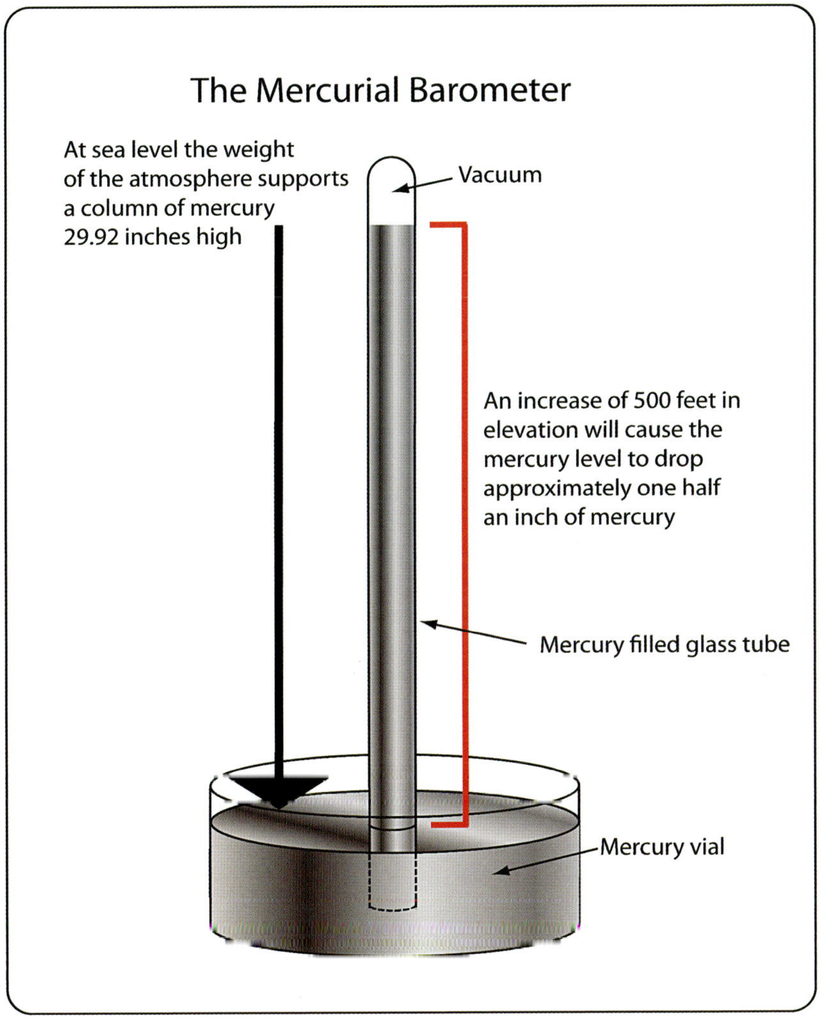

The Mercurial Barometer

At sea level the weight of the atmosphere supports a column of mercury 29.92 inches high

Vacuum

An increase of 500 feet in elevation will cause the mercury level to drop approximately one half an inch of mercury

Mercury filled glass tube

Mercury vial

Figure 9-24. *Mercury barometer* –Illustration by Philip Mobley

Figure 9-25. *A compensated aneroid barometer with scale.* –Photo by Philip Mobley

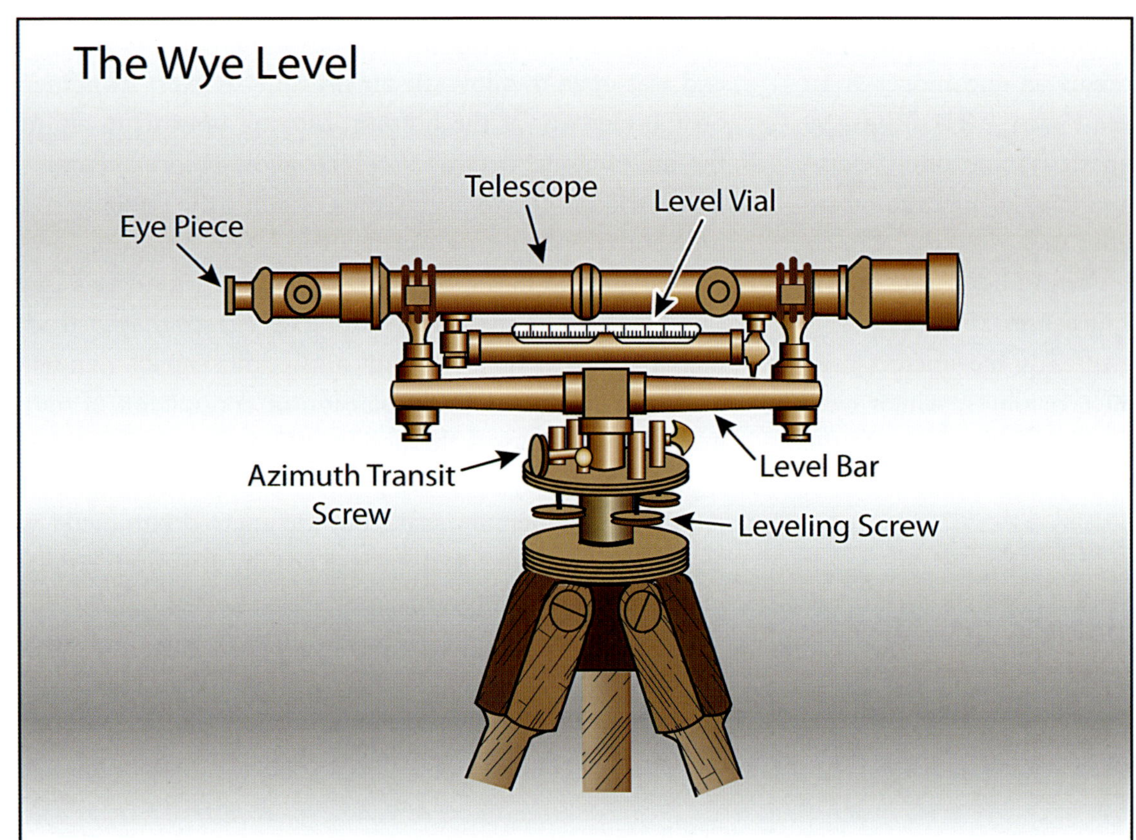

Figure 9-26.
A wye level
–Illustration by Philip Mobley

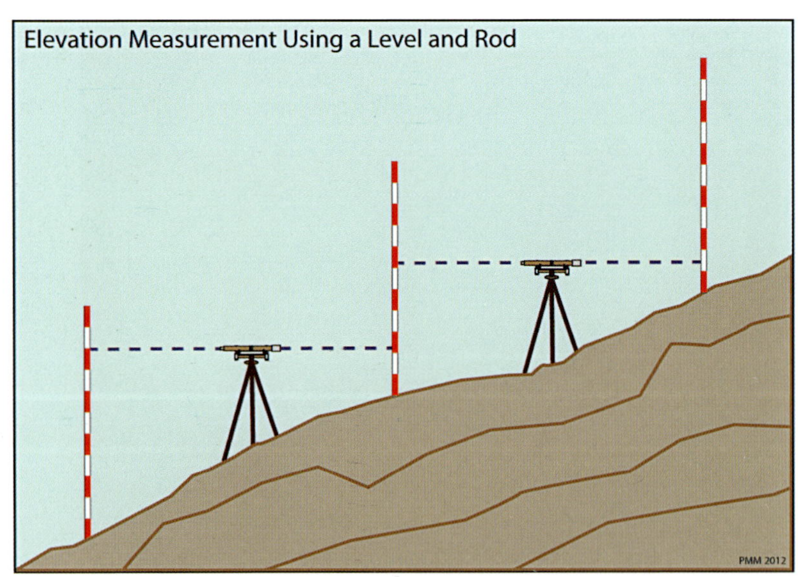

Figure 9-27. *How a rod and level are used to determine elevation*
–Illustration by Philip Mobley

Profile Notes

Station	Rod	Instrument Height	Rod	Point Elevation
Bench Mark				100.00
	6.25	106.25		
Angle 1			2.64	103.61
	2.40	106.01		
Angle 2			8.52	97.49
	6.14	103.63		
Angle 3			3.27	100.36
	7.25	107.61		
Angle 4			1.15	106.46
	5.83	112.29		
Angle 5			4.27	108.02
	2.79	110.81		
			1.86	108.95

Figure 9-28. *Profile table of elevation readings from a wye level and rod.*
–Table by Philip Mobley

needed during construction. Elevation changes also affected the degree of difficulty that wagons would encounter as they ascended or descended significant slopes. The expedition's other main consideration was railroad construction. They needed to be sure that the route they chose could be sufficiently graded to assure efficient railroad operation.

Making calculations more difficult for the surveyors was the fact that air pressure varies not only as the altitude changes, but also as different air masses pass over a region. Some masses are characterized by low pressure and others by high pressure, so those effects must be taken into consideration when calculating elevation.

Level: Mullan and the surveyors of his time collected elevation data using different techniques and instruments. Aside from the barometer, the most important device they had for measuring altitude was the level. Essentially an elongated telescope, the level was typically used with measuring rods (also called surveyors rods, leveling rods, or level staffs), a special rod or pole marked like a yardstick but often in increments of tenths and hundreths of a foot rather than in inches. Figure 9-26 shows a wye level, one of the earliest level designs.

To use a level, the surveyor first places a rod near the base of the hill or peak he wants to measure, in order to establish a benchmark point, then he adjusts the rod so that the zero mark is even with the instrument's eyepiece (or the surveyor's eye). He measures the benchmark's elevation by sighting the rod with the level and taking a reading. This reading is later used in the calculations, as is the elevation of the level's own location. The surveyor then places a second rod some distance away from the benchmark, rotates the level to sight the second rod, and takes that reading. After that, he moves the level to the top of the hill and takes another set of readings from there. Sometimes the process was repeated many times. Finally, the surveyor calculated the elevation of the hill based on these readings. Figure 9-27 shows how a level and rod are used in the field; Figure 9-28 is an example of a table showing the measurements used to calculate elevation.

Transit. Another instrument used to obtain elevation data in the nineteenth century was a transit, which provided a more direct and somewhat quicker means of estimating elevation

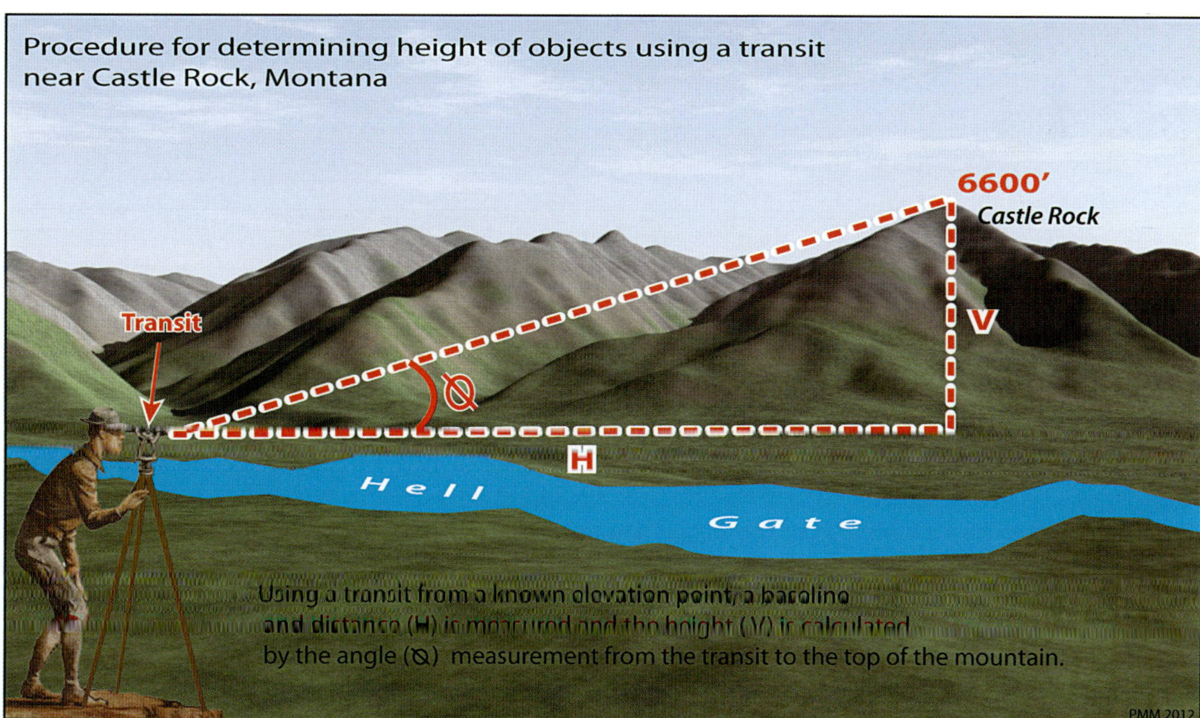

Figure 9-29. *How a transit is used to determine elevation* Illustration by Philip Mobley

than a level. Like a level, a transit is mainly a telescope and is used with rods or something similar, but the transit measures angles rather than height (see Figure 9-29). The method of measuring the elevation of a landform with a transit is similar to the method used with a level, but the elevation is calculated based on geometric angles, expressed in degrees, minutes, and seconds.

After the Surveys

Mullan's road project was completed in July 1862, by which time the Civil War was in full swing. Sadly, Isaac Stevens was killed at the battle of Ox Hill near Chantilly, Virginia, on September 1, 1862. It was under these chaotic circumstances that Mullan compiled his final report to Congress, thus it was a hurried project with only a handful of illustrations. He did include a number of new maps, however, although they were small in scale and details were essentially confined to the immediate region along the route.

Because the nation was so focused on the Civil War, there was very little interest in the West, so maps of the western United States were not immediately updated with the information derived from the Mullan surveys. Had it not been for the war and the resulting difficulty of obtaining sufficient funding for maintaining the wagon road, Mullan's work would perhaps have stimulated more progress in the Pacific Northwest than it did. Nevertheless, the road's impact on the region was significant.

Once the war was over, interest in creating a transcontinental railroad revived, producing an upsurge in demand for accurate maps of the West. Another, more urgent reason for this demand was the development of mining in Montana, Idaho, and other areas in the Northwest. Thus Mullan's maps garnered renewed attention. A few years later, during the construction of the Northern Pacific Railway, Mullan's maps of the region became vital, especially when an extension was built along the St. Regis River and across the crest of the Bitterroots into the Coeur d'Alene region. Later, the Milwaukee Railroad also traced a large stretch of the original Mullan Road. In the twentieth century, U.S. Highways 10 and 12 and, eventually, Interstate 90 were built along major sections of the road. In the end, many miles of roads, railroad tracks, and highways followed the route that Mullan and his men pioneered.

---

1 George B. McClellan, "Memoranda on Railways, Office of Pacific Railroad Surveys," in *Reports of Explorations and Surveys, to Ascertain the Most Practicable and Economic Route for a Railroad from the Mississippi River to the Pacific Ocean, made under the direction of the Secretary of War, in 1853–5,* vol. 1 (Washington, DC: Thomas H. Ford, Printer, 1855), pp. 115–30.

2 Edward T. Price, *Dividing the Land: Early American Beginnings of Our Private Property Mosaic* (Chicago: University of Chicago Press, 1955), pp. 7–9 and appendix 349.

3 Washington County land patents are examined in detail in a forthcoming book by Paul D. McDermott, *Backcountry: The Settlement of Western Maryland.*

4 Norman J. W. Thrower, *Maps and Man: An Examination of Cartography in Relation to Culture and Civilization* (Englewood Cliffs, NJ: Prentice Hall, 1972), p. 102.

5 Patricia Molen Van Ee, "The Coming of the Transcontinental Railroad: The Warren Maps," in Paul E. Cohen, *Mapping the West: America's Westward Movement, 1524–1890* (New York: Rizzoli, 2002), pp. 172–75.

6 In *Artists and Illustrators of the Old West, 1850-1900,* Robert A. Taft estimates that the publication of the thirteen-volume report cost over $1,000,000, while the surveys themselves cost $455,000. Taft, *Artists and Illustrators* (New York: Charles Scribner's Sons, 1953), pp. 5, 254–6.

7 Stevens survey cost figures from U.S. War Department, *Reports of explorations and surveys, . . .* (Washington, DC: Thomas H. Ford, Printer, 1855-60). See especially vol. 1, pp. 23–24.

8 Estimated cost of building the railroad as calcuated by the authors; various discussions concerning the estimated costs of building the proposed Pacific railroad routes can be found in U.S. War Department, *Reports of Explorations and Surveys*; see especially vol. 1, pp. 29–31 and 55, and vol. 12, part 1, pp. 348–51.

Maps of the Mullan Road

BY PAUL D. MCDERMOTT AND RONALD E. GRIM

Editor's note: An earlier draft of this paper appeared as "Maps of the Mullan Road" in *Proceedings of the American Congress on Surveying and Mapping, 36th Annual Meeting* (Washington, D.C.: American Congress on Surveying and Mapping, 1976). The Gustavus Sohon illustrations included here were originally published in John Mullan, *Report on the Construction of a Military Road from Fort Walla-Walla to Fort Benton,* S. Ex. Doc. 43, 37th Cong., 3rd sess., serial 1149 (Washington, D.C.: Government Printing Office, 1863).

The military wagon road from Fort Walla Walla to Fort Benton, commonly known as the Mullan Road, was one of numerous road-building projects sponsored by the federal government in the mid-nineteenth century. The Mullan project resulted from public demand for a new transportation system in the growing Pacific Northwest. The Democratic Congress of the time was initially opposed to projects of this type, but continuing pressure to provide military protection for settlers in the West persuaded legislators to support the construction of military roads. Mullan received his initial approval and appropriation in 1858.[1]

Officially, the function of the Mullan Road was to provide a link between two military posts–Fort Walla Walla in the Washington Territory and Fort Benton, then in the Nebraska Territory. It was anticipated that, upon completion, the road would not only facilitate the movement of troops and supplies, but it would also eliminate a missing link in a potential transcontinental transportation system, which included the navigable portions of the Columbia and Missouri Rivers. The importance of the two posts, one the head of navigation of the Columbia, the other the head of navigation of the Missouri, is revealed in illustrations of the two terminal points by expedition artist Gustavus Sohon, each of which depicts the movement of men and supplies at its respective post (see Figures 10-1 and 10-2).[2]

More significantly, the promoters of Mullan's road saw it as an essential prerequisite for the building of a transcontinental railroad. For one thing, the road expedition would supply vital topographical data for choosing a railroad route. Moreover, the completed road would provide the means by which men and materials would be moved to the railroad construction sites. Another hoped-for benefit, as evidenced in John Mullan's own comments relating to the project, was the road's potential as a major route for immigrants to the West.[3]

The road construction was scheduled to begin in 1858, but because of Indian hostilities near present Spokane, the project was postponed until 1859. The road was completed in 1862 at a site along the Hell Gate (now called the Clark Fork) River, southeast of Missoula, Montana. During those three years, some 624 miles of road were surveyed and constructed under the supervision of Lieutenant Mullan, assisted by an able group of cartographers, surveyors, soldiers, engineers, and laborers. In terms of map making, Mullan's key personnel included artist-interpreter Gustavus Sohon, topographer Theodore Kolecki, and civil engineer Walter Washington DeLacy. In one of Sohon's illustrations, several

Figure 10-1. *Sohon's lithograph of Fort Walla Walla, which appeared in Mullan's final report. This military post was constructed in eastern Washington in 1857 to provide protection for settlers in the region and travelers along part of the Oregon Trail and Mullan Road. Some of the original buildings still exist at the site of Fort Walla Walla.* –From John Mullan, *Report on the Construction of a Military Road from Fort Walla-Walla to Fort Benton* (1863)

men–assumed to be Theodore Kolecki, John Mullan, and W. W. DeLacy–are shown working on maps at Cantonment Jordan in the winter of 1859–60 (see Figure 10-3).[4] Described in detail in the next essay, "Illustrating the Mullan Road," this drawing is the only known image that shows nineteenth-century cartographers at work compiling data and drawing maps.

In this essay, we will examine some of the particulars of the Mullan Road project, with a focus on the cartographic products associated with it. In addition to describing and analyzing the maps, we will suggest several types of geographic applications for which they may be used by historians.

The Challenges of Building the Mullan Road

The road-construction process was difficult for a variety of reasons, including the team's insufficient knowledge of the region's climatic and hydrologic conditions, the complex

Drawn by C. Sohon. Bowen & Co. lith. Philada.

FORT BENTON;_HEAD OF STEAM NAVIGATION ON THE MISSOURI RIVER.

Figure 10-2. *This Fort Benton lithograph, created by Sohon in 1862, was published in Mullan's official report.*
–From John Mullan, *Report on the Construction of a Military Road from Fort Walla-Walla to Fort Benton* (1863)

Figure 10-3. *This 1860 illustration by Gustavus Sohon shows cartographers at work at Cantonment Jordan. The identities of the men pictured at the table, though uncertain, are likely John Mullan, Theodore Kolecki, W. W DeLacy, and Gustavus Sohon; the fifth man is possibly topographer Thomas Adams.*
–Courtesy Sohon family

physical nature of the terrain, and a lack of cooperation from the region's Indian inhabitants. All of these factors contributed to Mullan's accounting of the route, the reasons for his final location selections, and the types of construction problems that were encountered by the builders.

The route passed through portions of three physiographic regions, loosely defined as the Columbia Plateau, the Rocky Mountains, and the Great Plains. Of these, the builders found the mountain belt to be the most difficult. Subsequently, it was within this stretch that most of the construction efforts were centered. In the period between June 1859 and August 1860, 71 percent of the work was devoted to this section of the road. The proportion increased to 78 percent in the second construction phase, which took place during a ten-month period in 1861 and 1862, when the work was focused on the region west of Mullan Pass. (See Figure 10-4 below.)

Similarly, most of the cartographic work was concentrated on these more challenging regions. This assertion is supported by the index map (Figure 10-5), which shows the location and areal extent of the individual maps that were prepared during the course of the project. This index denotes the regions where most of the mapping efforts were focused, suggesting that those places were of the greatest interest or concern to the project's engineers. Here we see that the section through the Columbia River Gorge between The Dalles and old Fort Walla Walla and the section through the Rocky Mountains were the two locations where the bulk of the cartographic work was done.

Clearly, portions of the road were poorly chosen, either because of a lack of climatological and hydrologic data or because of faulty assumptions about seasonal conditions within a given region. The most serious problem was encountered near the south end of Lake Coeur d'Alene. It was not until after the road was constructed through this area, in the summer of 1859, that the builders discovered the severity of spring flooding there. Consequently, the road was rerouted north of the lake in 1861. The route change is delineated on the map in Figure 10-6.[5]

Contours and Form Lines

For centuries, mapmakers have struggled with the problem of depicting terrain on maps. At first, attempts to represent terrain were descriptive but imprecise. Early maps showed the general locations of hills or mountains but no specific data regarding heights, slopes, or detailed shapes. Beginning in the nineteenth century, however, terrain representation greatly improved. Topographers began to use hachures to show mountains and other significant landforms. These short parallel lines indicated the shape of the landform and sometimes, with changes in line weight, the steepness of slopes as well. Still, quantitative information, particularly elevations, could not specifically be ascertained from such marks.

TIME AND MILEAGE DEVOTED TO CONSTRUCTION OF MULLAN ROAD THROUGH THE MOUNTAIN SECTION

These percentages are approximations based on our estimates of the time and distance devoted to road construction in the mountain section.

Construction Phase	Percentage of Time in Mountain Section	Number of Miles Covered in Mountain Section	Percentage of Total Road Distance (624 miles)
1859-60 (14 months)	71%	308	49%
1861-62 (10 months)	78%	110	18%

Figure 10-4

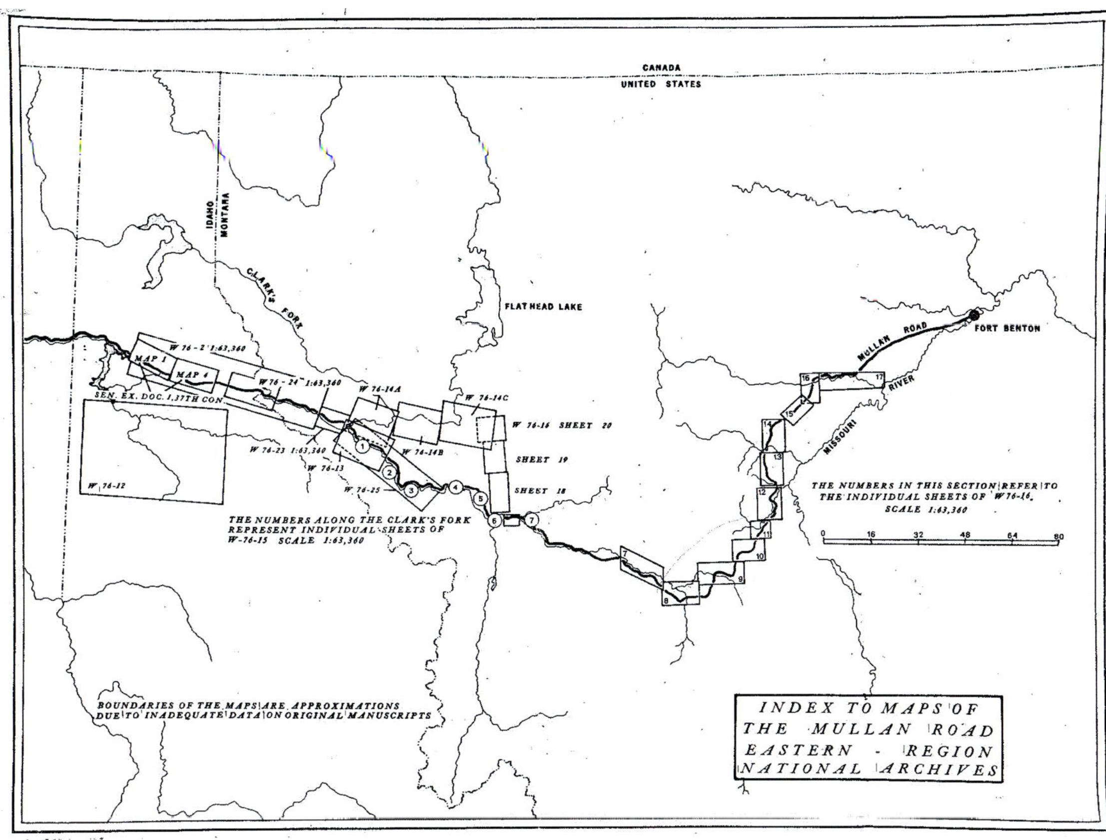

Figure 10-5. *Index of maps completed during construction of the Mullan Road. These manuscript maps can be viewed in the Cartographic Section of the National Archives in College Park, Maryland.* –Illustration by Paul D. McDermott

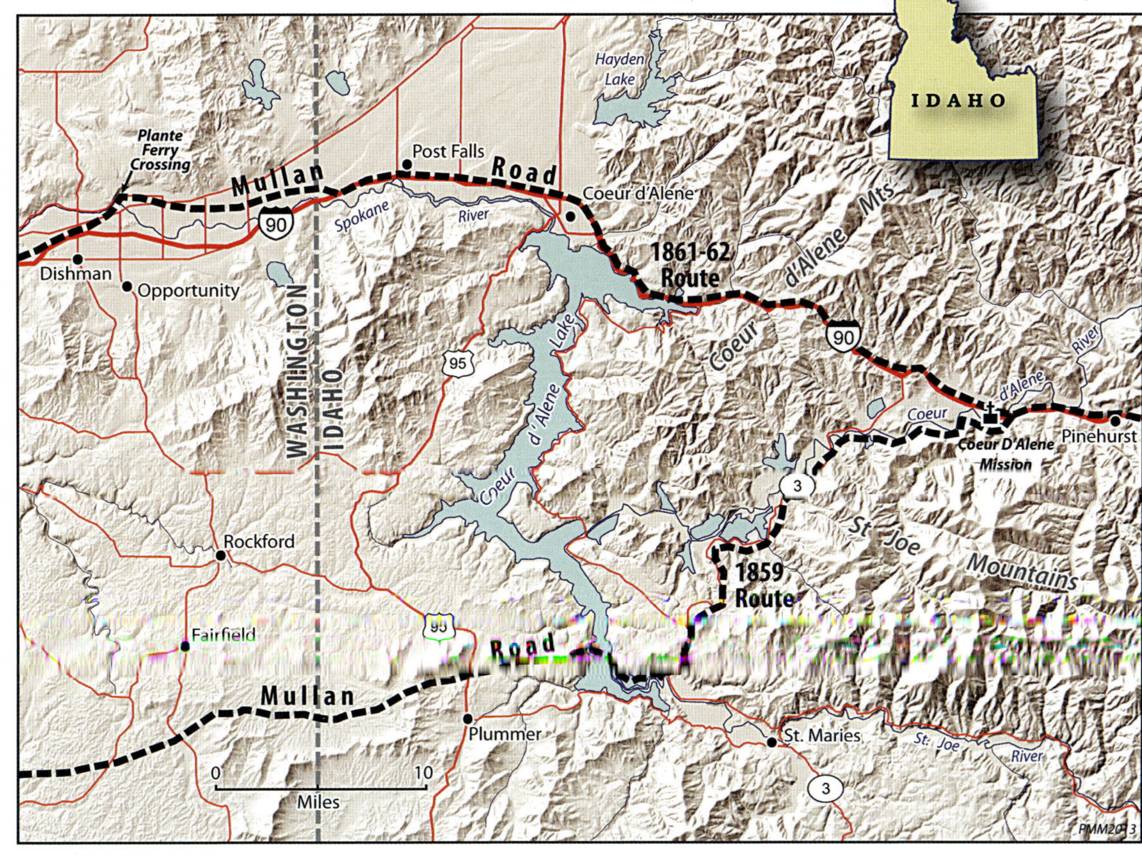

Figure 10-6. *This map shows the two different routes Mullan used around Lake Coeur d'Alene. The first, which traced the southern end of the lake, was abandoned because of flooding. The second followed the lake's northern end and crossed the Spokane River Plain.* –Map by Philip Mobley

As transportation evolved in the nineteenth century, it became essential to have more quantitative information in order to effectively construct roads, canals, and railroads. In response to this need, new mapping techniques and symbolization were developed. In this regard, the most important innovation in the nineteenth century was the creation of the *contour line*. Contours are lines connecting points of equal elevation above sea level. Generally, each contour represented a 5, 10, 20, 25, 50, or 100-foot change in elevation. The use of contours allowed the reader of the map to precisely calculate gradients, facilitating the planning of routes for roads, canals, and railroad beds.

The maps in Figure 10-7, created by two military officers, G. W. Whistler and J. A. McNeil, in 1822, illustrate the terrain at Salem, Massachusetts.[6] The map on the left uses contours to depict terrain, while the map on the right depicts the same area using hachures.

We know that Mullan and his topographers knew how to create contour maps, and that the party had the equipment to obtain elevation data–namely levels and barometers. The Mullan maps that were drawn with contours often used a 100-foot interval because the topographers calculated elevation with a mercury barometer, which measured changes in 100-foot increments. Not all of the Mullan maps used contour lines to depict terrain, however; many used simpler *form lines*, which showed only the approximate shape of landforms. The survey parties simply did not have the time to go up and down all the surrounding mountains to collect the necessary data to create contour maps. Understandably, they tended to record more-detailed elevation data in areas near the road, becoming less precise as they moved away from the road.

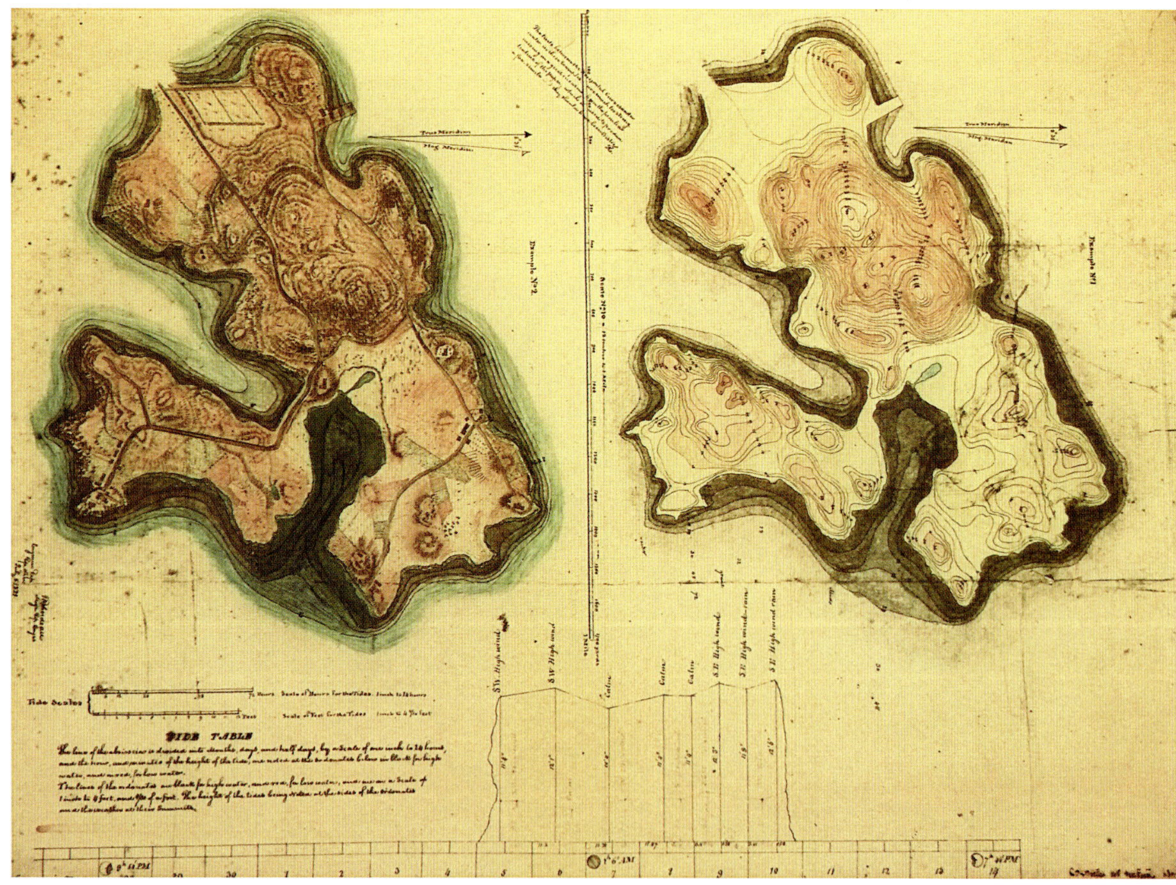

Figure 10-7. *Terrain maps of Salem, Massachusetts, by J. A. McNeil and G. W. Whistler* –Courtesy National Archives and Records Administration, Cartographic Section, College Park, MD

Historical Resources

The cartographic materials relating to the Mullan Road include not only the maps published in the official reports, but also manuscript maps found in the National Archives in Washington, D.C.[7] Specifically, the manuscript maps are located in the Headquarters Map File of the Office of the Chief of Engineers (Record Group 77). The Headquarters Map File, also known as the Civil Works File, was a central repository for maps compiled by various army units in the field and forwarded to the Chief of Engineers. The maps in this file reflect the activities of the army, particularly the Corps of Topographic Engineers and its successor units, during the nineteenth century. These maps document military campaigns, exploring expeditions, and construction projects such as forts, roads, railroads, canals, and material changes to improve river and harbor navigation.[8]

Within the Headquarters Map File are two groups of materials relating to the Mullan Road. These were assigned the file designations "W76" and "Rds 174." These alpha-numeric or mnemonic-numeric designations correspond to either a geographical area or a subject category. In this case, "W" refers to the geographic area west of the Continental Divide, and "Rds" refers to the subject category of road and railroad construction. The materials in W76 consist of fifty-one graphic items, including maps, landscape sketches, and road profiles. There are also fifty pages of textual documents, including copies of reports and letters pertaining to the operation of Mullan's expedition in 1858, 1859, and 1860. The second grouping, Rds 174, contains only six items, including two landscape sketches, a bridge plan, several plans of side cuts, and a copy of a map showing a portion of the road from the crossing of the Bitterroot River to Cantonment Wright.[9] A complete list of the materials in the Headquarters Map File can be found in its card catalog.

The cartographic records of Mullan's project reveal much about the construction process. The graphic materials can be classified in several different ways, based on such variable features as the amount of detail depicted, the relative scale employed, the type of drafting materials used, the time period in which they were created, the accuracy of the features, and the manner of reproduction. Here we will classify the items by type and subject, but we will also analyze them according to some of their other features.

The first group comprises reconnaissance sketch maps. Usually drawn in pencil, these displayed limited detail in terms of both symbolization and field notes. They were developed in two different time periods. The first were drawn by Mullan when he conducted exploratory surveys for the Pacific Railroad Surveys in 1853 and 1854. The others were drawn contemporaneously with the actual construction of the road, in the years 1858 through 1862. In most cases, these reconnaissance sketches note only the general course taken by the surveyor, the mileage from the base, and the approximate location of streams.

Figure 10-8 shows a typical reconnaissance map drawn by Mullan in 1854, which illustrates his route through the Flathead country. Drawn at a scale of 1:1,200,000, or nineteen miles to the inch, it depicts a large area in western Montana, extending from Cantonment Stevens north to the Kootenai River near the Canadian border. Although most reconnaissance maps failed to portray terrain, Mullan departed from convention here, depicting terrain, such as mountains, with crudely drafted hachures–a commonly used nineteenth-century terrain symbol.

The second group of graphics can be broadly defined as report manuscripts. These were maps plotted during the winter season, at either Cantonment Jordan or Cantonment Wright. The drawings vary in scale and draftsmanship; some were drawn in pencil, others in ink. Different paper types were used, ranging from vellum to rag. Perhaps the most distinctive characteristic of this group is the high degree of physical and cultural detail the maps depicted.

The field sketch of The Dalles shown in Figure 10-9, probably drawn in 1858, is a representative example of a map from the second group. This graphic, drawn in pencil at a scale of 1:62,500, is remarkable for several reasons. First, it contains an abundance of physical and cultural detail, including Indian trails, a fishing station, agricultural cropland, a mission, and a fort. Second, its symbolization for terrain

Figure 10-8. *This 1854 map by John Mullan is titled "Sketch of a Reconnaissance from Cantonment Stevens to the Kootenay River." The scale for this map was 1:1,200,000, or nineteen miles to the inch. Such maps were intended to provide a general picture of a region.* –Courtesy National Archives and Records Administration, Cartographic Section, College Park, MD

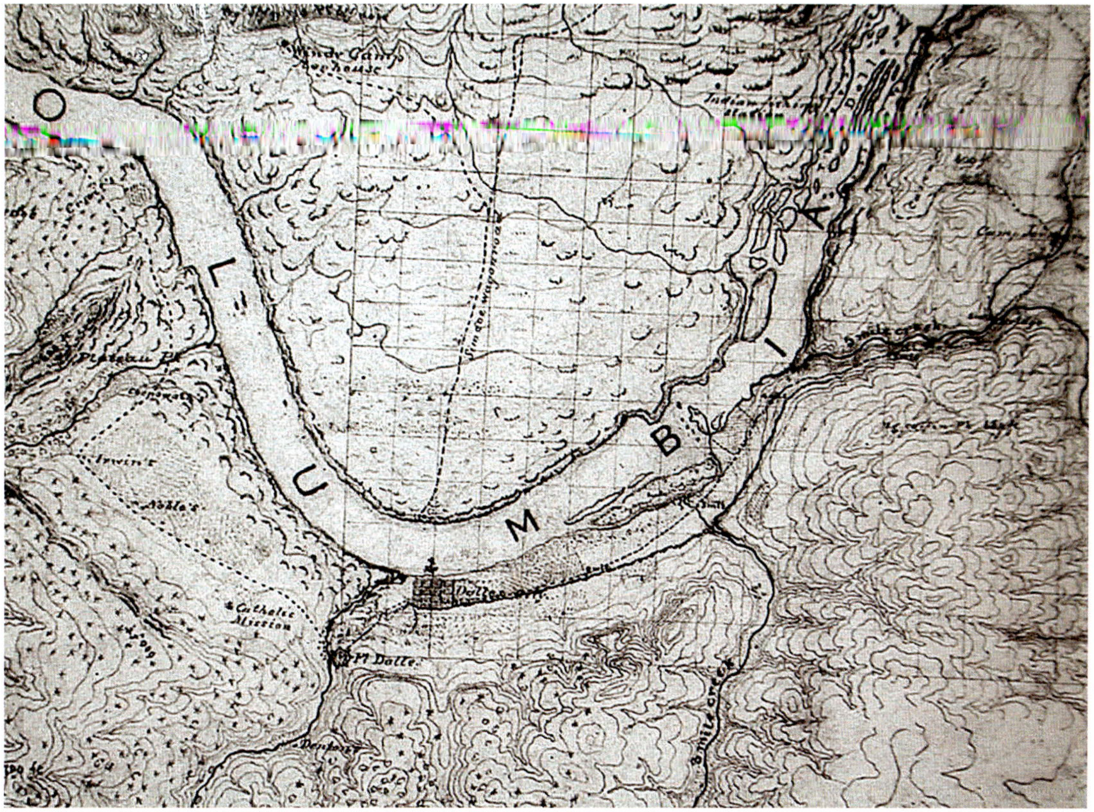

Figure 10-9. *This 1858 map of The Dalles by Theodore Kolecki and Gustavus Sohon, sketched in pencil in the field, was one of the first detailed maps of that area. It depicts the landscape using form lines. The course of the Columbia River was later altered by a hydroelectric dam (Dalles Dam).* –Courtesy National Archives and Records Administration, Cartographic Section, College Park, MD

representation was unusual for that time. These line symbols appear to be either standard form lines (approximate lines suggesting equal elevation based on general observation) or early examples of contour lines (lines of equal elevation based on precise measurements).

Based on the examinations of the maps we did as part of the Mullan Road project, arguments can be made for both possibilities. We found some clues in the map legends. Some of the maps specifically identify contour intervals. The most commonly designated interval is 100 feet, which corresponds to the change of elevation recorded by a barometer for each 1/10-inch change in the height of a mercury column. Surveyor Bill Weikel, however, provides an interesting argument defending form lines–it would take too much work to climb up and down the mountains to collect the extensive barometric data needed to create contours. But if they were only form lines, why would the cartographers specify precise contour intervals?

On the map of The Dalles (Figure 10-9), the lines appear to be form lines, while on Gustavus Sohon's map of Sohon Pass (Figure 10-10), they are clearly contour lines.[10] Consequently, it is our conclusion that some of Mullan and company's maps were among the first to apply the contour line as the primary form of terrain symbolization. As western historian William Goetzmann noted in 1966, the western frontier was frequently the site of experimentation in the application of scientific techniques and equipment developed elsewhere.[11]

The third category of maps and drawings are classified here as the published group. These were the graphics that appeared as printed illustrations in Mullan's official report,

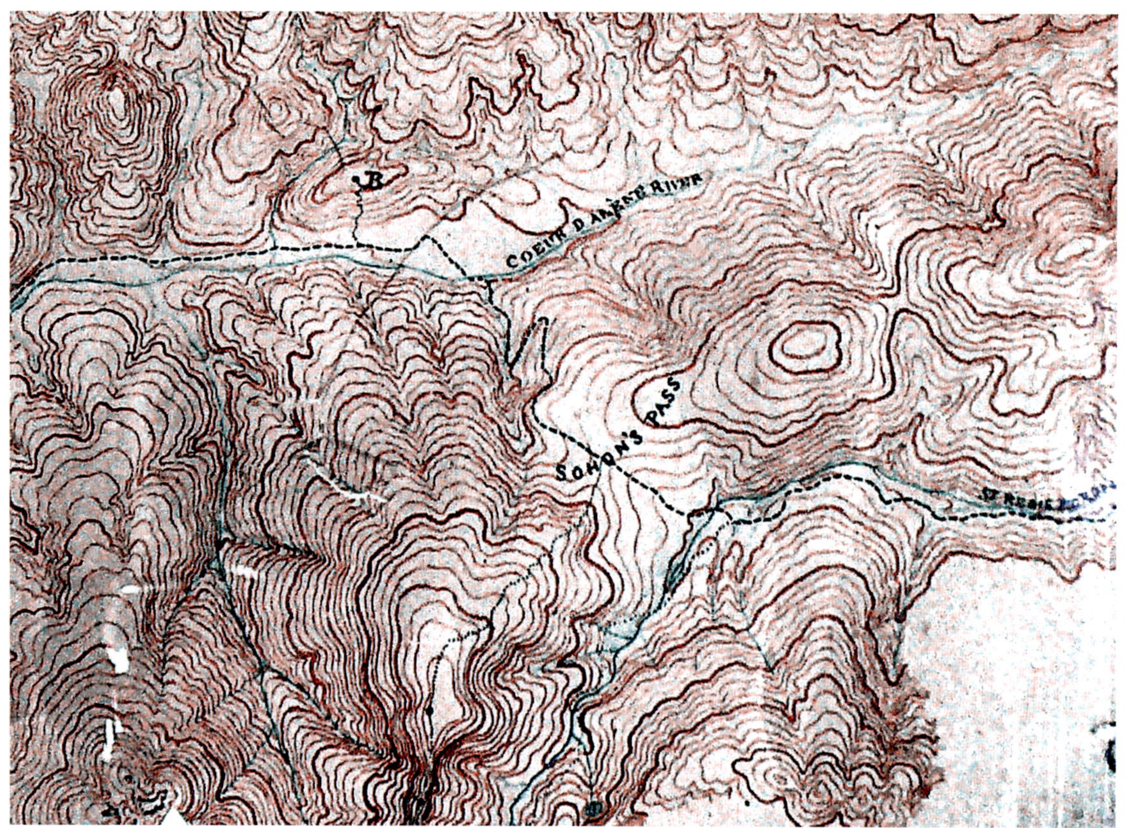

Figure 10-10. *Gustavus Sohon drew this detailed contour map of Sohon Pass in the winter of 1859–60, probably at Cantonment Jordan. It shows the rugged terrain around the pass, located near the Idaho-Montana border. The contour lines appear in red ink.* –Courtesy National Archives and Records Administration, Cartographic Section, College Park, MD

published as Senate and House Executive Documents. In this case, two basic maps documented the road-building project. The first was drawn by Theodore Kolecki at a scale of 1:300,000. (The entire map is not pictured here; see Figure 10-12 for a detail.) It is primarily concerned with the portrayal of mountain topography along the road from Lake Coeur d'Alene to the Dearborn River. On this map, Kolecki depicted the topography utilizing contour lines, thereby promoting this newly developing technique of terrain representation. In addition, it is evident that Kolecki attempted to show the location of settlers inhabiting the region at the time of the road construction, as well as the locations of missions, trails, and older wagon roads.

The second map was compiled by Edward Freyhold at a scale of 1:1,000,000 (Figure 10-11). It differs from the first in several aspects. First, it is a more comprehensive product in that it illustrates a larger region of the Northwest. Second, the cartographer has reverted to using the most common symbol for showing terrain in the nineteenth century–the hachure. Figure 10-12 is a side-by-side comparison of a detail from each of the two maps showing the same area, namely the Cantonment Jordan vicinity in northwestern Montana, including Sohon Pass, Steven Peak, and the St. Regis River. On the top map, created by Edward Freyhold, hachuring was used to depict the topography, while on the bottom one, by Theodore Kolecki, contour lines were used.

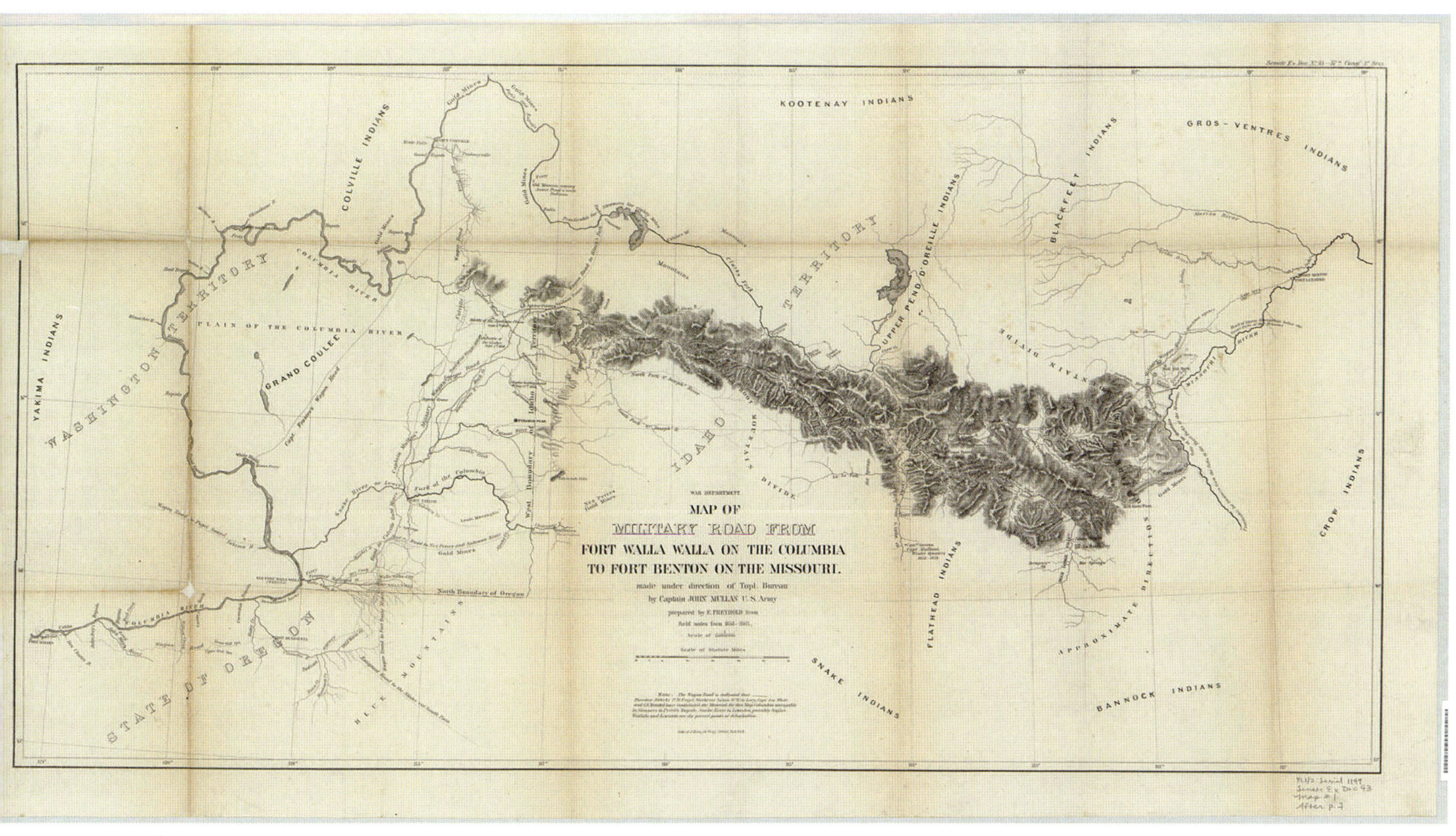

Figure 10-11. *This 1863 map by Edward Freyhold, titled "Map of the Military Road from Fort Walla Walla on the Columbia to Fort Benton on the Missouri," was based on field notes from surveys of 1858–63. It's a small-scale map (sixteen miles to the inch, or 1:1,000,000) and uses hachures, lines that represent slopes in the terrain.* –From John Mullan, *Report on the Construction of a Military Road from Fort Walla-Walla to Fort Benton* (1863)

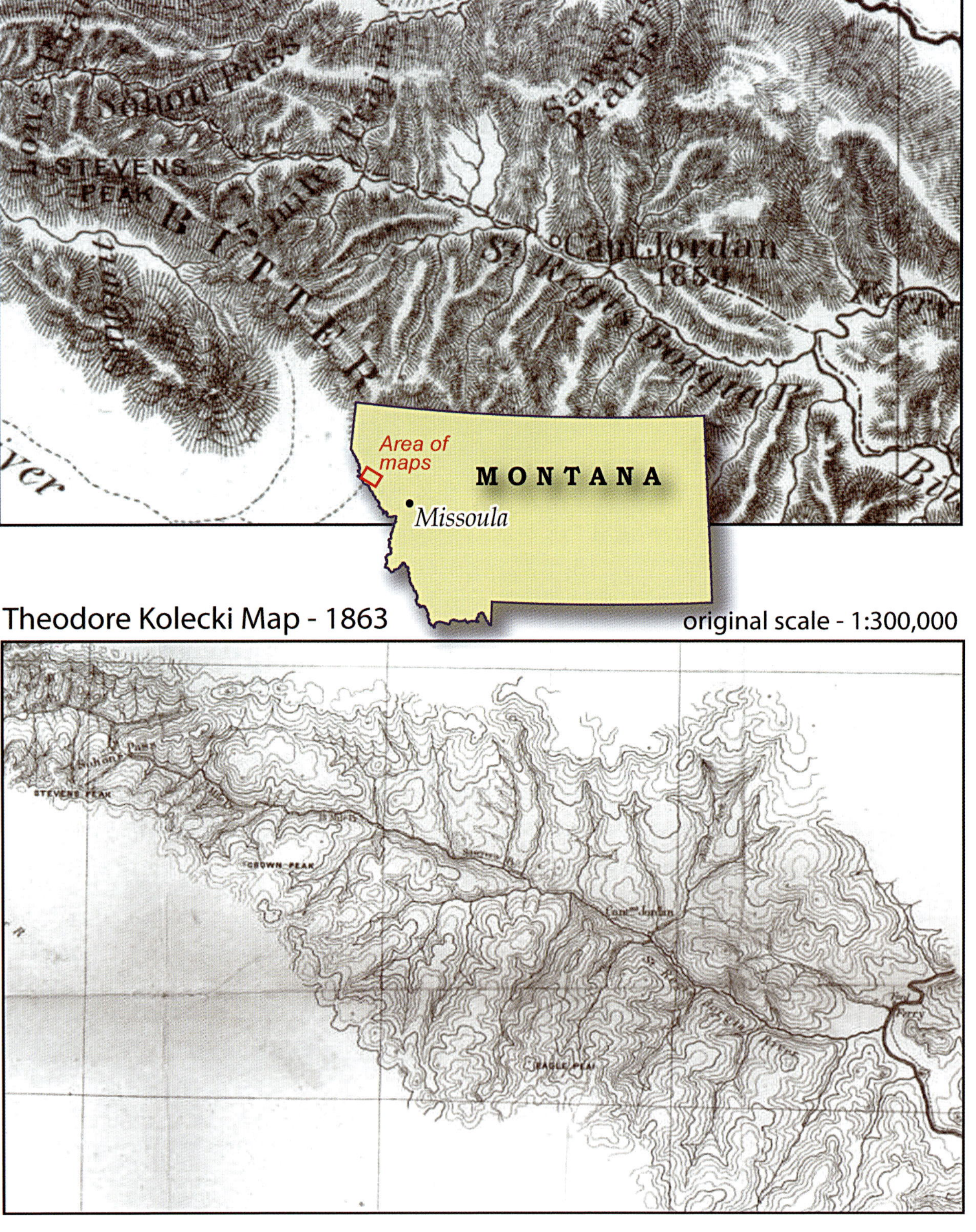

Figure 10-12. *Quantitative versus qualitative symbolization along the Mullan Road. These are details from two maps, one drawn by Edward Freyhold and the other by Theodore Kolecki, showing the same region along the Mullan Road using two different types of symbolization. The Kolecki map (bottom) is a more quantitative portrayal while the Freyhold map, certainly more visually interesting, is a qualitative representation.* –Maps courtesy Library of Congress, Geography and Map Division; comparison graphic by Philip Mobley

Sohon Pass Tunnel

Sohon Pass is a particularly interesting location on the road's route. It was depicted on the map using contour lines, representing 100-foot intervals. The key cartographer responsible for this map was Gustavus Sohon. One reason that this particular region was mapped with contours was to show the gradient, which would help to determine the degree of difficulty in building a railroad through the area. Ultimately, this pass through the Bitterroots, today known as St. Regis Pass, was deemed too difficult to cross over, and it was believed that a tunnel would be required.[12] Figure 10-13 shows the proposed plan for this tunnel. When the railroad was finally built, however, the tunnel plan was jettisoned. Instead, to reach the St. Regis Valley on the other side of the pass, a long grade with numerous switchbacks and timbered bridges was constructed.

Figure 10-13. *Location of the proposed railroad tunnel at Sohon Pass (now called St. Regis Pass), Montana. The tunnel was never built.* –Courtesy National Archives and Records Administration, Cartographic Section, College Park, MD

Research Potential

In accessing these maps for research value in historical geography, we have concluded that there are several topics for which these or similar records of other road-building projects offer potential utility. The subjects of greatest interest to us include the study of the development of the nation's transcontinental transportation network, the reconstruction of past landscapes, and the determination of local place names.

The construction of the road can be viewed on two levels in the context of transportation geography. On a micro level, the maps provide excellent documentation of the decision-making process used in selecting the road's final location.[13] There were numerous cases in which reconnaissance parties explored and mapped alternate routes, or construction crews had to reroute sections of the original road in order to avoid such obstacles as steep grades, spring floods, or excessive river crossings.

On a broader scale, this road-building project serves as an example of the federal government's involvement throughout the nineteenth century in financing and constructing various transportation projects. In this particular situation, the road provided a vital link in a major route across the northern part of the western territories by connecting two navigable rivers, the Missouri and the Columbia. Although the actual network that was developed emphasized the continued reliance on a combination of land and water modes of transportation, the fact that Isaac Stevens's Pacific Railroad Survey served as a prelude to this project also points to a transitional phase in which the route selected for the road was to become a major link in the development of a transcontinental railroad system.

It is also possible to use these records to reconstruct former landscapes, although their usefulness in this regard is mostly limited to the areas that were most extensively surveyed. Since the builders' attention was primarily focused on the places that offered construction problems, the maps emphasize the mountainous regions in the vicinity of Lake Coeur d'Alene and the Bitterroot Valley, to the exclusion of the plateau and plains at either terminus of the road. This point is illustrated by the index map discussed earlier, which shows the area of greatest concern to the developers as well as the routes identified for development. The temporal cross-section does not represent the white man's earliest entrance into the area; rather, it reflects the end of an era dominated by the activities of early explorers, major fur traders, and missionaries, as well as by the Indian wars. Its value lies in its portrayal of the landscape prior to the intensified economic exploitation of the region by miners, cattlemen, and farmers.[14]

A readily apparent feature of the maps is the occasional change in place names. Although this characteristic makes it difficult to ascertain the actual coverage of the maps, it does offer a potential source for studying the process by which local place names have been determined or modified through time, especially when the names were Indian in derivation. The changing of place names is particularly noticable in the nomenclature applied to the drainage systems (see Figure 10-14). For example, the river that is today known as the Bitterroot River was in the past called St. Marie's River.

In many cases, it appears that as the explorers and settlers increased their comprehension of the complex stream systems found in the mountain region, they changed the names of the rivers in order to better differentiate between tributaries. The river then called the Bitterroot is recognized today as part of the Clark Fork, a major tributary of the Columbia. In Mullan's time this river was identified as several different waterways with different names, including the Bitterroot, west of present Missoula, and the Hell Gate, south of Missoula. It was later discovered that the Bitterroot and Hell Gate Rivers were not individual tributaries of the Clark Fork but rather sections of the main stream itself.

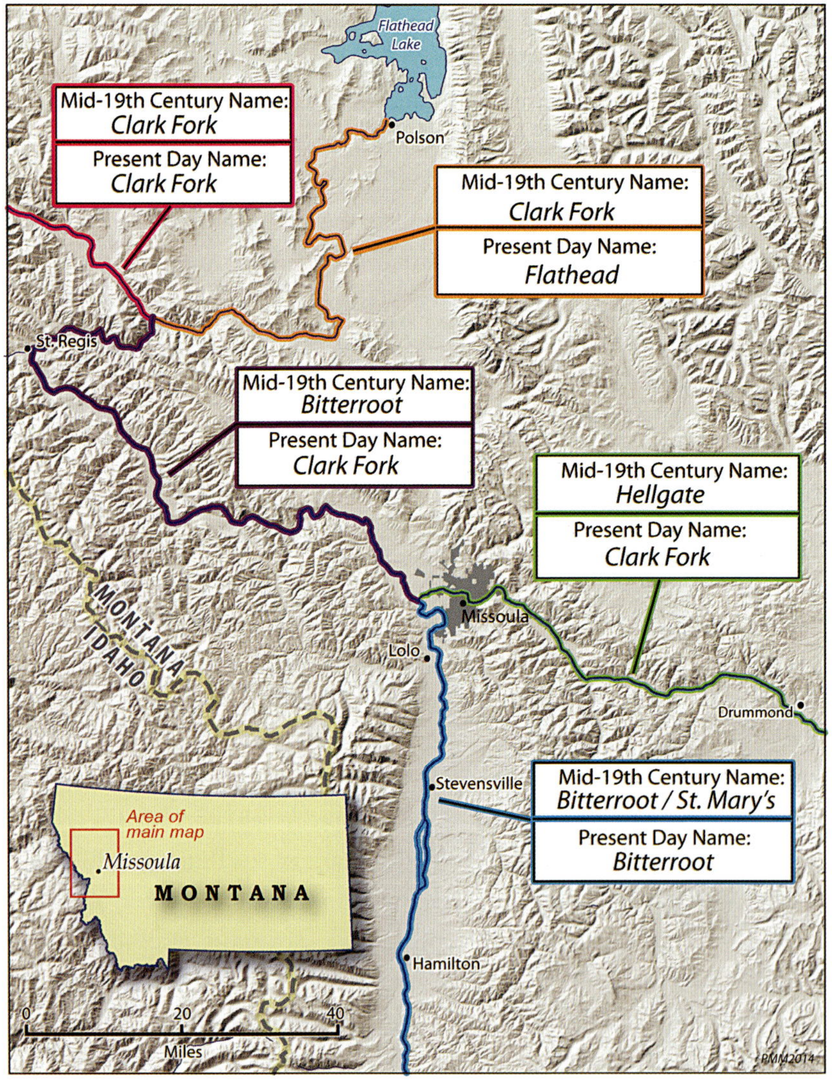

Conclusion

Our purpose in writing this essay was to illustrate the types of cartographic products associated with mid-nineteenth-century military road projects. Analysis of these maps not only reveals the existence and the location of cultural and physical features in the region traversed by the road, but also demonstrates how the route selection was made after a variety of possible routes had been explored and evaluated. From a cartographic perspective, the graphics provide some interesting clues as to how the use of the contour line evolved during the mid-nineteenth century; hopefully, as more maps of this period are examined, a clearer picture of the evolution of this important map symbol will emerge. Finally, we anticipate that as scholars continue to study these maps and others relating to the activities of the army, they will be able to reconstruct an ever fuller portrayal of the settlement process in the Pacific Northwest.

Figure 10-14. *Place-name changes on intermontane streams in western Montana. This map shows the changes in river names from Mullan's time to today.* –Map by Philip Mobley

1 Additional information relating to John Mullan and the military road is available in the following publications: Louis C. Coleman and Leo Reiman, *Captain John Mullan: His Life, Building the Mullan Road, as It Is Today, and Interesting Tales of Occurrences along the Road*, compiled by D. C. Payette (Montreal: privately printed for Payette Radio Ltd., 1968); Samuel F. Bemis, "Captain John Mullan and the Engineers' Frontier," *Washington Historical Quarterly* 14 (July 1923), pp. 201–05; Helen Addison Howard, "Captain John Mullan," *Washington Historical Quarterly* 15 (July 1934), pp. 185–202; and W. Turrentine Jackson, "Across the Northern Plains: The Mullan Road, 1853–1869," in *Wagon Roads West: A Study of Federal Road Surveys and Construction in the Trans-Mississippi West, 1846–1869* (Berkeley: University of California Press, 1952), pp. 257–78 and 369–72.

2 The graphic work of Gustavus Sohon is further discussed in our book *Eye of the Explorer: Views of the Northern Pacific Railroad Survey, 1853–54*, (Missoula, MT: Mountain Press Publishing, 2010). See also Ronald E. Grim and Paul D. McDermott, comp., *Gustavus Sohon's Cartographic and Artistic Works: An Annotated Bibliography*, Philip Lee Phillips Society Occasional Paper Series, no. 4 (Washington, DC: Geography and Map Division, Library of Congress, 2002); a PDF file is available for a small reproduction charge at MapMcD@aol.com).Sohon is also discussed in: John C. Ewers, *Artists of the Old West,* enlarged edition (New York: Doubleday, 1973), pp. 173–81; Ewers, *Gustavus Sohon's Portraits of Flathead and Pend d'Oreille Indians, 1854,* Smithsonian Miscellaneous Collections vol. 110, no. 7 (Washington, DC: Smithsonian Institution, 1948); and Robert Taft, *Artists and Illustrators of the Old West, 1850–1900* (New York: Charles Scribner's Sons, 1953), pp. 275–76.

3 One of Mullan's objectives for the road project was to attain data for a potential railroad route; this can be seen in the maps and terrain profiles drawn in conjunction with the surveying and construction work on the road. These illustrations frequently depict the best location for the railroad as well as proposed sites for tunnels and fills. Moreover, in his reports to Congress, Mullan provided additional commentary identifying where construction problems would be encountered as the road reached certain sections; see John Mullan, *Report on the Construction of a Military Road from Fort Walla-Walla to Fort Benton,* S. Ex. Doc. 43, 37th Cong., 3rd sess., serial 1149 (Washington, DC: Government Printing Office, 1863), pp. 2–3; this source is hereafter cited as Mullan, *Report*, 1863. Mullan's ideas for the use of the road for immigration are revealed in his promotional guidebook, *Miners' and Travelers' Guide to Oregon, Washington, Idaho, Montana, Wyoming and Colorado via the Missouri River and Columbia River* (New York: Wm. M. Franklin, 1865).

4 This unique image was provided by the Sohon family for our use. For further details regarding the image, see "Illustrating the Mullan Road" in this book.

5 The map shown in the text was based on a larger map titled "Map of Military Reconnaissance from Fort Taylor to the Coeur d'Alene Mission, Washington Territory," which Mullan made under the direction of Captain A. A. Humphreys, U.S. Topographic Engineers, assisted by civil engineers Theodore Kolecki and Gustavus Sohon while these men were attached to the 1858 military expedition of Colonel George Wright, 9th Infantry." It was later updated to include the original route of the military road, identified by the years in which it was built, 1859–60, and the 1861 rerouting. The revised map, scaled at 1:300,000, was published with Mullan's 1863 report.

6 A reproduction of this map, which is the first known contour map drawn in the United States, as well as some brief notes about the evolution of the contour line as a terrain model, can be found in Norman J. W. Thrower, *Maps and Man: An Examination of Cartography in Relation to Culture and Civilization* (Englewood Cliffs, NJ: Prentice Hall, 1972), p. 91–92. The 1822 map of Salem, Massachusetts, employed both hachures and contours to depict the terrain. Thrower further notes that "the battle for adoption of the contour (a quantitative method of terrain rendering as opposed to the qualitative hachure) in topographic mapping was not won as early as 1822. Actually this method was not officially approved for the British Ordinance Survey until 1849, and it was decades after this before most topographic sheets were contoured."

7 The specific sources for the published materials are: John Mullan, P. M. Engle, et al, *Military Road from Fort Benton to Fort Walla Walla: Letter from the Secretary of War . . .*, H. Ex. Doc. 44, 36th Cong., 2d sess., serial 1099 (Washington, DC: Government Printing Office, 1861), hereafter cited as Mullan, *Report*, 1861; Mullan, "United States Military Road Expedition from Fort Walla Walla to Fort Benton, W.T." in *Annual Report of the Secretary of War, 1861*, S. Ex. Doc. 1, 37th Cong., 2d sess., serial 1118, pp. 549–69, and serial 1120, "Report of the Surveyor General" (maps); and Mullan, *Report*, 1863. The utility of the Congressional Serial Set for the study of geographical exploration is examined by Herman R. Friis in "The Documents and Reports of the United States Congress: A Primary Source of Information on Travel in the West, 1783–1861," in John F. McDermott, ed., *Travelers on the Western Frontier* (Urbana: University of Illinois Press, 1970), pp. 112–27.

8 While there is no comprehensive published finding aid for the maps in the Headquarters Map File, general descriptions of many of these maps are provided in Ralph Ehrenberg, *Geographical Exploration and Mapping in the 19th Century: A Survey of the Records in the National Archives,* Reference Information Paper no. 66 (Washington, DC: National Archives and Records Service, 1973), and Patrick D. McLaughlin, *Transportation in Nineteenth Century America: A Survey of the Cartographic Records in the National Archives of the United States*, Reference Information Paper no. 65 (Washington, DC: National Archives and Records Service, 1973). The primary finding aids for locating individual maps within this collection are the original card file and a bound set of registers that were compiled as the maps were added to the files.

9 These two portfolios of maps are not the only records in the National Archives pertaining to the construction of the Mullan Road. For example, the noncartographic records of the Office of the Chief of Engineers contain a number of field survey books that pertain to the road's construction (Record Group 77, Field Survey Data Cupboard 1, Shelf 4, nos. 7, 8, and 9, and Cupboard 37, Shelf 5, nos. 56, 57, 58, and 59).

10 A number of maps drawn during the next few years utilized the same type of symbolization but with two additions: First, varying line weights appeared in order to differentiate main, or index, lines from intermediate lines (in most cases, these maps represented areas in the Coeur d'Alene region); second, the contour interval shown by the lines was now specified, making it logical to conclude that they were indeed contour lines.

11 For more on scientific experimentation on the western frontier, see William M. Goetzmann, *Exploration and Empire: The Explorer and Scientist in the Winning of the American West* (New York: Alfred A. Knopf, 1966).

12 This book's editors, Grim and McDermott, have recommended to the US Geological Survey Geographic Board of Geographic Names to change today's St. Regis Pass back to its original title–Sohon Pass. The appeal has been tabled until sufficient support is obtained from parties in Montana and Idaho.

13 John A. Jakle points out the need for investigating the decision-making process in historical geographic studies in "Time and Space, and the Geographic Past: A Prospectus for Historical Geography," *American Historical Review* 76 (October 1971), pp. 1097–98.

14 See Donald W. Meinig, "Isaac Stevens: Practical Geographer of the Early Northwest," *Geographical Review* 45 (October, 1955), pp. 542–58.

Illustrating The Mullan Road

BY PAUL D. MCDERMOTT AND RONALD E. GRIM

In 1863, when Captain John Mullan submitted his final report on the construction of his military road from Fort Walla Walla to Fort Benton, only ten illustrations were included. Most of these were created in 1862, as the road-building project was ending. All ten images were the work of Gustavus Sohon–artist, engineer, linguist, and cartographer. Because the United States was in the midst of the Civil War, Mullan's report was rapidly produced, and consequently the work was not profusely illustrated.[1]

Certainly Mullan was under considerable pressure to complete the report. Federal funding was limited, and the outcome of the Civil War was still unknown. On the personal side, Mullan had recently become engaged to Rebecca Williamson; his pending marriage also helped spur him to complete the report. Together, these factors account for the lack of illustrated material in the final document. Of the images that were included, several were previously published, derived from drawings originally used in Isaac Stevens's Northern Pacific Railroad report some ten years earlier. Among these were a lithograph of Cantonment Stevens and another of the Great Falls of the Missouri.

Though Mullan's report focused on the road, only three of the lithographs–those of Cantonment Wright, the Coeur d'Alene Mission, and Fort Walla Walla–actually showed the newly constructed thoroughfare. The other images depict either Native American activities or interesting physical landforms, subjects only indirectly associated with the road. Sohon did create two splendid images focused directly on the road's final appearance, but ironically, these were not used in the report, though we will discuss them here. The question remains: Why weren't these images used? Apparently Mullan was a man in a rush. There just wasn't time to convert all the new drawings into a lithograph format.

The following is a list of the illustrations used in the 1863 report:[2]

- *Pend d'Oreille Mission in the Rocky Mountains in 1862*
- *Military Post & City of Walla-Walla, W.T. in 1862*
- *Cantonment Stevens–Capt. Mullan's Winter Quarters 1853–54*
- *Great Falls of the Missouri, 2500 Miles from St. Louis*
- *Upper Falls of the Missouri River*
- *Mode of Crossing Rivers by the Flathead and Other Indians*
- *Cantonment Wright, Capt. Mullan's Winter Quarters in 1861–62*
- *Paloose Falls in Washington Territory*
- *Coeur d'Alene Mission in the Rocky Mountains*
- *Fort Benton–Head of Steam Navigation on the Missouri River*

Who Was Gustavus Sohon?

Three members of John Mullan's crew were capable of creating realistic landscape images–Gustavus Sohon, Theodore Kolecki, and Theodore Adams. We know relatively little about Kolecki and Adams. The greatest amount of information pertains to Sohon. He was born in Tilsit, Prussia, on December 10, 1825. He migrated to New York City in 1842. For the next ten years, he probably lived in Brooklyn. At the age of twenty-seven he joined the U.S. Army and was sent west, eventually stationed at Columbia Barracks–later known as Fort Vancouver. His first major assignment was a relief column under Lieutenant Rufus Saxton, transporting 2,500 rations to Fort Owen (near today's Stevensville, Montana) for Isaac Stevens's Northern Pacific Railroad Survey personnel.[3]

Figure 11-1. *This photograph of Gustavus Sohon, born in Prussia in 1825, was probably taken circa 1863, when he was in his thirties. Sohon died in Washington, D.C., in 1903.* –Courtesy Sohon Family

Sohon began working with John Mullan in 1853, when he was reassigned to Mullan's command during Stevens's railroad survey. Initially, Sohon was responsible for collecting meteorological data, but with the passage of time, Mullan began to recognize the young man's many talents. On this expedition, Sohon began drawing landscapes and portraits of Native peoples. He made his first such images when he accompanied Mullan and six other men on an exploration of the region near Flathead Lake and up to the Canadian border. More details about this trip can be found in "Journey to the Flathead Country" in this book.

In 1855 Sohon was assigned to assist Stevens, the Washington territorial governor, on a treaty expedition through Washington Territory. Stevens's goal was to negotiate treaties with the Nez Perces, Flatheads, Blackfeet, and other area tribes. On this expedition, which occupied much of 1855, Sohon continued making portraits of Indians and created images documenting the council meetings. The artist was assigned this task because Stevens wanted to publish a book about what he had accomplished on the expedition. In addition, Sohon functioned as an interpreter and linguist on this important journey. In all these roles, he proved invaluable. Yet he will best be remembered for the sixty-eight portraits he drew in 1854–55.[4]

After the Stevens survey, Sohon was stationed for a time at the army's Topographical Engineers headquarters in Benicia, California. He resigned from the military in 1857. One suspects that Mullan persuaded Sohon to work with him on his military road, a project that began in 1858. Most of the early work on the road was focused along the Columbia River upstream from Fort Dalles. Unfortunately, Mullan's project was delayed indefinitely when Indians attacked and soundly defeated the troops of Colonel Edward Steptoe on May 17, 1858, near present-day Rosalia, Washington. In August, Mullan volunteered his command, including Gustavus Sohon, to serve under Colonel George Wright in his subsequent campaign to punish Steptoe's attackers and to terminate Native aggression in the region. Several officers and a number of enlisted men were killed in the battle. While on the Wright expedition, Sohon drew eleven distinctive images documenting the campaign, including:

Crossing the Snake River at the Mouth of the Tou-kannon, August 25th & 26th, 1858

View of Snake River at Mouth of Tukañon Showing Fort Taylor and Peaks Taylor & Gaston, 20th August 1858

Battle of Col. Steptoe on the In-gos-so-man Creek, W.T. Fought 17th May 1858

Battle on the Spokane Plain—Col. G. Wright in Command and Against Forces of the Indians, 1858

Great Falls of the Spokane River, W.T., 30 Miles below the Coeur-d-alene Lake

Horse Slaughter Camp on the Spokane River, W.T, 8th, 9th, 10th Sept. 1858

Figure 11-2. *Sohon's re-creation of the Steptoe Battle. This engagement, which took place on May 17, 1858, in Rosalia, Washington, precipated the conflict between the U.S. Army and Native Americans in the summer of 1858.* –Courtesy Library of Congress, Geography and Map Division

Some of these works were in color, others simply drawn, albeit exquisitely, in pencil.[5] Sohon's artistic reconstruction of the Steptoe battle (Figure 11-2) is archived in the collection of the Geography and Map Division of the Library of Congress in Washington, D.C.

Road construction began in earnest in 1859, starting, appropriately, at Fort Walla Walla. At this time, Sohon did some artwork, rendering the fort and Paloose (Palouse) Falls. His skills were most vigorously exercised in the Palouse River corridor and later in the valley of St. Joseph's (now called St. Joe) River.

Late in 1859, Mullan took his road crew eastward to the Bitterroot Mountains in present Montana, where the men established their winter camp, dubbed Cantonment Jordan. En route, Sohon created a detailed contour map of the mountain pass named for him–Sohon Pass (today known as St. Regis Pass).[6] After passing the winter at Cantonment Jordan, the construction team moved across Camel's Hump, a hill near today's St. Regis, and across the Bitterroot River. Work proceeded rapidly, and in August 1860, the men arrived at Fort Benton. Sohon then returned to Fort Walla Walla with the first wagon train ever to travel on the newly completed route.

In 1861, while making repairs and improvements on his military wagon road, Mullan decided to alter the route. Instead of following his original southern path around Lake Coeur d'Alene, he rerouted the road north toward the Spokane Plain and around the northern shore of Lake Coeur d'Alene. It was during this period that Sohon created his image of Palouse Falls and also of Fort Walla Walla. Mullan assigned Sohon other responsibilities as well, making him the supervisor of one of three bridge-building crews. Obviously, Sohon was unable to do much artwork at such times. But by the time of the project's completion in 1862, Sohon had created two pencil sketches portraying the finished road–these will be discussed later in this chapter.

By 1863 Sohon was in Washington, D.C., with Mullan to aid him in creating the final congressional report. In April of that year, Sohon married Julia Groh, just days after Mullan married Rebecca Williamson. Immediately after the wedding, Sohon and his bride moved to San Francisco, where he set up an ambrotype studio. He would operate this studio for only two years. We speculate that his new wife was dissatisfied with life in the far West and persuaded her husband to return to Washington. Ironically, for the remainder of his working life, Sohon made his living in the shoe business. He died in 1903, at the age of seventy-eight.[7]

Theodore Kolecki

Besides Sohon, the other member of Mullan's party who produced many maps and topographic drawings for the road-building project was Theodore Kolecki. Very little is known about Kolecki's life, but we have located some examples of the work he did during his time with Mullan. As the road-building project began in 1859, Kolecki maintained a journal in which he documented the topographic character of the region between Fort Dalles and Fort Walla Walla.

Before joining Mullan's expedition, Koleski kept similar journals and drew maps for Colonel Wright's 1858 campaign. Many of these maps, drawn at a very small size within leather-bound journals, usually in pencil, have never been reproduced. Pieced together, however, these sketched maps document the route Wright followed and show some of the major events that occurred during the campaign. Particularly interesting are Kolecki's maps of the Steptoe Battlefield, the Battle at Four Lakes, the Battle of Spokane Plain, and the Horse Slaughter Camp (just east of present Spokane). During the Mullan Road project, Kolecki continued his practice of making topographic drawings in his journals.[8] Not all of the landforms are easily identifiable in these sketches, but they are intriguing to look at. One example of his topographic drawings made in the road-building project is illustrated in Figure 11-3.

Figure 11-3. *This topographic map was created by Theodore Kolecki in 1860, while Mullan was building the road in the Bitterroot Mountains.* –Courtesy National Archives and Records Administration, Old Military and Civil Records Branch, Washington, DC

Theodore Kolecki produced a number of topographic drawings in his journals. The locations at which he drew these works are difficult to determine. The journals are now housed in the National Archives.

The Illustrations

As previously mentioned, Gustavus Sohon created all of the illustrations submitted with Mullan's final report. In many of them–including those of Cantonment Stevens, Fort Walla Walla, and Fort Benton–a tall pole flying the U.S. flag dominates the scene. What is the significance of the recurring presence of the flagpole? It is our hypothesis that both the height of the flagpole and its prominence in each composition suggest the American army's claim and authority over this new territory.

Cantonment Stevens–Capt. Mullan's Winter Quarters 1853–54. To this day, the location of Mullan's winter camp of 1853–54 has not been precisely determined. Mullan reported that

Figure 11-4. *Sohon rendered several full-color studies of Cantonment Stevens; this view is toward the north-northwest.* –Courtesy Yale University Art Gallery

Figure 11-5. *Sohon lithograph of Cantonment Stevens, near present Corvallis, Montana, looking westward* – From John Mullan, *Report on the Construction of a Military Road from Fort Walla-Walla to Fort Benton* (1863)

he built the cantonment some fourteen miles south of Fort Owen, at the mouth of Willow Creek. Our thinking is that the cantonment was located on the southeast side of today's Corvallis, Montana. Our reasoning for this conclusion is based upon three factors. First, Sohon created his rendition from a high point looking northwest onto the camp, but there is no such perspective near the junction of Willow Creek with the Bitterroot River. Second, as Mullan used it, the term *mouth* may have meant the place where Willow Creek enters the Bitterroot Valley, not the river. Third, we know that six buildings existed in Corvallis in 1872, shortly after the first post office opened in the community. It is very likely that some of these buildings may have been erected by Mullan as part of the cantonment. Interestingly, the image shows a relatively large Indian settlement a short distance north of the cantonment. If archaeological investigations could pinpoint the location of the teepee circles, it might be possible to determine the true location of the cantonment.

Sohon's images of the cantonment show four buildings in the complex and a U.S. flag fluttering over the scene, denoting, perhaps, American supremacy. Figure 11-4 is from a collection now held by Yale University. Another Sohon study

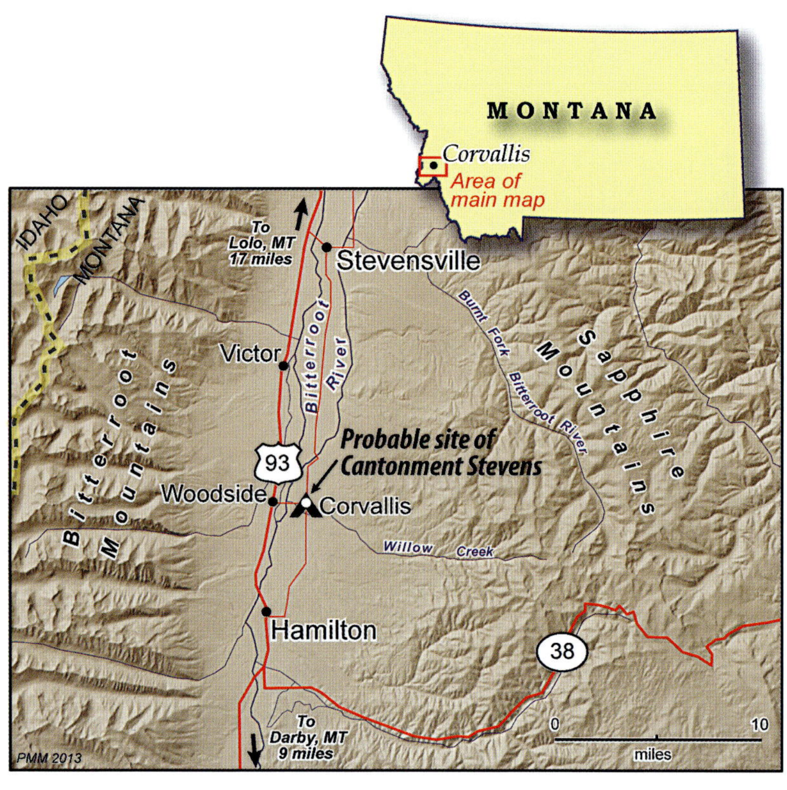

Figure 11-6. *Location of Cantonment Stevens* –Map by Philip Mobley

of Cantonment Stevens, archived at Georgetown University (Figure 11-7), also depicts the flagpole, as does the lithograph included in the Mullan report (Figure 11-5).[9]

From Cantonment Stevens, Mullan conducted numerous explorations of the intermontane region between the Canadian border and Fort Hall. One expedition traversed the Rocky Mountains to Fort Benton. His final expedition in this time frame was across the Lolo Pass toward the Columbia River and Fort Dalles.

One final question remains. Why did Mullan choose to include the picture of Cantonment Stevens (Figure 11-5) in his final report on the Mullan Road? It is strangely out of place, having had no role in the construction of the military wagon road. Apparently it was used only as filler.

Figure 11-7. *Another color study of Cantonment Stevens by Sohon* –Courtesy Georgetown University

Military Post & City of Walla-Walla, W.T. in 1862. Interestingly, Mullan included two drawings in his report that defined the terminal points of the Mullan Road—Fort Walla Walla (Figure 11-9) and Fort Benton. Fort Walla Walla was established in the late 1850s in response to increasing agitation in the region after the signing of the Walla Walla treaty in June 1855, a treaty not every Indian agreed with.

The fort evolved in three stages, moving from location to location until 1857, when its final site was determined. From this base, Colonel George Wright began his campaign to subdue the Natives responsible for the attack on Colonel Steptoe's troops.

In 1859, Mullan began building the road northward from Fort Walla Walla and then eastward toward Fort Benton. Later, in 1861, Mullan began a second road-building phase in order to make improvements and repairs along the route. One of these improvements included rerouting the road to the north side of Lake Coeur d'Alene. Another of his objectives

Figure 11-8. *This sketch by Sohon shows the same view of Fort Walla Walla seen in the lithograph used in the report (Figure 11-10) and was probably the basis for it.* –Courtesy Georgetown University

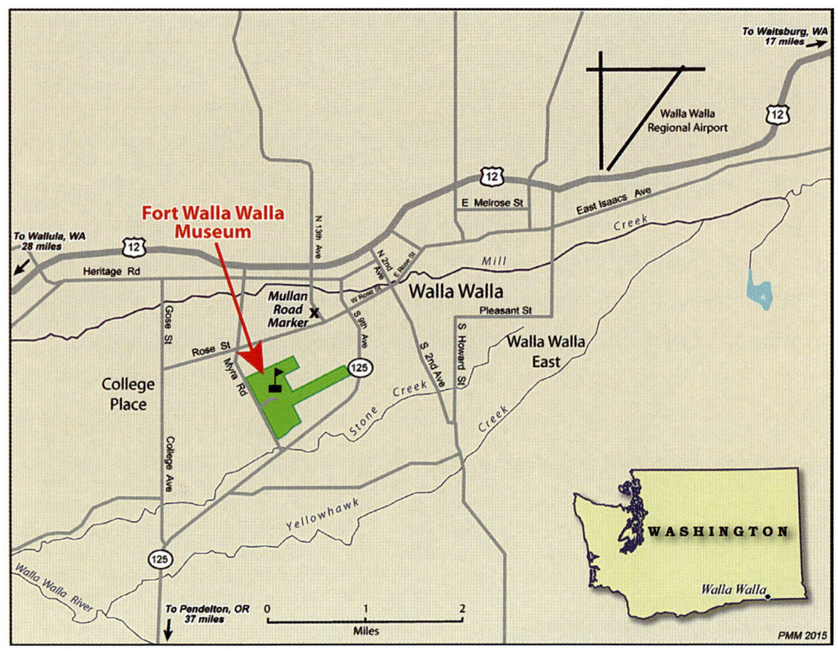

Figure 11-9. *Location of Fort Walla Walla* –Map by Philip Mobley

was to eliminate difficult and unnecessary river crossings–he was obsessed with this problem, earning him the nickname the "Duke of Bridgewater"!

Sohon's color sketch (Figure 11-8) and his lithograph (Figure 11-9) depict the newly constructed Fort Walla Walla. The fort stood at the southwest end of the young town of Walla Walla. In the background of the image toward the northeast, one can see the hazy outline of the Walla Walla settlement, and the Blue Mountains appear faintly in the background toward the southeast. Once again, an American flag dominates the composition. Especially interesting are the officer quarters, scattered on both sides of the parade ground. Today, some of the buildings from the original fort still adorn the complex; a number of the officer quarters are now private homes. The primary structure at the fort site today is a veteran's hospital with its many facilities.

Figure 11-10. *Lithograph of Fort Walla Walla as it appeared in 1862. Some of the buildings that comprised the complex are still standing. Like other artists of that time, Sohon incorporated human figures to provide scale.* –From John Mullan, *Report on the Construction of a Military Road from Fort Walla-Walla to Fort Benton* (1863)

Pend d'Oreille Mission in the Rocky Mountains in 1862. Sohon frequently used ovals—or a variation thereof—to frame his images. Such was the case in his lithograph depiction of Pend d'Oreille Mission nestled at the base of the Mission Mountains in present St. Ignatius, Montana (Figure 11-11). Founded in 1844 by Jesuits, the original mission was called the Pend d'Oreille Mission, as it was located on Pend d'Oreille River in today's Washington state. In 1854 it was moved to its current location in northwestern Montana and renamed the St. Ignatius Mission. The mission still exists, and its 1893 church actively serves area parishioners.

Sohon's lithograph of the mission is one of the few images used in the Mullan report for which we are able to locate the original painting, which can be seen at Georgetown University in Washington, D.C. (Figure 11-13). It is very colorful, detailed, and vibrant, although Sohon's depiction of the Mission Mountains is not as spectacular as the reality. The perspective of this image faces northeast, showing the mission in the foreground and the mountains, with their steep peaks projecting skyward, in the background.

W. W. DeLacy described the St. Ignatius Mission in a report he made on April 4, 1860. The sanctuary encompassed

Figure 11-11. *This lithograph of the St. Ignatius (Pend d'Oreille) Mission was used in Mullan's report to Congress. The view, which faces north-northeast, shows considerable human activity. The image depicts the mission in 1854, when Sohon visited the region, but whether or not this much activity existed is questionable.* —From John Mullan, *Report on the Construction of a Military Road from Fort Walla-Walla to Fort Benton* (1863)

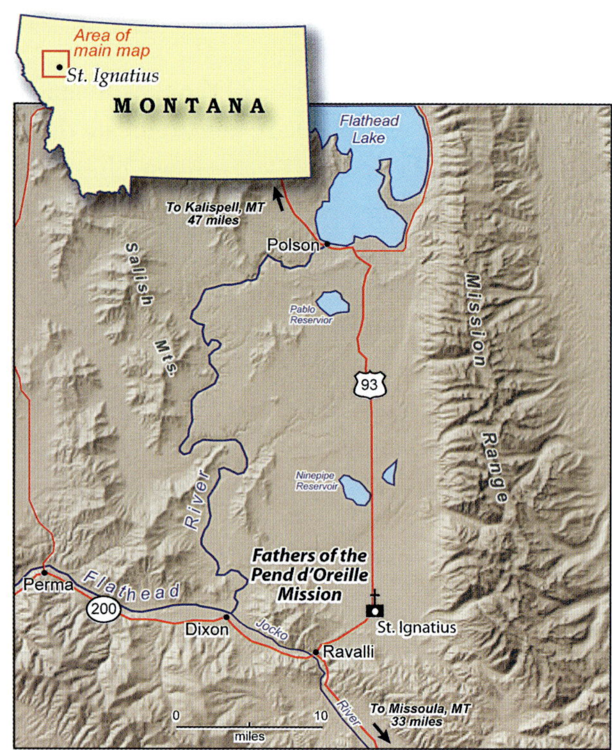

Figure 11-12. *Location of the Pend d'Oreille Mission* –Map by Philip Mobley

"... fields and gardens, cultivated by the Indians under the direction of the fathers. They have a saw and gristmill, several log houses, and a church.... Scattered around the settlement were the usual conical wigwams of the Indians ... the greater portion of the tribe being settled there."[10]

Figure 11-13. *St. Ignatius (Pend d'Oreille) Mission, original painting by Sohon* –Courtesy Georgetown University

Fort Benton–Head of Steam Navigation on the Missouri River. Sohon's 1862 lithograph of Fort Benton (Figure 11-16) depicts three steamboats anchored in the river adjacent to Fort Benton. The scene stands in marked contrast to John Mix Stanley's view of Fort Benton (Figure 11-14), which was published in Isaac Stevens's report on the Northern Pacific Railroad survey of 1853–54.[11] Stanley did not show any steamboats in his 1853 rendition, suggesting that steamboat traffic to Fort Benton was not yet a common occurrence. Sohon's lithograph, however, created nine years later, suggests the beginning of a new era of trade on the Missouri River and of settlement in Montana. With the road completed and the Columbia and Missouri Rivers connected, relatively efficient transcontinental travel, using a combination of wagons and steamboats, was now possible.

Like Stanley's image of Fort Benton, Sohon's was rendered from a perspective facing west. Assumedly the artist stood on the high bluff that stands east of Fort Benton. In Sohon's image, Indian tepees are scattered along both banks of the Missouri. An American flag symbolically unites both the steamboats and the fort, suggesting that American soldiers and traders now controlled the commerce on this portion of the Missouri River.

Figure 11-14. *This lithograph of Fort Benton by John Mix Stanley was created in 1853, before the fort became a major steamboat port.*
–From Isaac I. Stevens, *Narrative and Final Report of Explorations for a Route for a Pacific Railroad, . . .* (1860)

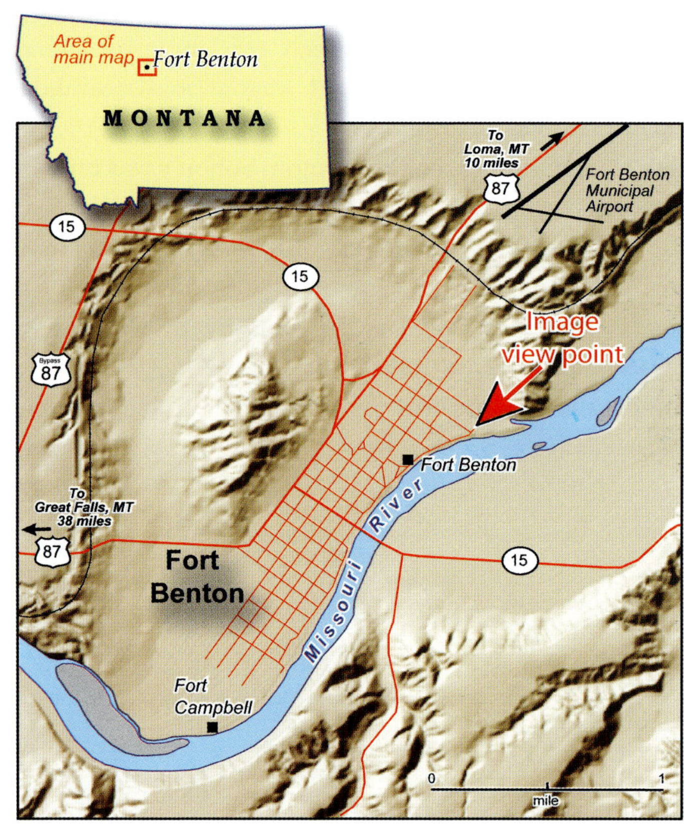

Figure 11-15. *Location of Fort Benton; this map also identifies the hill where Sohon made his drawing. The fort and town were built adjacent to a former meander in the Missouri River. Today, the site contains the remains of the original fort and several museums.* –Map by Philip Mobley

Figure 11-16. *This Sohon lithograph of Fort Benton, created in 1862, appeared in Mullan's report. The fort, built in 1847 as a fur-trading post, became the head of steamboat navigation on the Missouri River and the terminus of the Mullan Road in 1862. This rendition shows three steamboats visiting the port.* –From John Mullan, *Report on the Construction of a Military Road from Fort Walla-Walla to Fort Benton* (1863)

Cantonment Wright, Capt. Mullan's Winter Quarters in 1861–62. Mullan's crew spent two very cold and snowy winters in the Rocky Mountains. In the winter of 1859–60, the men were confined at Cantonment Jordan, sixteen miles northwest of today's St. Regis, Montana. Two years later, the expedition wintered near the banks of the Big Blackfoot River, just above its junction with the Hell Gate River (now called the Clark Fork). Mullan named this camp Cantonment Wright.

The main task for Mullan's crew during the winter of 1861–62 was to build a bridge across the Big Blackfoot River. The men constructed five triangular piers on top of the frozen Blackfoot and filled them with rock, then they cut holes in the ice and sank the piers to the river bottom—an ingenious solution to an otherwise very complex type of bridge construction.

The lithograph shows three buildings by a river terrace close to the bridge across the Big Blackfoot (Figure 11-17). One structure served as a root cellar, another as a stable, and the third as the men's quarters. But these three buildings could not accommodate Mullan's entire force—others were situated on the upper terrace, near the now-abandoned sawmill in present Bonner, Montana.

The image is interesting because it shows the Big Blackfoot River completely frozen over while the Hell Gate (Clark

Figure 11-17. *Sohon made at least two images of Cantonment Wright. This lithograph is the most well-known, created in the winter of 1861–62. Here, Mullan was obligated, in the dead of winter, to build a bridge across the Big Blackfoot; the bridge is prominently featured in the scene.* –From John Mullan, *Report on the Construction of a Military Road from Fort Walla-Walla to Fort Benton* (1863)

Figure 11-18. *Location of Cantonment Wright, at the confluence of the Big Blackfoot and Hell Gate (Clark Fork) Rivers (at present-day Bonner, Montana, southeast of Missoula)*

Fork) is only fringed with ice. This scene is dramatic due to the large number of men scattered throughout the picture–some on the bridge, others carrying game back to camp, and the remainder doing work about the three buildings by the bridge. Another group of at least six buildings can be seen in the background; presumably these functioned as additional housing. To the right, the Mullan Road ascends a terrace along the Hell Gate River into the distance.

The other Sohon images of Cantonment Wright is a pencil sketch drawn in the summer or early spring of 1862 (Figure 11-19). This study was donated to the Smithsonian Institution by Sohon's daughter, Dr. Elizabeth Sohon.

Figure 11-19. *This sketch of Cantonment Wright was made when the weather had warmed–probably in April of 1862. Originally owned by the Sohon family, the drawing was given to the Smithsonian in 1947 by Sohon's daughter Elizabeth.* –Courtesy National Anthropological Archives, Smithsonian Institution

Palouse Falls in Washington Territory. Most of the images Sohon created for both Isaac Stevens's Northern Pacific Railroad Survey report and Mullan's Military Road report were done in a landscape, or horizontal, format. One exception was Sohon's lithograph of Palouse Falls, which has a portrait, or vertical, orientation (Figure 11-20). These falls create a wondrous scene as they cascade over a large lava precipice into a plunge pool some 180 feet below. From there, the Palouse River meanders through a very deep canyon for approximately six miles until it reaches its juncture with the Snake River to the south. A series of lava flows define the valley walls on both sides. Sohon drew this image from a lava-flow terrace

Figure 11-20. *Lithograph of Paloose (Palouse) Falls, created by Gustavus Sohon in either 1859 or 1861. The artist was captivated by the falls of the Palouse River, which tumble over massive lava walls into the canyon below.* –From John Mullan, *Report on the Construction of a Military Road from Fort Walla-Walla to Fort Benton* (1863)

Figure 11-21. *An early oil painting of Palouse Falls by Gustavus Sohon. This is Sohon's only known image done in oils.* –Courtesy Sohon family

on the valley's west side, his view to the northeast. In the background, the gently rolling hills of eastern Washington overlook a cathedral-like structure that protrudes dramatically above the river's left (west) bank. Gustavus Sohon made an oil painting of the falls (Figure 11-21), which is still owned by the Sohon family. Compare these images with a modern photograph of the same falls (Figure 11-23).

Figure 11-22. *Location of Palouse Falls, about four miles upstream from the confluence of the Palouse and Snake Rivers* –Map by Philip Mobley

Figure 11-23. *Palouse Falls today, with its magnificent 180-foot waterfall* –Photo by Philip Mobley

Mode of Crossing Rivers by the Flathead and Other Indians. The specific river Sohon portrays in this lithograph (Figure 11-25) is not known. Our best guess is the Clark Fork or Kootenai River, or possibly the Bitterroot River. Here several Native American women and children crowd the interior of a small round vessel of skins stretched over a frame of tree branches; this type of craft is often called a bull boat. Similar boats can be seen in the background. The bull boats are being pulled across the river by horses tethered to the boats and guided by men swimming alongside. Most likely, the idea of bull boats diffused westward across the mountains from the Great Plains. Contact between tribes on both sides of the Rockies promoted the spread of this interesting innovation. The original pencil study for this lithograph (Figure 11-24) is owned by Georgetown University.

Figure 11-24. *Sohon's "Mode of Crossing Rivers by the Flathead and Other Indians"* –Courtesy Georgetown University

Figure 11-25. *Lithograph entitled "Mode of Crossing Rivers by the Flathead and Other Indians" by Sohon. The artist captures on paper the type of vessels Native Americans used to cross streams in the Northern Rockies.* –From John Mullan, *Report on the Construction of a Military Road from Fort Walla-Walla to Fort Benton* (1863)

Great Falls of the Missouri, 2500 Miles from St. Louis. Originally, there were five major waterfalls on the Missouri River between present-day Great Falls, Montana, and Fort Benton. So hazardous were these falls that Lewis and Clark were forced to portage around them for eighteen miles. This portage was not only difficult but it also caused a significant delay in their journey westward. Most of the grandeur of these magnificent cataracts has been lost over time, due to the construction of four hydroelectric dams in the late 1800s and early 1900s.

Sohon created intriguing lithographs of two of these falls for the Mullan report, one of the main cascade, the Great Falls (Figure 11-26), and the other of the Upper Falls (now called Black Eagle Falls), seen in Figure 11-29. While the title

Figure 11-26. *Lithograph of the Great Falls of the Missouri by Sohon. These falls originally had an eighty-foot drop.*
–From John Mullan, *Report on the Construction of a Military Road from Fort Walla-Walla to Fort Benton* (1863)

of the former image indicates that the Great Falls of the Missouri were 2,500 miles from St. Louis, the actual distance is shorter, approximately 1,843 miles. Sohon also made a sketch of the main falls, preserved at the Yale University Art Gallery (see Figure 11-28).

In Mullan's time, the Great Falls dropped some eighty feet, but its once-surging waters are now diverted by Ryan Dam (first called Volta Dam), a concrete structure built in 1915 to create electrical power for the town of Great Falls and its nearby copper smelters. The same fate befell three of the other four waterfalls; today, only Crooked Falls remains in its natural state.

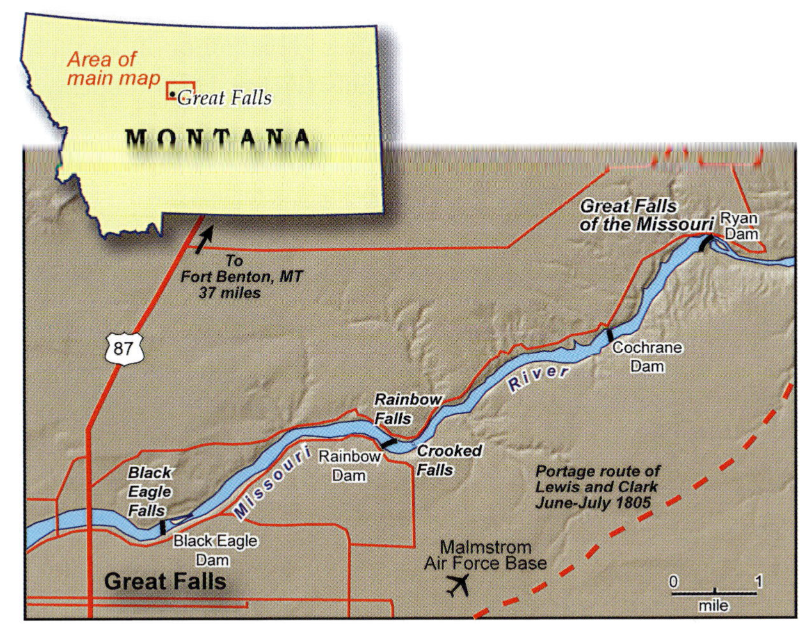

Figure 11-27. *Location of the five falls of the Missouri River* –Map by Philip Mobley

Figure 11-28. *Great Falls of the Missouri, pencil sketch on three-color paper by Sohon. The viewpoint is from the northern bank of the Missouri, looking toward the west, to replicate Lewis and Clark's vista of the falls in the summer of 1805.* –Courtesy Yale University Art Gallery

Upper Falls of the Missouri River. This Sohon lithograph of the uppermost of the Great Falls (Figure 11-29) appeared in the report. Today the Upper Falls are called Black Eagle Falls. Lewis and Clark also made a sketch of the Upper Falls.

Figure 11-29. *Upper Falls of the Missouri River lithograph by Sohon. Today this cascade is known as Black Eagle Falls.*
–From John Mullan, *Report on the Construction of a Military Road from Fort Walla-Walla to Fort Benton* (1863)

Coeur d'Alene Mission in the Rocky Mountains. This 1850s mission, also known as the Cataldo Mission or Mission of the Sacred Heart, still stands along Interstate 90 in northern Idaho as a state park and National Historic Landmark. The mission's Sacred Heart Church is Idaho's oldest intact structure. Construction began under the direction of Father Antonio Ravalli in 1850. The church is a magnificent, hand-built edifice fashioned from huge wooden timbers complemented by waub-and-daub filling. As materials were expensive, much ingenuity was used in its construction and decoration. Its beautiful interior was created using such items as old newspapers, tin cans, and homemade stain made from huckleberry juice.

Sohon made several different illustrations of the mission. One, in pencil, was done during Colonel Wright's expedition in 1858 (see Figure 11-31). The image used in the Mullan report, a distinctive lithograph (Figure 11-30), was probably created in either September 1859, when Mullan and his men first visited the mission, or in the summer of 1861, when they returned there. In this depiction, the mission is viewed from the northwest looking southeast. The mission buildings stand prominently on a small hill behind the meandering river. Over the years, the course of the Coeur d'Alene River has changed—it now flows south of the hill where the mission stands. Today, the interior of the mission church appears much as it did when it was built (see Figures 11-33 and 11-34).

Figure 11-30. *Sohon's lithograph of the Coeur d'Alene Mission, also known as the Caltado Mission, on the Coeur d'Alene River east of Coeur d'Alene Lake. Officially Idaho's oldest standing structure, the mission was built over three years, 1850–53, and many travelers on the Mullan Road stopped here. This image was created in 1859 or 1861.* –From John Mullan, *Report on the Construction of a Military Road from Fort Walla-Walla to Fort Benton* (1863)

Figure 11-31. *A detail of Sohon's pencil drawing of the Coeur d'Alene Mission, rendered in September 1858* –Courtesy Library of Congress, Geography and Map Division

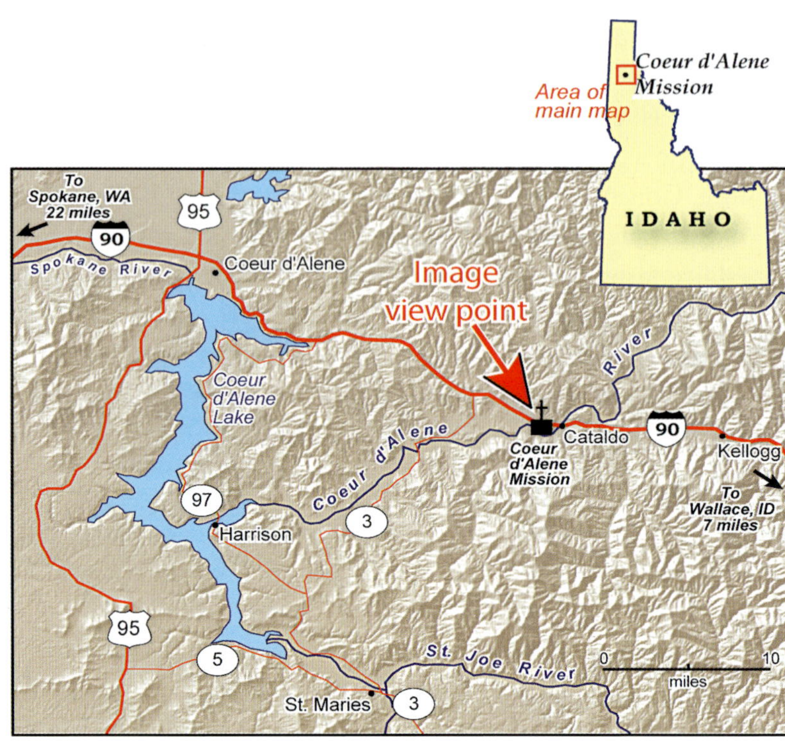

Figure 11-32. *Location of the Coeur d'Alene Mission in northern Idaho.* –Map by Philip Mobley

Figure 11-33. *Interior of Sacred Heart Church in Coeur d'Alene, Idaho. Much of the original charm and craftsmanship still remains in the structure's architecture.* –Photo by Paul D. McDermott

Figure 11-34. *Hand-carved ceiling panel, Sacred Heart Church* –Photo by Paul D. McDermott

Other Illustrations

Although the images discussed above were the only illustrations published in Mullan's official report, Gustavus Sohon rendered two other noteworthy illustrations during the road's construction. Both are pencil sketches, dated April 1862, that show the road as it appeared when the work was nearly completed. They were both drawn on small sheets of paper, measuring approximately 8 x 10 inches.

The first drawing shows a place known as Rocky Point, located approximately ten miles southeast of today's Missoula (see Figure 11-35). Here the road was narrow, perched above the north bank of the Hell Gate (Clark Fork) River until it cut in to get around Rocky Point–a major obstacle. Part of this hill has since been removed, and the surrounding area has been altered by the construction of Interstate 90. Missoula journalist Kim Briggeman has provided a photograph of the highway near Rocky Point and annotated where the spur was cut during the construction of the highway (Figure 11-36).

Figure 11-35. *Sohon's drawing of Rocky Point* –Courtesy National Archives and Records Administration, Cartographic Section, College Park, MD

Figure 11-36 *This modern photograph shows Interstate 90 moving toward the Rocky Point spur. The line in red indicates where the spur was truncated during the excavation for the highway.* –Photo by Kim Briggeman

Sohon's second fascinating image documents the Three Mile Hill cut constructed by Mullan in 1861–62 near Turah, Montana (see figure 11-38). The cut can still be seen on aerial photographs of the region. Both of Sohon's sketches are significant not only because they show the road itself just after its construction but also because they identify the sites of some of the most difficult construction the crew encountered in building the Mullan Road. At both Rocky Point and Three Mile Hill, Mullan was attempting to minimize the number of crossings of the Hell Gate River, as illustrated in Figure 11-40. He also sought to keep the road elevated above the river to prevent flood damage to the road.

Figure 11-37. *Location of Rocky Point and the Three Mile Cut* –Map by Philip Mobley

Figure 11-38. *Sohon sketch showing the Three Mile cut, between the first and second crossings of the Hell Gate River.* –Courtesy National Archives and Records Administration, Cartographic Section, College Park, MD

Figure 11 39.
Modern photo of the Three Mile cut –Photo by Kim Briggeman

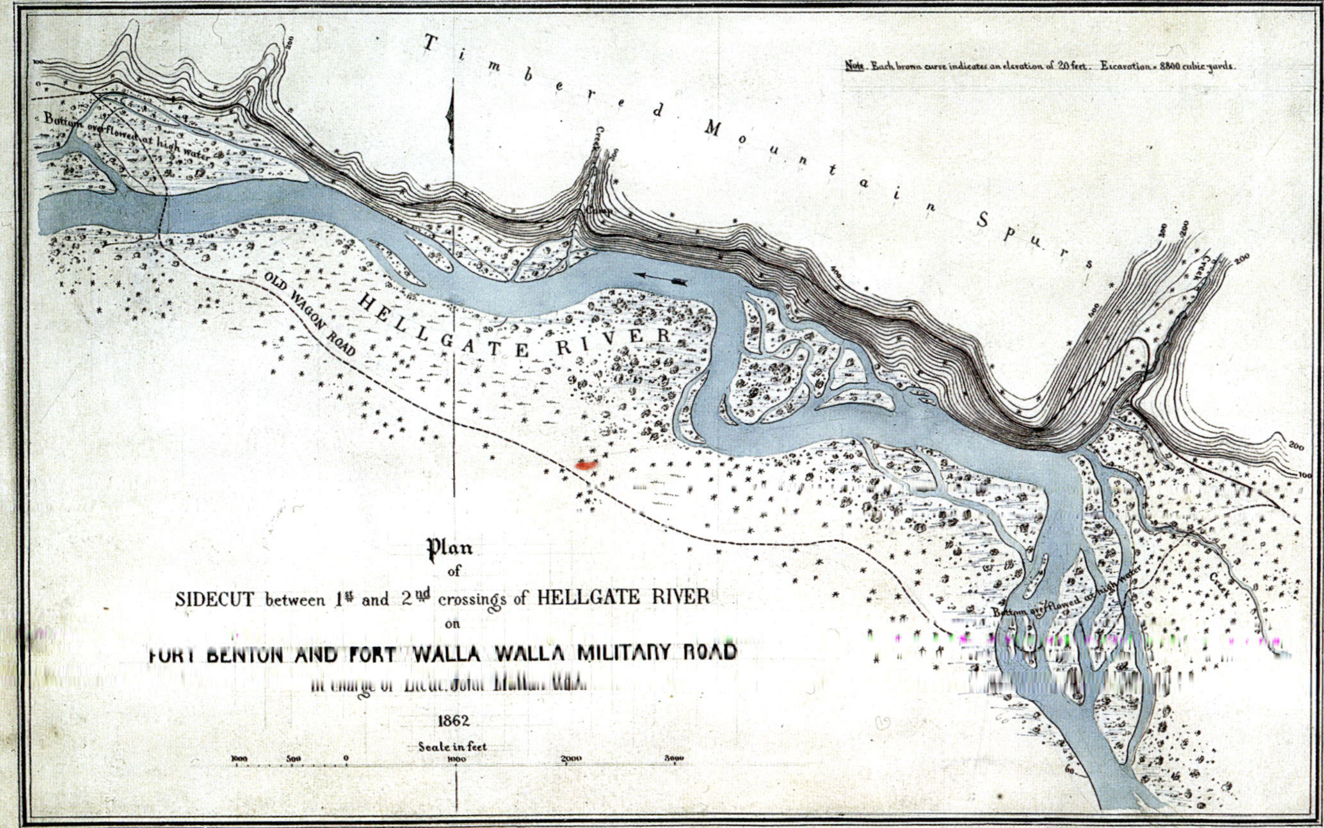

Figure 11-40.
Map of plans for the side cut at Three Mile Hill –Courtesy National Archives and Records Administration, Cartographic Section, College Park, MD

Cartographers at Work

"Cartographers at Work" is our own title for this illustration by Sohon (Figure 11-41). Significantly, it shows Mullan and his staff working at plotting maps in an office at Cantonment Jordan, a winter camp built in late 1859 near today's De Borgia, Montana. Considering all the exploration conducted in the western United States during the nineteenth century, it seems strange that this is the only known image that shows cartographers actively working on maps.[12]

This illustration is important for several other reasons as well. For one thing, it shows the effort Mullan and his men made to provide good lighting for drawing maps—notice the skylight cut into the ceiling, permitting natural light to fall on the drafting table below. This creative use of natural light was unusual for the time. Secondly, the tags on the walls mark where rolled maps were stored, revealing the meticulous care the cartographers took in storing and preserving their work—they even built shelves for their reference books and field survey notes, some of which remain in the collection of the Old Military Branch at the National Archives in Washington, D.C. Perhaps most important, through this image, we have likenesses of those who were presumably the most important

Figure 11-41. *Sohon's illustration of cartographers at work at Cantonment Jordan, Mullan's winter camp of 1859–61. It pictures the cartography team working on the maps in an office building especially constructed to facilitate map drawing—note the skylight in the ceiling.* –Courtesy Sohon family

surveyor-cartographers of the Mullan Road project—although we cannot be sure who the men in the picture are. We suppose them to be John Mullan, Gustavus Sohon, Theodore Kolecki, and W. W. DeLacy, but a positive identification has not been made. Using facial-recognition software to compare known portraits of these men with the faces in the illustration may someday help solve the problem.

Summary: The Meaning Behind the Art

An analysis of the illustrations created as the Mullan Road was being constructed prompts various thoughts. The most obvious is the recognition that, because the artwork documents how the road and surrounding landscape appeared at the time of construction, we can fairly easily compare today's landscapes with the way they appeared circa 1860. In many of these comparisons, significant cultural progress can be observed. For instance, the simple one-lane wagon road winding around Rocky Point in 1862 has been replaced by a multilane interstate highway. Another example is the Great Falls of the Missouri. Originally, the course of the Missouri River was punctuated here by five waterfalls, each with a different form and vertical drop. Today, most of these falls are no longer natural, having been altered by the installation of hydroelectric dams. Replacing the sound of rushing cascading water is the hum of electrical power lines and generators, forever muting a priceless part of the great Missouri River's history.

Likewise, at the sites of Mullan's winter camps, virtually all of the original features have disappeared. The three buildings that once comprised Cantonment Jordan are gone, as are the numerous cabins that once defined Cantonment Wright. At the latter spot, where two rivers once peacefully converged, the waters now meet in the midst of a huge Superfund cleanup site (see Figure 11-42).

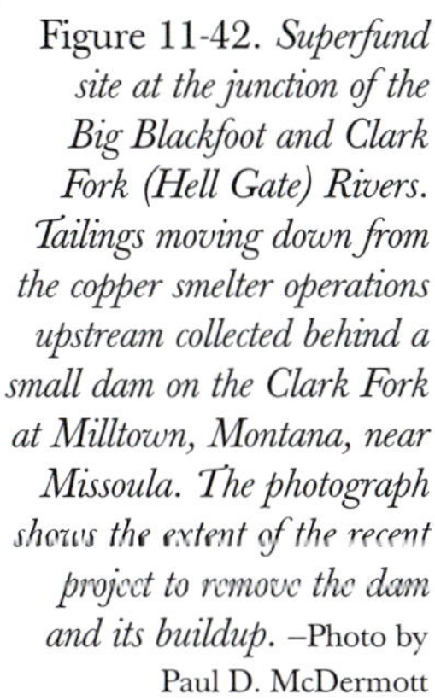

Figure 11-42. *Superfund site at the junction of the Big Blackfoot and Clark Fork (Hell Gate) Rivers. Tailings moving down from the copper smelter operations upstream collected behind a small dam on the Clark Fork at Milltown, Montana, near Missoula. The photograph shows the extent of the recent project to remove the dam and its buildup.* –Photo by Paul D. McDermott

Similarly, where Mullan's narrow wagon road once traced the South Fork of the Coeur d'Alene River through northern Idaho, a number of notable mining towns, including Wallace, Kellogg, and Mullan, still exist. Dotting this region's once-verdant mountainsides are the remains of some of the numerous silver and lead mines that emerged in the late nineteenth century and left many slopes stripped of their wonderful trees.[13] Some efforts to restore the original environment have been made, but the process is costly and takes many years.

Certainly the appearance of towns and cities along Mullan's route fulfills the promise of progress that Mullan had hoped for. The most outstanding of these is Spokane, Washington. What started as a humble fur-trading post is now the second-largest city in the state of Washington. Correspondingly, at Fort Benton, rather than a motley collection of fur trappers, soldiers, Indian families, riverboat captains, and frontiersmen mulling around a few simple log buildings, one will find a thriving modern town bustling with tourists, housewives, farmers, shopkeepers, and professionals. Quite a change!

Other transformations are also evident. In eastern Washington, the once-barren plains, peppered with lava-flow formations and rock outcroppings, prairie grasses and sagebrush, prairie dogs and rattlesnakes, have become fields of wheat, corn, and hay. What was then brown and bleak is now a rich green—at least for part of the year—although if today's irrigation systems failed, the landscape would return to its original austere state.

What other messages are imbedded in these images from the past? A sense of peace and serenity pervades many of the scenes. At the Coeur d'Alene Mission, a man and his horse drink water at the river's edge while a nearby fisherman tests his skills in the placid waters. In contrast, Sohon's portrayal of Fort Benton conveys a different message—one of progress. The three steamboats show that commerce has reached the far West and portend more economic development to come. The repeated presence of the American flag suggests another idea as well—that the government of the United States is in control. Strangely, however, only one illustration—that of Fort Walla Walla—includes figures of military personnel in the scene. Here the message, subliminally transmitted, is this: The U.S. Army is here, and its function is to protect you.

---

1 John Mullan, *Report on the Construction of a Military Road from Fort Walla-Walla to Fort Benton*. S. Ex. Doc. 43, 37th Cong., 3rd sess., serial 1149 (Washington, DC: Government Printing Office, 1863). Although only ten illustrations were included in this published report, there were other manuscript landscape sketches that were prepared during the process of constructing the Mullan Road. These graphic documents are filed in the Headquarters Map File (under the designations W79 and Rds 174) of the Office of the Chief of Engineers (Record Group 77), Cartographic Section, National Archives and Records Administration (NARA), College Park, MD. The associated manuscript maps, also included in these archival files, are discussed in this book's preceding essay, "Maps of the Mullan Road."

2 These illustrations are described more fully in Paul McDermott and Ronald Grim, *Gustavus Sohon's Cartographic and Artistic Works: An Annotated Bibliography*, Phillip Lee Phillips Society Occasional Paper Series, no. 4 (Washington, DC: Library of Congress, Geography and Map Division, 2002), pp. 45–54.

3 See Isaac I. Stevens, *Narrative and Final Report of Explorations for a Route for a Pacific Railroad, near the 47th and 49th Parallels of North Latitude from St. Paul to Puget Sound*, supplement to U.S. War Department, *Reports of Explorations and Surveys, to Ascertain the Most Practicable and Economic Route for a Railroad from the Mississippi River to the Pacific Ocean*, vol. 12 (Washington, DC: Thomas H. Ford, Printer, 1860), pp. 107–11.

4 The portraits created on the 1855 expedition are featured in David Nicandri, *Northwest Chiefs: Gustave Sohon's Views of the 1855 Treaty Councils* (Tacoma: Washington State Historical Society, 1986). These images were originally retained by Isaac Stevens's family, and Stevens's son, Hazzard, used much of his father's work in the biography he wrote: Hazzard Stevens, *The Life of Isaac Ingalls Stevens* (Boston: Houghton Mifflin, 1900).

5 Sohon's images related to the 1858 Wright campaign are described in McDermott and Grim, *Gustavus Sohon's Cartographic and Artistic Works,* pp. 39–43.

6 When the railroad was constructed through this part of western Montana in 1900, the names of several passes were inadvertently changed. Sohon Pass is now called St. Regis Pass, while the original St. Regis Pass was renamed Lookout Pass. Due to a lack of local support, an attempt to change the name of St. Regis Pass back to Sohon Pass has yet to be successful.

7 For more about the life of Gustavus Sohon, see McDermott and Grim, *Gustavus Sohon's Cartographic and Artistic Works*, pp. 1–2 and 54–55.

8 Kolecki's field survey journals are found in the National Archives in Washington, DC. They are filed among the non-cartographic records of the Office of the Chief of Engineers (Record Group 77, Field Survey Data Cupboard 1, Shelf 4, nos. 7, 8, and 9, and Cupboard 37, Shelf 5, nos. 56, 57, 58, and 59).

9 Some of the original source material for the lithographs used in Mullan's official report are now in the collection of Georgetown University, Washington, D.C. Previously, they were in the possession of Mullan's great-grandson–a Catholic priest who taught at Georgetown in the mid-twentieth century.

10 Mullan, *Report on the Construction of a Military Road*, p. 116.

11 Paul McDermott, Ronald Grim, and Philip Mobley, *Eye of the Explorer: Views of the Northern Pacific Railroad Survey, 1853–54* (Missoula, MT: Mountain Press Publishing, 2010), pp. 76-79.

12 The original is owned by the Sohon family, who gave us permission to use it here.

13 While mining played a significant role in forest desecration in the interior Northwest, a massive forest fire swept over the landscape in 1910. See Timothy Egan, *The Big Burn: Theodore Roosevelt and the Fire that Saved America* (Boston: Houghton Mifflin Harcourt, 2009).

Merchants, Mule Packers, and a Rugged Path to Success

Walla Walla and the Mullan Road, 1860–83

BY ALEXANDER C. MCGREGOR

In memory of my father, Sherman, who showed me the old trails and told me of the pioneers he had known.

And to Robert Whitner and Vernon Carstensen, who helped me hone my skills so I could share the story of the remarkable people of the bunchgrass prairies.

"They were all men who did not know the meaning of the word fear and they risked their lives and property to serve the public. . . . It was the packers and traders that were always foremost in blazing the trails and preparing the way for civilization."

—Ben Burgunder, 1907[1]

"The whole packing business called for courage, hardihood, and endurance; that wasn't all, you had to have skill, daring, initiative, gumption and loyalty. . . . All in all, a rough, lawless and profane bunch of men; but they were brave, hardy, and extremely loyal and trustworthy."

—James Watt, 1928[2]

Mule packers, with their long trains of work animals laden with supplies for distant mining camps, left in their wake trails still visible along the rough military road blazed across prairie, scab rock, timber, and rugged mountain terrain by John Mullan and the men under his command. For a quarter of a century, between the defeat of the Indian tribes of the inland Northwest by Colonel George Wright in 1858 and the arrival of transcontinental rail service in 1883, mules provided the energy to move an endless supply of goods along what became a vital transportation corridor.

Mule packing was an old trade, practiced for more than three centuries in Mexico and the Southwest, later employed by fur traders and gold seekers across the West. Some of the men who worked the mules were of Mexican heritage, recruited from California and "employed on account of their special experience and skill in such work," as one pioneer put it. Others were gold seekers and immigrants from east of the Rockies. The tools of the packers' trade included *aparejos*—Spanish pack saddles made of leather and stuffed with moss, dry hay, or grass, that protected the back of the mule—and *caronies*, fancy Spanish embroidered blankets laid over the "sweat" blanket and under the "bed" blanket, which were made of the best wool.[3]

Old packers vividly remembered the hard work of their job, as seen in this quote by James W. Watt:

> It was some work, I'll tell you. The average time was 1½ to 2 minutes for two experienced men to pack a mule. . . . If it was 400 pounds of flour . . . each man had to swing his half of the heavy pack up from the ground on to the animal's back, and hold it there on one side of the pack saddle with one arm, despite the kicks and bites and protesting lunges of the mule, until he and his partner had it securely lashed there by means of a "diamond" or other hitch. To round up, saddle and load twenty five such packs a day and to meet all the emergencies of the trail was a real man's job.[4]

Pioneer cattleman A. J. "Jack" Splawn claimed that he and his brother Billy could reach a mule-per-minute pace, loading fifteen mules in fifteen minutes. He remembered that it was "quite an imposing sight" when the *mulara* (the bell mare, which led the train) started out on the trail followed by the *caballada* (outfit) of thirty to a hundred loaded pack animals, with the *arrieros* (packers) riding alongside, watching for shifted or unbalanced packs. "When all the packs were satisfactorily adjusted the arrieros would break out in song, the music blending with the rattle of the bell on the old mulara," Splawn recalled. "With a well-trained mule and a well-trained packer, there is nothing imaginable from a bag of oats to a load of crockery that cannot be securely fastened on a mule's back."[5] James Watt remembered that it took "a

Two mule packers cinching aparejos, getting ready to head out on the trail –From "Pack-Mule Train" in *Harpers Weekly*, October 12, 1878; negative provided by Lawrence L. Dodd, Archivist Emeritus, Whitman College

"The Patient Pack-Mule" by Frederic Remington –From *The Century Magazine*, March 1891; negative provided by Lawrence L. Dodd, Archivist Emeritus, Whitman College

young muscular man with a stout back and good arms and shoulders" to lift an "8 cask," a twenty-eight-gallon liquor keg, packed one on each side of the mule. Mules were reported to have carried at one time or another a safe, heavy vats of quicksilver, a pool table, and parts for a steamboat.

John Mullan, in his *Miners and Travelers Guide* of 1865, advised men never to maltreat their mules "but govern them as you would a woman, with kindness, affection, and caresses, and you will be repaid by their docility and easy management." Veteran farmer Carl Penner echoed this sentiment as he described working with big mule teams: "You couldn't beat the meanness out of 'em. It just don't work. But if you'd treat 'em halfway right, and pet and curry 'em and give 'em plenty to eat, they soon learned who was boss. They was just some life." Mules and those who handled them had close working relationships. As Penner put it, "You know you get attached to a mule, same as you do people."[6] Whether packer, freighter, or skinner, each man had his own way of working with his animals. "Whispering Thompson," one of the freighters, had five span of work animals and could be heard from a mile away shouting at them.

Where mountains could be avoided and prairie trails were "worn down" enough to make it practicable, "mule skinners" and "bull whackers" drove cayuse horses, mules, and oxen to pull freight wagons with big payloads. Wagons could not pass through the roughest terrain, however. Packers, on the other hand, "penetrated into the most remote and inaccessible places; over all kinds of country and in all kinds of weather conditions," according to James Watt.[7]

"The Bell Mare" by C. M. Russell –From Montana: *The Magazine of Western History*, April 1967; negative provided by Lawrence L. Dodd, Archivist Emeritus, Whitman College

Such rugged travel was not anticipated in 1858, when Congress allocated the first $30,000 of what would eventually total $230,000 to build a wagon road connecting the Missouri and Columbia Rivers. The road was to be a quicker, more practicable alternative for immigrants, troops, and future railroad builders to reach the Pacific Northwest—better than the old Oregon Trail overland from Missouri, or an ocean voyage around Cape Horn or portage across the Isthmus of Panama from the East Coast up to California. Thus did Lieutenant John Mullan, with one hundred laborers and a team of engineers and talented cartographers and artists, set out in 1859 to traverse sand and prairie, scab rock and timber, and the rugged Bitterroot Mountains to build a 624-mile military wagon road across the Northwest.

In 1862 the editor of the *Washington Statesman*, in the tiny village of Walla Walla, expressed "feelings of the most lively anticipation" about the opening of a "great highway." That same year, Portland's *Oregonian* predicted 30,000 immigrants would travel Mullan's new road. "Much of this immigration may be poor, but it will possess the material—muscle, energy, principle, and intelligence—to hew out homes on this coast and add to our wealth, business, and prosperity."[8]

Though emigrants did use Mullan's road, heavy snows, spring flooding, fallen trees, and lack of maintenance soon made attempts to traverse the mountain sections with wagons so "horrible . . . that no word in the English language could convey an idea of its condition," according to one traveler quoted in the *Oregon Argus* in 1866. Between mountains, "where the road is all torn in pieces," emigrant Henry Cummins wrote to the *Argus* editor in September 1863, there were swamps "black with mosquitos of prodigious size with enormous powers of suction."[9]

Mule packers, however, undaunted by rocks, fallen timbers, and hungry insects, were frequent travelers on the road, using it to reach remote gold camps in Montana, Idaho, Washington, and British Columbia. Charley Broughton recalled traversing the most treacherous section of the Mullan Road, a thousand feet above the St. Regis River, and the gruff comment of trader Pat McGraw (or McGrath, as he was sometimes known) upon the train's arrival:

> Once, when Pat McGrath's pack train was going in, loaded, when passing over the point, one of the mules on the back of which was packed two twenty gallon kegs of whiskey, got to bucking, bucked off the road, plunged over the cliff, to the river below. Of course, that was the end of both mule and pack. When the rest of the train reached McGrath's store in Forest City, he was appraised of the mishap and was heard to say: "I don't givc a damn for thc mulc, he was a damned outlaw any way, but I sure do hate to lose that whiskey."[10]

Despite its limitations, the new road, which provided a route through the mountains, was probably the largest factor contributing to the Montana gold rush. The wealth extracted from the ground was staggering—an estimated $150 million from roughly five hundred gulches in western Montana alone during the great era of placer mining there, from 1862 through 1876.[11] This new prosperity rippled through the Northwest. Within a few years, Walla Walla, a town of but nine buildings in 1859 (including an Indian trading post, a boardinghouse, a general store, a saloon, and a liquor store) exploded as it became winter quarters for teamsters and miners. When the Mullan Road was built from Walla Walla, supplementing traffic from Oregon Trail to the south and the Nez Perce Indian Trail from the Clearwater River to Celilo Falls, which bisected the town, Walla Walla became a lively (and somewhat notorious) place in short order.[12] It hardly resembled the quiet and prosperous town it is today.

Randall Hewitt, writing of his experiences traveling with a pack train in *Across the Plains and Over the Divide*, described the newly hatched boomtown of Walla Walla in 1862: "The dirty streets were crowded with freighting wagons and teams and pack animals and a considerable army of rough men. A straggling, disorderly looking place." T. C. Elliot called it "a rambling collection of wooden shacks . . . morally quite low down on the scale of civilization, overrun with thieves, gamblers, and women of the demi-monde." Pioneer Nesmith Ankeny recalled it as a place of "many saloons, gambling dens, and dance halls. Brawls were a nightly occurrence. Often a miner was 'rolled' and robbed of his gold dust." When the local vigilance committee put to use the "hanging tree" along what would become South Second Avenue, settlers claimed it "purified the air for months." Rough and gritty though it may have been, the town was expanding at a feverish pace, enough so that the editor of the *Statesman* declared in 1862,

MAIN STREET, WALLA WALLA, 1859

Building No. 1, McClinchey & Lew McMorris, Indian traders; No. 2, Jim Galbraith's saloon; No. 3, I. T. Reese, store; No. 9, Martin's boarding house; No. 8, Ball & Stone's liquor store; No. 7, Baldwin Bros., licensed traders; No. 6, Dick Rackett's tin shop; No. 5, part of old Cantonment.

"Main Street, Walla Walla, 1859" –From Nesmith Ankeny, *The West as I Knew It*, 1953

"To-day the eyes of the Pacific and Eastern slopes are turned to what was yesterday the little trading post of Walla Walla."[13]

By the time Mullan and his associates completed work on their military wagon road, a full-scale gold rush was in the works throughout the Inland Northwest, with fortune seekers pouring in and scattering like quicksilver, following rumors of big strikes. The Boise Basin rush and a strike in Orofino, Idaho–accessible from Walla Walla by the old Oregon Trail and the Nez Perce Trail–were followed by larger strikes in the Kootenay (also spelled Kootenai) region of British Columbia (about fifty miles north of the U.S. border) and in western Montana. A correspondent from the Kootenay mines wrote to the *Statesman* of the "gold excitement" in his district in March 1863, noting that the "Jesuit fathers have discouraged miners. But the adventurous disposition of the American people and their desire to explore every nook and cranny of our vast territory, has placed it beyond the power of any to allow the mineral wealth of the country to longer be dormant." John Mullan, in his report to the Senate later that year, feared that "the location of our road and the swarms of miners and emigrants that must pass here year after year" would harm the Indian population the Jesuits had hoped to protect. "Now we propose to invade these mountain solitudes, to wrest from them their hidden wealth, where in heavens can the Indian go?"[14]

In 1863, the *Statesman* also predicted a "general stampede" to Colville, Washington, where would-be miners had earlier stirred discontent with the Indians by intruding upon tribal lands before Colonel Wright's 1858 campaign subdued them. With the land now open, "great amounts of gold dust" were brought to Walla Walla from Colville, including $35,000 in Columbia River dust brought by John Hofstetter "and several other gentlemen." In 1862 Marcus Oppenheimer and his two brothers opened a store in Colville, with supplies brought in via the Mullan and Colville Roads.[15]

Samuel W. "Wild Goose Bill" Condit started his long career as a packer in Walla Walla, delivering 13,000 pounds of freight to Pinckney City (Colville) and collecting $1,950 for his efforts. He became a regular on the route, leading sixty to one hundred pack animals per trip from Walla Walla to Fort Colville and occasionally beyond, along the Dewdney Trail to the newly developing mines in British Columbia. The pace was sometimes so hurried that the pack animals were not unloaded along the way—at night they were hobbled in pastures with their packs and saddles still on their backs. Trade with the Colville district helped make 1862 "a year of unparalleled prosperity" for Walla Walla. The editor of the *Statesman* predicted that "Walla Walla trade will continue to double every year."[16]

By the fall of 1863, mines farther north, in the Kootenay region, had also caught the eye of Walla Walla merchants and packers. According to Ned James of the local office of Wells, Fargo & Co., as reported by the *Statesman* in October 1863, "gold from the Kootenai mines has been brought down here in considerable quantities." The new market brought new-year predictions of "immense travel this coming season," the *Statesman* enthused in January 1864. "This will open up a new avenue of trade and, added to the long lines of pack trains and wagons that must necessarily be employed in some of the old mines, we may with certainty expect to see an equally busy one in transit for the new El Dorado, laden with the products of our valley and the goods of our merchants." With the usual contingent of "a large number of miners [who] have returned to Walla Walla" in the fall, and plans under way for the new year, the *Statesman* editor reflected in the same article that "Those who have witnessed the changes that have taken place here within the three years past—who have seen the thousands of people passing to and from the mines—will know how to appreciate this and prepare for the coming harvest."[17]

For several years after it was built, the Mullan Road was improved here and there to accommodate the additional traffic—a new bridge across the Touchet River and improved ferryboats for pack trains bound for the various Northwest mines. In May 1864, William McWhirk announced a new "swing" ferry on the Snake River, near the mouth of the Palouse, claiming he had spared no pains or expense "to make this ferry a safe, expeditious and convenient crossing for animals or vehicles of any description." McWhirk asserted that his ferry could handle a hundred pack mules per hour, and he offered "liberal" discounts for pack trains. Moreover, he offered a $1,000 reward "to any man who can find as good or direct a Wagon Road to the Kootenai mines or Colville as the route crossing the Snake River at McWhirk's ferry . . . the most direct and only really practicable WAGON ROAD from Walla Walla to the Kootenai, Columbia, or Beaver Head mines or to Ft. Benton or Colville. Much the shortest to Kootenai; plenty of grass and good water."[18] Two to five large pack trains crossed the Snake on McWhirk's ferry every day, a *Statesman* correspondent reported a few months after it opened. A second ferry a couple of miles upstream "on the Mullan Road as laid out in 1859" was kept busy with the additional traffic, too, he added.[19]

Dozens of pack trains loaded with merchandise left Walla Walla throughout the spring and summer of 1864 for the "upper country." In early April, a hundred pack mules reached McWhirk's ferry and two hundred left Walla Walla for the Kootenay mines. During a single week in May, eleven pack trains of forty animals each loaded up on the docks at Wallula, and in August, two parties from the Kootenay arrived in Walla Walla.[20]

The takings were impressive. One Walla Walla packer, a Mr. Gray, brought home "40 oz. of dust" in July of 1864. Later that month, the Walla Walla to Kootenay expressman, Mose, returned with several "big lumps," the largest of which was assayed at $78. In September another packer, a Mr. Day, came back from Wild Horse Creek, in the Kootenay region, with gold "which assayed 900 fine, worth $18.60 to $46 an ounce."[21]

At the height of the boom, freighter Dan Drumheller and his partner, Charley Allenburg, left Walla Walla for the Kootenay with a pack train, setting up a store there to sell supplies to the miners. Walla Walla merchants profited from supplying the other merchants, too, particularly those in Colville, some 250 miles to the north. On three consecutive days in the fall of 1864, "pack trains, freight teams, expressmen &c." came to Walla Walla from the Spokane country and the Colville Valley, according to the *Statesman*. "The six large mule teams belonging to Messers. Ferguson & Co., merchants in the Colville valley, arrived from the Touchet Bridge en route to

Wallula [a port on the Columbia River about thirty miles west of Walla Walla] to receive supplies and goods for the market at Colville. The train has made six trips during the past summer, furnishing the Colville merchants with some 50,000 pounds of goods."[22]

In February 1865, the editor of the *Statesman* predicted a "general stampede" of placer-mining hopefuls to the Kootenay diggings from Portland and beyond, with others moving north from the Boise Basin, bringing "promise of large amounts of business for teamsters in the upper country."[23] He was right. Beginning in March 1865, a Mr. Nye and a Mr. Johnson, both of Walla Walla, started an outfitting service, offering a riding horse, equipment, and supplies for trips to Fisherville, B.C., in the Kootenay mine district, for a fare of $80. Walla Walla merchants offered "every article of mining supplies" at their stores, advertising "everything that the most fastidious could desire for an outfit."[24]

This illustration, titled "The Romantic Calzada or 'Shod' Mule Pack," portrays packers in camp, loading their mules for another long day on the trail. –From *McClure's Magazine*, October 1905; negative provided by Lawrence L. Dodd, Archivist Emeritus, Whitman College

In June 1865, thirty-five pack trains, carrying 151,000 pounds of flour, departed Walla Walla, bound for the Kootenay mines. A Wild Horse Creek correspondent reported in the fall of 1866 that "we have used none but Walla Walla flour here this season. I think that next season Walla Walla will have to supply about 1,000 to 1,500 people with flour." Walla Walla merchants made an extra effort to attract packers with low prices. At least one store, Elias & Bros., promised to sell its wares at "San Francisco cost prices."[25]

The rush did not subside. Walla Walla packer George Dacres wrote favorably in 1867 of what the *Statesman* editor called the "new and rich mines in the Kootenai," the news of which created "intense excitement" in town and inspired "a large number of persons" to leave "this Valley for the land of promise." A regular Kootenay correspondent to the *Statesman*, known as "Otter," wrote in February 1868 that "I consider you part and parcel of us, having the claim of adding to your valley and commerce one hundred thousand dollars for produce &c." When the season "opened up" in June, the *Statesman* reported, "a number of pack trains started out from Walla Walla . . . for Kootenai country. Their cargoes are comprised principally of the three great necessaries of life in a mining camp–flour, bacon and whiskey. The trains numbered 180 packs in all."[26]

Meanwhile in the 1860s, rich strikes along the Little and Big Blackfoot Rivers of western Montana meant "a continuous stream of miners leading pack horses or burros or driving an occasional oxcart . . . over the Mullan Road from California, Oregon, Washington, and Idaho."[27] Walla Walla merchants and packers had done business in Montana for several years. In 1860, Frank Worden, later considered to be one of the founders of the city of Missoula, and his partner, C. P. Higgins, left Walla Walla with seventy-five horses, $7,000 worth of goods, $4,000 in livestock, and a large safe to set up a store at Hell Gate (later Missoula), Montana, 450 miles east of Walla Walla. They would return for more supplies later.[28]

Pioneer Montana settler Granville Stuart, in his journal, mentioned making a large purchase from packers who'd arrived at Alder Gulch from Walla Walla:

> May 5, 1863. Two pack trains arrived from Walla Walla. Bought from them the following:

52 lbs. tobacco @ $4.00	$208.00
168 lbs. bacon@ $ .40	$ 67.20
241 lbs. sugar @ $ .60	$144.60
17.5 lbs. soap @ $ .50	$ 8.75
	$428.55[29]

With the Montana strikes, traffic on the Mullan Road soon hit flood stage. "Nearly all of the vast crowd of gold seekers came over the Coeur d'Alene mountains by way of the military road constructed by Lieut. John Mullan," noted Frank H. Woody in his history of early western Montana. "During the whole of the summer and the fall of 1865, the road was literally lined with men and animals on their way to the new El Dorado."[30] Supplies were freighted to the mines from Walla Walla by hundreds and even thousands of pack mules. On September 1, 1865, the *Statesman* reported the departure of five hundred mules on the Mullan Road for the mines, and on September 8, the paper predicted that "the general line of the Mullan Road will be taken by those who contemplate visiting the Blackfoot and Prickly Pear mines." The following March, a veteran packer quoted in the *Statesman* estimated that "seventy pack trains, numbering from 15 to 90 animals each, left Walla Walla last fall for Montana." In the same issue, Cronly & Co. announced that they'd purchased and upgraded a crucial link for the Montana trade—McWhirk's Ferry. A few years earlier, McWhirk had said his ferry could handle one hundred pack animals per hour; now Cronly & Co promised it could handle a thousand animals an hour.[31]

The wealth being extracted in the Montana placer mines attracted interest from merchants spread halfway across the country, from St. Louis and Chicago to San Francisco and Portland, Oregon. By the mid-1860s, more than thirty light-draft steamboats laden with merchandise from St. Louis had arrived in Fort Benton, up from six a few years earlier.[32] In an 1866 *Statesman* article, a San Francisco merchant estimated that "over 100 pack trains, averaging 50 animals each, with three hundred pounds to the animal, making an aggregate of 750 tons" were sent from the Columbia River docks to Montana. The cost of transportation was estimated to have been $240,000, and the value of the goods, about $1.2 million. Fifty would-be prospectors arrived in Walla Walla on a single night "for the purpose of buying animals and generally procuring outfits."[33]

In the spring of 1866, the *Statesman* editor speculated that "in the history of mining excitement, we doubt whether there has ever been a rush equal to that now going to Montana. From every point of the compass, they drift by the hundreds and thousands all bound for [the Blackfoot mines]. Men of enterprise have placed passenger trains on the route from Walla Walla to Blackfoot with these trains going out daily with full passenger lists." One of those enterprising men came by the next week and reported he'd brought from the Blackfoot "quite a liberal installment of dust."[34]

Not every strike would "pan out." In June 1865, acting upon the reports of one Charles Wilson, there was "great excitement on account of the Coeur d'Alene mines." Wagons loaded with flour, bacon, and beans from Walla Walla stores were hauled by cayuse horses and mules along the Mullan Road. When Wilson was unable to relocate his find, irate miners began "to look rather more after that personage than for the new diggings." He was "only saved from being hung by the intercession of the priest at the [Coeur d'Alene] Mission."[35]

In spite of the uncertainty, most people remained inexorably optimistic. Teamsters and packers in Walla Walla reportedly had smiles on their faces as they made arrangements for freighting to the Idaho and Montana mining regions each spring. Spring was the season for travelers to purchase work animals and "lay in" supplies. The *Statesman* editor wrote in March 1866 that "the spring trade can scarcely be said to have commenced, and yet there is an evident revival of all branches of business. What with receiving new stock and attending to the demands of customers our merchants are in a constant stretch." In the fall came the second busy season of the year, as miners and packers came to town for the winter and the city presented "a lively and animated appearance." It was the beginning of what local merchants and saloon keepers called their "winter campaign." Main Street was "thronged each day by strangers, mostly from the different mining camps, and as a consequence, quite a 'run' has been made on hotel proprietors for board and lodging."[36]

The Washington territorial legislature estimated in 1866 that the young town at the foot of the Blue Mountains had six flour mills, six sawmills, two planing mills, two distilleries, one foundry, and fifty-two "thrashing, heading, and reaping machines." With a population of nearly 1,400 within its borders, and more than 5,000 in the surrounding county, Walla Walla was by 1870 the largest city in Washington Territory.[37]

By 1866, the Mullan Road had deteriorated into a rough pack trail, particularly through the Bitterroots. In March of that year, "farmers, merchants, mechanics and capitalists" met in Walla Walla to urge Congressional action to fix the road. The Montana legislature also petitioned Congress, and the governor of Idaho joined in as well. Also in March, public meetings were held in San Francisco, whose merchants supplied goods for stores in Walla Walla. That same week, the Washington Territorial Legislature petitioned Congress for $100,000 to open a wagon road "of the greatest, most vital importance to the people of Washington, Idaho, and that portion of Montana lying west of the Rocky Mountains." The legislators, using pack-train information from "reliable sources," catalogued traffic from Walla Walla to western Montana from January through mid-November 1866: 1,500 horses purchased by miners at Walla Walla horse markets, 5,000 cattle driven from the valley across the mountains, 6,000 mules loaded with freight, 20,000 travelers, and $1 million in gold hauled over the road. In 1867 Walla Walla resident Philip Ritz spent six months in Washington, D.C., to help plead the cause.[38]

Congress, in the aftermath of a devastating Civil War, Lincoln's assassination, and the trauma of Reconstruction, had other, more pressing problems. "It is greatly to be regretted," the editor of the *Statesman* wrote, "that Congress could not drop the 'negro' for a few moments and attend to a question on which the people of the Pacific Slope are so greatly interested." Thus the pleas to improve the road fell on deaf ears: the Mullan Road would remain a trail for mule packers.[39]

With the Mullan Road left in disrepair, higher value items could be shipped less expensively to the mines via river routes from St. Louis to Fort Benton. The *Statesman* described the Montana trade as anything "of the character that can be shipped and sold in that market at lower figures than can be furnished from the Eastern states–mostly now flour, bacon, and whiskey." There was also competition from merchants and "capitalists" of Portland, who had up until now been aligned with Walla Walla merchants and packers. The mining trade had an impact in what Walla Walla residents called "webfoot country" (Oregon)–nearly $1.4 million in gold was assayed in Portland during three months in the fall of 1864. The Oregon Steam Navigation Company (OSN) "fitted out" three riverboats to move cattle upriver and serve the mining trade–as many as 46,000 head in a single year were shipped upriver from the Willamette Valley to be grazed on the interior bunchgrass prairies. More than 60,000 tons of freight was transported to the docks at Wallula over a three-year period, from 1861 to 1864.[40]

Simeon G. Reed of the OSN took note that more than a thousand pack animals had left Walla Walla in a single week, bound for the mines of Montana by way of the Mullan Road. Alert to opportunities, the firm transported a steamboat, piece by piece, on the backs of mules from Walla Walla to Lake Pend Oreille and reassembled it there, offering pack trains passage by water–an alternative that Walla Walla packers sometimes used en route to the Kootenay or around some of the worst of the Mullan Road mountain terrain. Walla Walla–based packer George Dacres was among those who boarded the OSN boat, the *Mary Moody*, on his way to the Kootenay mines.[41]

More of a threat to Walla Walla merchants was the attempt to bypass Walla Walla and Wallula in favor of docks the OSN had established at White Bluffs, farther north on the Columbia. The firm offered preferential shipping rates there and promoted a path through the desert sagebrush and sand from their docks to the lake. The editor of the *Statesman* railed against the company's "heartless" attempt to "induce the miner to travel out of his way." It was "the greatest humbug," which the "Oregon Steam Navigation Co. has begotten for purposes vile as the prostitution of its own offspring," the paper spouted with characteristic hyperbole. But the editor had been correct in his earlier prediction that "the cream of the mining trade"—flour and bacon sales—would remain with Walla Walla, "notwithstanding every sand bar on the Columbia be dignified with the title of town-site." The OSN also promoted a potentially shorter path, the "Wastuckna Wagon Road," which ran from Wallula to an island on the Snake River and past Kahlotus Lake.[42]

The merchants of Walla Walla needn't have worried. Most packers preferred that town—"a considerable city where they could have a 'good time,'" according to historian Otis Freeman. "Most of the large packing outfits had their headquarters at Walla Walla," mule packer James Watt remembered, "but they packed out of Wallula, Umatilla, Walla Walla, or Lewiston, wherever they could get the freight." John Mullan, comparing Walla Walla to the sandy desert of Wallula (White Bluffs had similar terrain), contended that Walla Walla would remain the chief emporium "so long as [men] shall desire pleasant homes, where the eye is as [desirous] of drinking draughts of pleasure and beauty as the pocket is of accumulating wealth, where mills, farms, gardens, and pleasant enclosures can be had, where the product of the fields are garnered with a short transportation to a ready market." In April 1866, the *Statesman* carried an advertisement signed by twenty-five teamsters and freighters living in the Walla Walla Valley, who, "being engaged in the transportation of goods from the Columbia River inland" wanted it known that they preferred "freighting from Wallula to Blackfoot, Boise, Colville than from any other point on the Columbia River." Attempts to reach the Blackfoot River via Lolo Pass were similarly scorned. "On the Mullan Road a man would be laughed at if he should be found traveling on foot and leading a stout American horse," the *Statesman* asserted.[43]

When the muddy Missouri was low, steamboats became stranded, creating "fine prospects" for Walla Walla mule packers. With the boats beached in the spring of 1868, for instance, the *Statesman* reported that everything on the Mullan Road was quickly readied and put in "an active state of preparedness for the upcoming rush. . . . The proprietors of the Spokane Upper Bridge—'the Pioneer'—have worked hard and faithful and spent considerable money to merit the continuance of public patronage." By March, the departure of pack trains from Walla Walla was "an almost hourly occurrence. On Wednesday we noticed the trains of Dan Hayes, James McAuliff, and Jim Buckley start out almost simultaneously."[44]

With trails smoothed out across the prairies by heavy traffic, some of the hauling across the lowland was done by freight wagons, saving the aparejo saddles, the caronie blankets, and the two-hundred-pound double packs for the rougher trails beyond. Arrangements were made with merchants and packers "for having their goods teamed up to the Spokane" and "packed from there to the mines either by [steamboat on] Lake Pend Oreille or [via] Coeur d'Alene pass." P. M. Lynch's teams, for instance, "loaded up this week [September 1869] with 22,000 pounds of flour and merchandise for [Montana merchant] Charles Buck, to be freighted to the Spokane Bridge and then packed to Blackfoot."[45]

A poor wheat crop in western Montana in 1869 meant food had to be brought in from elsewhere, once again creating "full employment to the packers, who are scarce allowed to return from one trip before they are summoned to load up for another." The tide was turning in favor of "the producers, manufacturers, and tradesmen of our valley." Parties bound for the mines could find "horses and everything else required to complete the miner's outfit in full supply at Walla Walla and at very low prices." In September 1869, the *Deer Lodge Independent* reported the arrival of 30,000 pounds of Walla Walla flour by pack train and enthusiastically added, "More is coming!" Ten or twelve pack trains loaded up in less than a week, the paper noted. "Our packers are . . . constantly kept busy and our merchants actively employed."[46]

By the late summer of 1869, the *Statesman* related that "not a day passed" that did not "witness the departure of one or

"Ready to Be Packed: Rocky Mountain Mules" –From *Scribner's Monthly,* April 1880; negative provided by Lawrence L. Dodd, Archivist Emeritus, Whitman College

more trains to the mountains. As a rule these trains are mainly laden with flour and provisions–the products of the valley, thus bringing us a constant stream of treasure." Weeks later, the paper noted that freighting activity had not slowed down: "The business of packing and teaming continues active and our stores generally are crowded with customers." In April 1870, two Walla Walla stores–Adams Bros. and Schwabacher Bros.–reported having pack trains leaving every day.[47]

Detailed weekly reports from two other firms–I. T. Reese and Jordan & Co.–listed the names of the packers, the size of mule teams, the destinations, and the cargo in the "City Trade" column of the *Statesman* from July 30 through December 4, 1869. Traffic was substantial at the two stores, as this list for the week of July 30 attests:[48]

Packer's Name	Number of Mules	Destination	Cargo*
J. F. Galbraith	44	Kootenai	F, W, GM
P. Cobabley	70	Blackfoot	F, W, GM
Pardee & Sprague	39	Blackfoot	F, W, GM
C. P. Rutherford	20	Susanville, OR	F, W, GM
James McGinney	21	Olive Creek, OR	F, W, GM
F. Freer & Co.	60	Bear Gulch	F, W, GM
P. Kent	55	Helena	F, W, GM
Dan Hayes	48	Bear Gulch	F, W, GM
Leeman	38	Helena	F, W, GM
Davey Bros	40	Idaho City	F, W, GM
Louis Eskone	16	Jefferson City, MT	F, W, GM

*(Flour, Whiskey, General Merchandise)

The cargoes reported on July 30 were "comprised in part of 700 barrels of flour and 400 gallons of whiskey," carried in pouches on the backs of trusty mules.[49] Four of the packers—Perry Kent, Frank Freer, Dan Hayes, and a man named Leeman—were back in town just over a month later, loading up once more with supplies for the mines. Other packers and mules were also leaving town, most headed for Montana mines: 8 packers with 301 mules were recorded in the August 13th column, and 15 packers with 667 mules were listed on September 3rd.[50] During the fall and winter, a total of 49 mule packers and 1,993 mules loaded up at Walla Walla stores, as did 18 horse and ox teams.[51] On September 3, 1869, the *Statesman* reported, "Trains from all parts of the country arrive here daily and leave with the least possible delay, going out loaded mainly with domestic produce for the different mining camps. At no time during the past five years has business in Walla Walla been as healthy as present."[52]

Mule packers celebrated their ties to the community with a special event one winter evening. The "Packers' Social Party," which the *Statesman* called the "Ball of the Season," was held on January 12, 1869, at the Bankers Exchange Hall. "Decidedly one of the most successful as well as pleasant affairs of its kind it has ever been our good fortune to attend," the newspaper gushed. "The packers, numbering 60, were all dressed in blue shirts and black pants, and certainly were as fine a looking body of hearty, robust men, as could be gathered out of the same number of men anywhere. . . . [T]he Hall was densely packed with the elite and fashion [*sic*] of our valley. . . . At one time there [were] fully five hundred persons in the Hall." Upstairs, the organizers decorated the rooms to highlight the mule-packing trade—the place was "covered with evergreens, and made as nearly as possible to resemble a 'packers camp.'" The supper was "'pork and beans,' got up in packer's style, and greatly relished." Dessert was much fancier; "One of the cakes was surmounted by an apparahoe . . . another was surmounted by a pack mule, with the packs on its back, and all ready for the trip to the mines." Overall, "mirth and good humor prevailed." The affair's success, the article noted, was "eminently creditable to the liberality and public spirit of the packers who thus devoted their time and means to the entertainment of their fellow citizens."[53]

The "[illegible]" of the valley had [illegible] new brick buildings appeared along Main Street in 1869 and were open for business. Among them was the new home of the Brechtel Bakery, which advertised for "merchants, packers and others, in want of *crackers* to ship to the mines." Another was an assay office established by Dr. J. N. Day, a former druggist, in partnership with A. H. Reynolds. Day, it was said, contributed his stake in the form of a forty-pound bar of gold he'd molded into one large brick. The Day and Reynolds company would eventually become the First National Bank of Walla Walla. Merchants Dorsey S. Baker and John F. Boyer, whose Walla Walla store had served miners and packers from the earliest gold-rush days, sold that business in 1869 and used the proceeds to set up Baker Boyer Bank, a venerable firm that remains today.[54] Their successor on the main floor was the mercantile firm Paine Brothers & Moore. Frank Paine, one of the proprietors, had made a handsome profit hauling supplies to the Blackfoot mines. Another partner, Miles C. Moore, a former Blackfoot merchant, later served as territorial governor.

The growing prosperity of Walla Walla owed much to the mines of Montana, four hundred miles away and connected only by the rugged pack trail first opened by John Mullan. Walla Walla Valley farmers found markets for their produce in the distant placer mines along the Mullan Road. The Bergevin brothers were among several Walla Walla farmers who mule-packed apples to the mines each fall. Another settler, Philip Ritz, hauled "fruit trees and machines" by mule team from Walla Walla to establish his flour mill and home in the Hell Gate Valley.[55]

Although mines would begin to play out in the 1870s, the change was not immediate, and the economic growth in Walla Walla continued. In a ten-day period in October 1870, sixty two freight wagons, with a capacity of 3,000 pounds of cargo each, were loaded at Wallula, in addition to "a great number of pack trains." When miners grew weary of holding their pans in chilly water as fall advanced, they'd be seen "flocking to Walla Walla from all mining sections" with "dust pouring in lively."[56]

The "floating population" of Walla Walla, ever alert for word of a new El Dorado, again caught the "mining fever" in early 1870, upon learning of the new diggings at Cedar Creek, in today's Mineral County, Montana, near today's Frenchtown. By April, the papers in Walla Walla were reporting that "every day witnesses the departure of parties. . . . They start with all conceivable outfits–some with teams of oxen, horses, and mules; some go on horseback, with pack animals; some go on foot with a cayuse to carry blankets and grub, others go on foot and pack their blankets on their backs."[57] News from Cedar Creek–perhaps "the best strike yet"–included reports that the camps along those canyons and creek beds could receive "goods, provisions, & c." cheaper from Walla Walla than from the nearer Helena, Montana. With the Cedar Creek discoveries, packers were "constantly kept busy." They returned to the lower country in droves to load up, "all bring[ing] dust with them to pay for their cargoes."[58]

Newly established merchants in the Cedar Creek camps were "actively employed." C. F. Buck, who packed a 125-mule train into Cedar Creek district, brought in "a lot of whiskey," and with a partner set up a store in Frenchtown, where they sold apples and pears from the lower country and offered miners boots for $18, bacon for 75 cents, and beans for 50 cents. His teenage cousin, Charley Broughton, who had ridden the bell mare and cooked for the packers on the trip from Walla Walla, became the storekeeper. The mining camp boasted two thousand inhabitants, "quite a few of whom are from Walla Walla, Washington Territory." Dan Hayes and Pat McGraw "both well-known Walla Walla packers . . . opened stores and are doing well." Walla Walla was "well represented . . . in saloons, and in every other branch of business." The rush to Cedar Creek did not last long, however. In October of 1871, the *Helena Herald* reported that "A year ago everything was bustle and excitement, now everything wears the aspect of decay and dilapidation."[59]

By 1873 the Cedar Creek mines were "played out," but only after a total of $4 million in gold had been extracted by the placer miners. Most of the money ended up in the hands of saloon keepers (fifty were listed in the 1870 census) and merchants. The mines in the Kootenay started to play out at about the same time, but there was still substantial business to be done selling "flour, tea, sugar, tobacco, bacon, lard, gum boots and good liquor" to prospectors. Liquor could be picked up along the way to the mines–traders at each watering place had "a keg of 500-yard whiskey and a sack of beans which they are ready to dispense with at about ten times The Dalles prices."

Packer "Wild Goose Bill" Condit set up a a ferryboat and way station on the Upper Columbia, where he carried guns, ammunition, cooking utensils, knives, and axes, which he sold for what he called "a one percent margin," which, to Bill, meant buying an item for fifty cents and selling it for a dollar. Liquor, Bill told friends, was the only profitable item he carried–he diluted it to half strength with water, then doubled the price per gallon for good measure.[60]

To reach the Kootenay mines, and newer ones in the Peace River country to the north, the Walla Walla *Union* advised in 1871 that "all intelligent miners" knew to use "the route via the Mullan Road to Snake River and so across 'Spokane Jimmy's' bridge and so on to Colville," where ferryboats crossed the Columbia. The Peace River district might be a flash in the pan, said the *Union* in the spring of 1872–it was not "generally believed that there are any very rich mines . . . but as there are a great many people going there, packing will no doubt pay." Although Chinese miners were "only allowed to work the claims that have been abandoned by white men," miners returning from the Coeur d'Alene diggings in the winter of 1870 were "discouraged," complaining that "'Mongolians' have captured many of the mines."[61]

The Kootenay mines continued to produce, though, if at a more modest pace. Dispatches sent south to Walla Walla papers by correspondents calling themselves "Syntax," "Sluice Fork," and "Otter" told of "good clean gold dust" and "the same old thing–'ounce dust' and good freight." Assay offices counted $20,000 of gold in a month and reported "miners are daily flocking to Walla Walla from all mining sections." The Kootenay correspondents wrote on January 1, 1870: "To you of Walla Walla we must look for our flour, bacon, fruit &c., but we can return you gold dust worth $18 per ounce. Treat the boys kindly who visit you, and in every grey shirt who comes to your country recognize the bulwark that sustains your greatness and whose assistance has raised your valley to its present cash basis and general prosperity."[62]

Nevertheless, times were changing. In a report from Montana in July of 1871, the *Union* noted that "in anticipation of the railroad many have quit the mines and taken up farms." In October of that year, the *Statesman* carried a column titled "Passed Away": "[T]here is no doubt that the flush times of Montana have passed away. The decline in surface mining, the growing necessity for capital, the economy in conducting deeper operations and the gradual absorption of claims into the hands of companies, or large owners, necessitates a change in the methods of business."[63]

Though teamsters had "anxiously looked forward to the packing season" in 1875, Adam McNeilly returned home with news from the Cariboo gold mine in British Columbia that "the packing business is very dull there, as well as everywhere." Frank McMahon packed only two mule trains to the Kootenay that year, one which carried "a hundred pairs of blankets" to barter with the Indians.[64] By 1880, the mines were all but dead. In the fall of that year, the *Statesman* reported on "good diggings" with a sense of nostalgia: "We met Mr. L. W. Meyers, of Colville, on Thursday in Schwabacher's store. He was 'weighing out' in a style that reminded us of a good 'clean up' of early days."[65]

When one of the last Walla Walla packers, George Dacres, headed for the Kootenay in 1878 with thirty-five pack mules and "an assorted cargo," a reporter explained that "he has been doing business with that camp for the last twelve years, and has run fifty odd packs every year. There are several of our citizens who have brought home large sums of money from these mines in the last decade." Dacres was one of those citizens; having prospered at the packing trade, he later bought and operated the Dacres Hotel in Walla Walla.[66]

Despite the decline of the mining trade, as early as 1873 the editor of the Walla Walla *Union* anticipated promising days ahead for the region: "Never before, since the commencement of the settlement of this valley, have the prospects of our whole people looked so bright. In early days there were 'flush' times but these were of short duration, they were created by the discovery of gold in the interior and lasted just as long as we held the key of their trade and travel." It was becoming clear that farming and ranching, not "dust" from mountain ravines, would provide lasting prosperity across the lowland prairies. Washington's territorial governor, Isaac Stevens, had once predicted that "the great body of the country" stretching eastward from the Cascades–the twenty-two million–acre Columbia Plateau–"promises, in its cattle, its horses, and above all, its wool, to open up a vast field to American enterprise."[67]

Miners had provided the early markets for livestock–as the *Statesman* put it in 1868, "the mining districts beyond depend on us [in the Walla Walla Valley] almost exclusively for their supply of fat cattle." Packer William McEnery "brought from southern Oregon 300 head of cattle which he intends to 'winter' at his place in the upper Walla Walla Valley and then plans to drive to the Blackfoot mines." Sheep, too, were wintered on the prairie–in the spring of 1866, the *Statesman* reported that "ten thousand head of sheep have already been driven from the Walla Walla Valley this spring and are now on the way to Montana." Five years later the *New Northwest* of Deer Lodge, Montana, noted the sale of 2,500 sheep by Walla Walla sheepman William Harkness, "which he drove to Montana last fall. . . . Mr. Harkness will soon return to Walla Walla to spend the winter."[68]

Some innovative livestock ranchers met miners' seemingly insatiable demand for bacon by moving pigs to the camps on foot. "Large numbers of hogs" were driven out of the Willamette Valley "for Boise, Kootenai, and Blackfoot," enough to cause a bacon shortage in Walla Walla. Moving those unruly animals was a considerable feat, as anyone who ever showed pigs at a county fair could attest. The flatboat at Lyons Ferry was transporting a load of hogs to the upper country when a thunderstorm struck and the hogs stampeded and jumped overboard. Fortunately, their owner got help from Chief Old Bones at the nearby Palus village, and all but one were saved.[69]

Livestock drovers faced various hazards along the way, not the least of which were rustlers. In 1871 William "Virginia Bill" Covington, whose name regularly appears in the big ledger books kept by Daniel Lyons at his Snake River ferry, had "his entire band of pack horses, numbering nineteen head, stolen from his place on the upper Columbia River." Horse and cattle rustlers had a hideout in Big Cove, in rugged scab-rock canyon country north of Winona, Washington. The outlaws set up shop around the time the Mullan Road was constructed. "Here trading took place with rustler gangs from Montana and Canada. Drives were made at night in constant exchange with bunches coming and going," according to the recollections of one Washington settler. Vigilantes in the western Montana mining camps took care of some of the criminals, lynching fifty-seven "cut-throats" and robbers in 1864 and 1865 alone.[70]

By the mid-1870s, as strikes became less frequent, the livestock market in the placer mines was in decline, but booming demand in the Great Plains made up for it. "Scarcely a day [passed] but that bands of stock arrive from the Willamette" on their way east, the Walla Walla *Union* noted in 1870. The *Statesman* echoed these reports: "Large bands of cattle . . . from the 'web-foot' country" were driven through town, the paper said. The herds were driven through Walla Walla via the Mullan Road to Montana, then to Dakota. In one week, four or five "purchasing companies" were reportedly in the market for 10,000 head. The livestock drives from the Pacific Slope and across the Columbia Plateau to Montana and the Great Plains were massive in scale–several hundred thousand cattle and fifteen million sheep, which is five times the size of the fabled cattle drives from Texas after the Civil War.[71]

Sheep herds, with a few skilled herders and "bell wethers"–castrated buck sheep–in the lead, trailed as far as Kansas and Nebraska, where they were fed in lots and marketed. Pioneer sheepman Arthur Cox remembered that the sheep trails east "usually followed . . . the old Mullan Road route into Montana Territory, or as close to it as possible."[72] Thus, in addition to helping to open a placer-mine market in Montana in the 1860s, the Mullan Road assisted in the development of a new enterprise–open-range livestock grazing–on the prairies east of the Rocky Mountain diggings.

With the wane in mining, some prospectors abandoned their pans, claimed land, and took up the plow, anticipating the eventual arrival of rail service and better access to distant markets. Packers, too, sometimes settled down to become farmers or ranchers. These mule skinners had learned a lot about the land, eating alkali dust in the lowlands in the summer, getting teams mired in mud after heavy rains in the spring. But they also saw lush valleys and hills and prairies covered with bunchgrass. Many of them liked the land enough to stay and put down deep roots.

Among the miners and freighters who had come to the Northwest diggings and decided to stick around were Damase Bergevin and several other former Hudson's Bay Company employees. Early settlers in Frenchtown, Washington Territory, just west of Walla Walla, they packed, grew crops, and became well established. Francis M. Lowden, Sr., Philip Ritz, William Reser, Milt and Newt Aldrich, and other miners and packers who had come north from the California diggings to the strikes of the Northwest also stayed in the valley and broke the land for farming, raising large crops of wheat and planting successful orchards.[73]

George Lucas, an Irish immigrant who had prospected in Idaho and Montana, set up a "half-way house" near the junction of the Mullan and Colville Roads, where he raised cattle and horses. Freighters Dan and Jesse Drumheller became ranchers and farmers in a big way–they and their descendants developed a 110,000-acre sheep ranch near Ephrata and a wheat farm just east of Walla Walla. A postcard from the early twentieth century showed five combines pulled by 165 horses at work on a steep hill on the Drumheller place. Freighter Charley Broughton and his descendants became wheat growers and cattlemen on a large ranch near Dayton, and Broughton's uncle C. F. Buck now relied on the Mullan Road to move his sheep to and from their summer range. Another freighter, "Uncle Jim" Kennedy–a rough, profane, hard-working man with a rollicking sense of humor–became a cattleman near Cow Creek. Kennedy continued to use the old pack trail he'd once traveled with mule trains, now with cattle.[74]

Livestock raisers on the Columbia Plateau found the land well suited to sheep and cattle. Pioneer cattleman Virgil Bennington remembered, "the bunchgrass was tall and just marvelous. It grew up to 18 or 20 inches high and would just wave like golden grain." Or, as another area rancher put it, "These hills reach from hell to heaven, with bunchgrass from top to bottom." North Yakima stockman A. J. Splawn remembered that "cattle were fat the year around. Cattle could be seen grazing the white sage in the coldest weather absolutely shaking with fat." Federal surveyor Thomas W. Symons conveyed an even more effusive optimism and enthusiasm for the land in his 1882 report to Congress: "'Bunch-grass' has become the synonym for things good, strong, rich and great: the bunch-grass country is the best and finest country on earth; bunch-grass cattle and horses are the sweetest, fleetest, and strongest in the world; and a bunch-grass man is the most superb being in the universe."[75]

Alfred and D. M. Holt, young settlers from the South, wrote back home from their fledgling ranch on Rebel Flat, four miles west of modern-day Colfax, Washington: "I think this country is perfection in the stock raising line. . . . You'd love this Palouse country. Our sheep, cattle, and horses will do splendid." In letters to their family back East, the Holt boys wryly suggested that there was still something missing from their prairie paradise, though. As one of them put it, "We would like some of your spare girls as they are rather scarce out here. Whenever you find a girl who is good looking, smart, agreeable, and will furnish money for me, let me know."[76]

The Mullan Road became a convenient access route to this ranch and farm country north of the Snake River. According to the *Union*, "Large droves of stock cattle" passed through Walla Walla in the spring of 1871 "for the country beyond Snake River. The country between the Snake and Spokane rivers seems now to be the favorite region with stock raisers. . . . Not less than five thousand cattle have been taken across Snake River at the different ferries since last fall." Sheep raisers used the road when moving to and from summer ranges. By 1882 a reporter claimed that "the country northeast of Crab Creek is well occupied by sheep. On a radius of fifteen to twenty miles there are upward of 30,000 head." Among the owners were "Messrs. McGilvery, Neace, and Drumheller," all former packers. Dan Lyons's ferry records for that year show that McGilvery crossed the Snake River ten times, each time with several teams of horses and a few hundred sheep; Neace crossed four times, escorting a total of 9,600 ovine passengers; and Drumheller crossed twice, herding 4,900 animals. In the late 1870s and early '80s, more than 30,000 sheep annually were herded onto that old cable ferry.[77]

Many ranchers moved cattle on the Mullan Road, herding them from ranges north of Snake River to Walla Walla. Some opened meat markets and butcher shops in Walla Walla to meet local demand. Rancher Chris Ennis set up a meat market in town, as did John "King Cattle" Dooley and his partner William Kirkman, who opened the Pioneer Market, claiming in advertisements that "being practical stockmen, largely engaged in the business, we have very superior facilities for carrying on a meat market."[78]

Packers, too, still used the Mullan Road and the Lyons Ferry, but they were now heading for farming settlements rather than mines. Regional packers "Wild Goose Bill" and "Virginia Bill" crossed back and forth on the Lyons Ferry many times each year with cayuse pack horses and sometimes with Chinese workers. In 1874 Wild Goose Bill crossed eleven times, each time with twenty to twenty-eight horses heading "up" or "down," and twice with five "Chinamen." He hired these Chinese laborers to hew out a road from the bedrock down to his ferryboat landing on the Upper Columbia. That same year, Virginia Bill crossed a dozen times, also heading "up" or "down," usually with ten to seventeen horses. One day in 1880, Virginia Bill crossed with six teams of horses and twelve "Chinamen."[79]

Walla Walla merchants outfitted the immigrants who came to the Columbia Plateau to take up farming. In 1870 there was "quite an emigration to the Union Flat country, which is beyond the Snake River on the right hand side of the Mullan Road." The next year, Walla Walla did "a very large trade with the Palouse country. A single store in town reports having loaded five teams for the Palouse within the last week." Almost every day the next spring, emigrants were seen with "one to half a dozen wagons loaded with farming implements, household goods and women and children bound either for Stevens or Whitman counties." As a result, the Mullan Road between Walla Walla and Coeur d'Alene enjoyed brisk traffic. In 1872 a local reporter called the stretch "a splendid wagon road . . . over which immense quantities of merchandise have been transported from below on heavy freight wagons for the last six or seven years." In 1880, as settlement moved farther north to the Big Bend and Spokane Rivers, the *Union* carried "Information for Immigrants," advising that "All persons who wish to go to Spokane Falls, Crab Creek, Four Lakes, or Fort Colville should leave Main Street near the Stine House and follow the Mullan Road." According to the paper, "The Road is an old one, a good one, and has plenty of guide boards, wood, water and grass on it."[80]

Those who took up farming began with machinery that was crude at best. Reflecting upon his career "as the youngest 12-year-old freighter in Washington Territory," Robert Cummins remembered that when his family settled down and started raising crops "we had to cut the grain with those old cradles. We'd haul it in and tramp it out and then wait for a windy day, because we didn't own a fanning mill." Mules often provided the motive power for the walking plows aptly known as "foot burners."[81]

For the miners and packers who had taken up farming, new markets were badly needed for their food products. While the mules that had packed pouches of flour to the mines had been reliable, the miners themselves were fickle, the markets unsteady and, in the long run, ephemeral. In search of a new market for Walla Walla wheat, Philip Ritz, a longtime advocate for the Mullan Road, shipped fifty barrels of flour to New York in 1868. Demand in Liverpool, England, 18,000 miles distant (a five-month ocean trip around Cape Horn), was booming. By 1890 nearly eighty ships per year carried Northwest wheat to Great Britain. A considerable portion of the grain shipped overseas came from the Willamette Valley, with an ever-increasing portion from the Columbia Plateau.[82]

As immigrants "settled up" the prairies north of the Snake River and ranchers claimed the ancient flood channels of Washington's scablands, rural communities were born. Merchants set up shop in these fledgling towns, and as they did, the settlers in the area made fewer and fewer trips south along portions of John Mullan's old trail to get supplies in Walla Walla. Still, the town endured. No longer a raucous carnival of miners, packers, and saloon keepers, Walla Walla evolved into a supply center for Washington's fertile farming country south of the Snake. It became a good place to raise a family, to get an education, and to be part of a community. Successful wheat farmers, sheepmen, and cattlemen–many of whom had first arrived in mule-packing days–built substantial homes in Walla Walla to retire there after long years of hard work.

In 1906 a writer for the *Pacific Monthly* shared with local readers a story that had already faded into the past: "Along the Mullan Road," the author reminisced, "were going the fortune seekers bound for the new gold fields of Washington and Idaho; the woodsman, the gambler, and other human flotsam and jetsam, for the Mullan Road stretched north and eastward" to the rough-and-tumble worlds of the placer mines.[83]

Into the late 1870s, the Mullan Road was still occasionally used by troops traveling between Forts Colville and Walla Walla. On September 3, 1878, for instance, Dan Lyons' ferry ledgers listed a crossing by the quartermaster with 235 mounted men, two mule wagons, forty-five mules, six horses, nineteen wagons, and six kitchen wagons. That same year, General William Tecumseh Sherman, "having heard a great deal said about the old 'Mullan Road' across the mountains," decided to inspect it. The road was, of course, in terrible shape in the mountains, passable by pack mules only. Sherman became convinced that, had the road been functional, the retreat of Chief Joseph and his Nez Perce band through the snow toward the Canadian border in the recently ended "war" could have been avoided. "Had the Mullan Road been open," he believed, "[General] Howard could have reached Missoula before the Nez Perces and would not have been forced to follow them across the mountains from Lapwai."[84]

In 1880, at Sherman's urging, Congress allocated $20,000 to improve the road. Troops cleared fallen trees and repaired the road from Fort Missoula to Fort Coeur d'Alene. In the six weeks that followed, observers reported thirty-nine wagons, several loaded with fruit from the "lower country," and 30,000 sheep crossing the rebuilt mountain pass. But these efforts were much too late to make a lasting difference. Progress on alternative transportation had marched on. The Northern Pacific Railway reached Spokane Falls in 1881, and the transcontinental line was completed to the Pacific two years later.[85]

When the 1880 census was tabulated, Walla Walla was still the largest city in Washington Territory and was still growing, with 3,588 residents—nearly double the last count. Since 1875 Walla Walla had had its own pioneer "strap iron" railroad, built by an innovative and determined local banker, Dorsey Baker. The line, which ran the thirty miles to the wharf at Wallula, was used to move more than 27,000 tons of wheat flour annually. It later became a branch of the Oregon Railway and Navigation line, its original thin strands of metal replaced with standard-issue rails. Some of the old "straps" were purchased by packer-turned-cattleman "Uncle Jim" Kennedy to build a sturdy metal-girdled corral, which is still in use.[86]

Although Walla Walla remained an important trade center into the 1900s, its prominence gradually faded. In its heyday, the town had served as the main regional transportation hub, but as the nineteenth century drew to a close, long-distance rail service became the order of the day. By 1890 upstart Spokane, with but 350 residents in 1880, had a population of 19,222, four times more than Walla Walla, owing to its transcontinental rail service and its many "feeder" lines across the Palouse and Big Bend regions.[87]

The impact of the rough old Mullan mule trail was more significant than many Americans realize—most notably, it had set the stage for the other transportation routes that followed. It was also one of the first, if not the first, scientifically engineered roads in the Pacific Northwest and across the entire trans-Mississippi West. It was recognized as such in 1978 when the American Society of Civil Engineers named it a National Historic Engineering Landmark. By then, Interstate 90 followed the Mullan route through the mountains via Fourth of July Pass—where trail builders had carved that date into a tree trunk—and Mullan Pass.[88]

John Mullan had recommended that transcontinental rail service be routed though the mountains via the Clark Fork River and Lake Pend Oreille—a route he'd once considered for his wagon road. Mule packers had sometimes gone that way, too, crossing the lake with their mule teams aboard the steamer *Mary Moody* as a shortcut. The Northern Pacific rail builders followed Mullan's route; today it is a heavily used corridor, nicknamed "the funnel" by railroaders, that carries Montana Rail Link and BNSF freight trains and Amtrak passenger service. The Union Pacific and the old Spokane, Portland, and Seattle rail lines followed close to the wagon road's ruts across many miles of scab rock and prairie. Much as Mullan and the road's other planners had hoped, the wagon road had laid the foundations for future transportation routes in the Northwest.

Father Joseph M. Cataldo, S.J., who came to the Northwest as a missionary in 1865 and knew the road firsthand, once wryly said, "The Mullan trail wasn't much of a road. It was a big job, well done, but we used to say, 'Captain Mullan made just enough of a trail so he could get back out of here.'"[89] It was a ragged path, laboriously constructed but never regularly maintained, yet it crossed the rugged mountain terrain that provides challenges even today for highway and railroad maintenance crews. Mullan's link between two great river systems and over the Rockies transformed the economy of the Northwest, profitting merchants in Portland, San Francisco, and other western cities; benefitting the cattle and sheep businesses up and down the Pacific coast and on the Great Plains; and, especially, expanding a once-tiny southeastern Washington village into a major commercial and social center. New settlers as well as former mule packers and miners began growing wheat in a region that had earlier been dismissed, a land that would become, with the Dakotas and Kansas, one of the three major wheat belts in the United States, producing dry-land yields unmatched anywhere in the nation. These pioneers and their descendants would grow other crops, too—apples, sweet onions, and recently, wine grapes. Not bad for what had once been a "straggling, disorderly looking place" where mule packers, bullwhackers, miners, and wayfarers caroused.

"In my enthusiasm," John Mullan once wrote of the road project that he invested seven years of his life scouting, planning, and building, "I saw the country thickly populated, thousands pouring over the borders to make homes in the far western land."[90] We can suppose that Mullan would be pleased to see this land now claimed, settled, extraordinarily productive, and still as beautiful as it was when his own associate, Gustavus Sohon, made his immortal sketches. This remarkable country of farm, range, and forest interspersed with cities and towns still bears many of the old ruts and skid marks of the pack trains and wagons that traversed it so long ago, marks on the land that inspire today's young westerners to learn more about their unique heritage.[91] This is John Mullan's legacy. It is a story that dates back a century and a half, when rough, hardy, and dedicated packers drove teams of twenty, fifty, even seventy mules carrying heavy payloads of whiskey, flour, bacon, and gum boots atop grass-stuffed saddles. It is to these remarkable pioneers that we owe the transformation of this remarkable land.

---

1 Ben Burgunder, quoted in the foreword by Lawrence Dodd of James W. Watt, *Journal of Mule Train Packing in Eastern Washington in the 1860s* (reprint, Fairfield, WA: Ye Galleon Press, 1971).

2 Watt, ibid., pp. 46–47.

3 Mexican packers "employed on account," ibid., p. 46; aparejos and caronies, ibid., p. 39; see also ibid., pp. 41–42.

4 Ibid., p. 41, 45.

5 A. J. Splawn, *Ka-mi-akin: The Last Hero of the Yakimas* (Portland, OR: Kilham Stationary & Printing Co., 1917), p. 371.

6 Watt, *Journal of Mule Train Packing*, pp. 38–47; John Mullan, *Miners and Travelers Guide to Oregon, Washington, Idaho, Montana, Wyoming and Colorado via the Missouri and Columbia Rivers* (1865; reprint, New York: Arno Press, 1973), p. 9; Carl Penner, interview, 1973, audiotape in "Horse Era Tape Series," Whitman College and Northwest Archives, Walla Walla, WA.

7 Whispering Thompson shouting, Ben Burgunder, "The Recollections of Ben Burgunder," edited by J. O. Oliphant, *Washington Historical Quarterly* [*WHQ*] 17, no. 3 (July 1926), p. 205; "penetrated into," Watt, *Journal of Mule Train Packing*, p. 47.

8 "Feelings of the most lively anticipation," [Walla Walla, WA] *Statesman,* July 19, 1862; 30,000 immigrants, [Portland] *Oregonian,* August 2, 1862; "much . . . may be poor," ibid., July 19, 1862.

9 "Horrible," [Oregon City] *Oregon Argus,* May 7, 1866, in Teakle Papers, vol. 31, Whitman College Archives, Walla Walla, WA, p. 245; "road is all torn in pieces," *Oregon Argus*, September 28, 1863, in Teakle Papers, vol. 31, pp. 164–69. For more about emigrant travel, see Alexander C. McGregor, "The Economic Impact of the Mullan Road on Walla Walla, 1860–1883," *Pacific Northwest Quarterly* [*PNQ*], July 1974, p. 119.

10 Richard Hobbs, *The Broughtons of Dayton: Family and Business in the Northwest Heartland* (Cambridge, MA: Winthrop Group, 2010), pp. 31–32.

11 William S. Greever, *Bonanza West: The Story of the Western Mining Rushes, 1848–1900* (Norman: University of Oklahoma Press, 1963), p. 215.

12 Nesmith Ankeny, *The West As I Knew It* (Lewiston, ID: R. G. Bailey Printing Co., 1953), p. 149, illustration of Main Street; ibid, pp. 55–57.

13 Randall Henry Hewitt, *Across the Plains and Over the Divide: A Mule Train Journey from East to West and Incidents Connected Therewith* (New York: Broadway Publishing Co., 1906), quoted in Teakle Papers, vol. 31, pp. 157–61; T. C. Elliott quoted in Louis C. Coleman and Leo Rieman, *Captain John Mullan: His Life, Building the Mullan Road, As It Is Today and Interesting Tales of Occurrences Along the Road* (Montreal, QC: Payette Radio Ltd., 1968), p. 72; Ankeny, *The West as I Knew It*, p. 57; "to-day the eyes," *Statesman,* December 13, 1862.

14 "Gold excitement . . . Jesuit fathers," *Statesman,* March 7, 1863; John Mullan quoted in Richard Scheuerman, "John Mullan and the Northern Overland Road: Through the Indian Country," *PNQ,* Fall 2010, p. 27.

15 "General stampede," *Statesman,* August 15, 1863; Hofstetter and "other gentlemen," ibid., September 12, 1863; Oppenheimer brothers, Burgunder, "Recollections of Ben Burgunder," pp. 190–21.

16 "'Wild Goose Bill' and His Ferry," *The Pacific Northwesterner* 14, no. 1 (Winter 1970), p. 1; see also *Statesman*, February 21, 1863; "unparalleled prosperity," *Statesman*, December 6, 1862; "double every year," ibid., February 21, 1863.

17 Ned James quote, *Statesman*, October 24, 1863; "immense travel," ibid., January 16, 1864; "those who have witnessed," ibid.

18 Touchet bridge, ibid., March 17, 1865; McWhirk quoted in ibid., May 20, 1864.

19 Ibid., September 16, 1864. The original Mullan Road, as developed in 1859, crossed the Snake River and followed the east side of the Palouse northward; it was subsequently moved to the west side of that river.

20 Pack mules at ferry in April, ibid., April 2, 1864; pack trains in May, ibid., May 20, 1864; two parties in August, ibid., August 26, 1864.

21 Mr. Gray, ibid., July 1, 1864; Mose, ibid., July 22, 1864; Mr. Day, ibid., September 16, 1864.

22 Drumheller and Allenberg, Daniel Drumheller, *"Uncle Dan" Drumheller Tells Thrills of Western Trails in 1854* (Spokane, WA: Inland American Printing Co., 1925); three consecutive days, "six large mule teams," *Statesman*, September 23, 1864.

23 "General stampede," "promise of large amounts," ibid., February 10, 1865.

24 Nye and Johnson, ibid., March 31, 1865; "every article," ibid., February 24, 1865; "most fastidious," ibid., June 9, 1865.

25 Thirty-five pack trains, ibid., June 30, 1865; "none but Walla Walla flour," ibid., October 26, 1866; "San Francisco cost prices," ibid., June 23, 1865.

26 "New and rich mines," ibid., August 30, 1867; "Otter" quote, ibid., February 21, 1868; "three great necessaries," ibid., June 26, 1868.

27 Helen Addison Howard, *Northwest Trailblazers* (Caldwell, ID: Caxton Printers, 1963), p. 161.

28 Frank H. Woody, "A Sketch of the Early History of Western Montana, Written in 1876 and 1877," in *Contributions to the Historical Society of Montana* 2 (1896), pp. 98–99; Worden and Higgins, Albert J. Partoll, "Frank L. Worden, Pioneer Merchant, 1830–1887," *PNQ* 40, no. 3 (July 1949), pp. 189–204, and "Worden Reached West by Way of Cape Horn," *Missoulian* (Missoula, MT), May 14, 1960.

29 Granville Stuart, *Forty Years on the Frontier as Seen in the Journals and Reminiscences of Granville Stuart, Gold-Miner, Trader, Merchant, Rancher and Politician,* ed. Paul C. Phillips (Cleveland: Arthur H. Clark, 1925), pp. 213–17, 239.

30 Woody, "A Sketch of the Early History," pp. 98–99.

31 Departure of 500 mules, *Statesman*, September 1, 1865; "general line," ibid., September 8, 1865; "seventy pack trains," Cronly & Co. ad, ibid., March 9, 1866.

32 "Steamboat Arrivals at Fort Benton, Montana, and Vicinity," in *Contributions to the Historical Society of Montana* 1 (Helena: Montana Historical Society, 1876), pp. 317–21.

33 "Over 100 pack trains," cost of transportation, *Statesman*, March 9, 1866; fifty would-be miners, ibid., March 2, 1866.

34 "In the history of mining excitement," *Statesman*, April 13, 1866; "liberal installment of dust," ibid., April 20, 1866.

35 "Great excitement," ibid., June 23, 1865; "to look rather more," ibid., June 30, 1865; "only saved from being hung," ibid., July 21, 1865.

36 Smiling faces, ibid., February 15, 1867; "the spring trade," ibid., March 9, 1866; "lively and animated appearance," ibid., September 9, 1864; "winter campaign," Jerry Bryant, "Daniel Lyons and His Ferry," *Bunchgrass Historian* 23, no. 2 (1996); "thronged each day," *Statesman*, September 9, 1864.

37 "Memorial Relative to the Mullan Road," Statutes of the Territory of Washington, 14th Sess., 1866–67 (Olympia: T. F. McElroy, 1868), p. 237; population statistics, Calvin F. Schmid and Stanton E. Schmid, *Growth of Cities and Towns, State of Washington* (Olympia: Washington State Planning and Community Affairs Agency, 1969), p. 40, 107.

38 "Farmers, merchants" meet in Walla Walla, *Statesman*, March 23, 1866; Montana legislature, ibid., June 22, 1866; governor of Idaho, ibid., December 29, 1865; San Francisco meeting, ibid., March 9, 1866; Washington legislature, "greatest, most vital importance," ibid., March 9, 1866; packing statistics and petition to Congress, "Memorial Relative to the Mullan Road," p. 234, 237; Ritz in Washington, *Statesman*, May 17, 1867.

39 "Greatly to be regretted," *Statesman*, May 17, 1867.

40 "Flour, bacon, and whiskey," ibid., April 9, 1870; gold assayed in Portland, Leslie M. Scott, "The Pioneer Stimulus of Gold," *Oregon Historical Quarterly* (*OHQ*) 18, no. 3 (September 1917), pp. 147–66; three OSN riverboats for cattle, 60,000 tons of freight, Earl K. Stewart, "Transporting Livestock by Boat Up the Columbia, [illegible]," *OHQ* [illegible], no. [illegible] (December [illegible]), pp. 251–59.

41 Reed noting a thousand pack animals, Simeon G. Reed to J. W. Ladd and D. F. Bradford, September 4, 1865, quoted in Dorothy Johansen and Frank Gill, "History of the Oregon Steam Navigation Company," *OHQ* 38, no. 2 (March 1937), p. 29; steamboat transported by mules, *Statesman,* November 16, 1866; George Dacres on boat, Watt, *Journal of a Mule Packer*, p. 31.

42 "Heartless" attempt, "greatest humbug," *Statesman*, April 6, 1866; "notwithstanding every sand bar," ibid., [illegible] ibid., February 22, 1867.

43 Otis W. Freeman, "Early Wagon Roads in the Inland Empire," *PNQ* 45, no. 4 (October 1954), pp. 125–30; Watt, *Journal of a Mule Packer*, p. 44; John Mullan, "From Walla Walla to San Francisco," quoted in Donald W. Meinig, "The Walla Walla Country: A Century of Man and the Land, 1805–1910," PhD diss., University of Washington, 1953, p. 142; teamsters' ad *Statesman*, April 6, 1866; "a man would be laughed at," ibid., October 13, 1865.

44 "Fine prospects," ibid., May 22, 1868; "active state of preparedness," ibid., January 31, 1868; "an almost hourly occurrence," ibid., March 27, 1868.

45 "Goods teamed up to the Spokane," ibid., November 1, 1867; Lynch's teams "loaded up this week," ibid., September 10, 1869.

46 "Full employment to the packers," ibid., August 20, 1869; "producers, manufacturers, and tradesmen," ibid., April 9, 1869; "horses and everything else," ibid., February 26, 1869; 30,000 pounds of flour, "constantly kept busy," *Deer Lodge Independent,* quoted in ibid., September 18, 1869.

47 "Not a day passed," *Statesman*, August 13, 1869; "the business of packing and teaming," ibid., September 25, 1869; Adams Bros. and Schwabacher Bros., ibid., April 2, 1870. The Schwabacher brothers–Louis, Abraham, and Sig–were originally San Francisco merchants who became prominent Northwest traders, first in Walla Walla, then later in the farm communities of Dayton and Colfax, and finally in Seattle, where they sold supplies to prospectors headed for the Klondike (see Hobbs, *The Broughtons of Dayton*, pp. 36–37).

48 "City Trade" column in *Statesman*, July 30, 1869.

49 "700 barrels of flour," ibid.

50 Four packers, 8 packers with 301 mules, ibid., September 3, 1869.

51 Totals of packers and mules are compiled from "City Trade" column in *Statesman,* July 30; August 6 and 13; September 3, 10, and 25; October 2, 9, and 23; November 13; and December 4, 1869.

52 "Trains from all parts," *Statesman,* September 3, 1869.

53 "Ball of the season," all quotes from *Statesman*, January 15, 1869; article provided by Lawrence L. Dodd, Archivist Emeritus at Whitman College and great-grandson of pioneer packer Francis M. Lowden, Sr.

54 New Walla Walla businesses, Robert A. Bennett, *Walla Walla: Portrait of a Western Town, 1804–1899* (Walla Walla, WA: Pioneer Press, 1980), pp. 71–99; new Brechtel Bakery, *Statesman*, December 18, 1869; Day and Reynolds, Bennett, *Walla Walla*, p. 84; Dorsey Baker and John Boyer, ibid., p. 73; and bakerboyer.com/about us/our founders.

55 For more on the Mullan Road connection between Walla Walla and Montana, see Merrill G. Burlingame, *The Montana Frontier* (Helena, MT: State Publishing Co., 1942), p. 131; Bergevin brothers' apples, *Statesman*, October 15, 1870; Ritz flour mill, ibid., October 26, 1866.

56 Sixty-two freight wagons, *Union*, October 18, 1870; "flocking to Walla Walla . . . dust pouring in lively," *Statesman*, September 24, 1870.

57 "Floating population," *Union*, February [illegible], 1870; "every day witnesses," ibid., April 9, 1870.

58 Goods cheaper from Walla Walla, ibid., May 21, 1870; "constantly kept busy," ibid., August 13, 1870; "all bring dust," ibid., July 23, 1870.

59 "Actively employed," ibid., August 13, 1870; C. F. Buck and Charley Broughton store, Hobbs, *The Broughtons of Dayton*, pp. 31–32; "quite a few . . . from Walla Walla," *Statesman,* March 26, 1870; Hayes and McGraw, "well represented," ibid., July 2, 1870; "decay and dilapidation," *Helena* [MT] *Herald,* October 28, 1871.

60 $4 million in gold, 50 saloons, Hobbs, *The Broughtons of Dayton*, p. 32; "flour, tea, sugar," *Statesman*, April 8, 1871; "a keg of 500-yard whiskey," Teakle Papers, vol. 31, p. 246; "Wild Goose Bill," John F. Weber and Eugene H. Wyborney, "'Wild Goose Bill' and His Ferry" *The Pacific Northwesterner* 14, nos. 1 and 2.

61 Route advice, *Union*, July 1, 1871; Peace River district not "generally believed," ibid., March 23, 1872; Chinese miners, *Statesman*, December 7, 1866; "Mongolians have captured," *Union*, November 26, 1870.

62 "Good clean gold dust," *Statesman*, February 3, 1872; "same old thing," ibid., September 28, 1972; "miners are daily flocking," ibid., September 24, 1870; "to you of Walla Walla," ibid., January 1, 1870.

63 "In anticipation of the railroad," *Union*, July 22, 1871; "Passed Away," ibid., October 21, 1871.

64 "Anxiously looked forward," [Walla Walla, WA] *Spirit of the West*, April 16, 1875; McNeilly news from the Cariboo, ibid., November 5, 1875; "a hundred pairs of blankets," ibid., August 27, 1875.

65 *Statesman*, November 6, 1880.

66 George Dacres "doing business," *Union*, April 20, 1878; Dacres Hotel, Watt, *Journal of Mule Train Packing*, p. 32. The daily logbooks at Lyons Ferry show that Dacres was a regular customer, year after year; he crossed once with forty packs in 1873, twice with fifty-two packs in 1874 and 1875, and so on (from Lyons Ferry Record Books, privately held by the descendants of the last ferry operators, Nae and Ruth Turner).

67 *Union,* June 8, 1872; Isaac Stevens quoted in James G. Swan, *The Northwest Coast, or Three Years' Residence in Washington Territory* (1857; reprint, Seattle: University of Washington Press, 1972), p. 398.

68 "Depend on us almost exclusively," *Statesman*, September 25, 1868; McEnery's 300 cattle, Todd Vernon Boyce, "A History of the Beef Cattle Industry in the Inland Empire," master's thesis, Washington State College, 1937, pp. 25–35; 10,000 head of sheep, *Statesman*, April 13, 1866; Harkness sheep, [Deer Lodge, MT] *New Northwest*, quoted in *Union*, December 16, 1871.

69 "Large numbers of hogs," *Statesman,* November 6, 1869; Lyons Ferry hog incident, W. F. Fletcher, *The Era of Chief Old Bones* (privately printed, n.p., 1994), p. 149; "Horses, too, were moved," *Union*, October 28, 1871.

70 Covington's "entire band of pack horses" stolen, *Union*, April 27, 1878; Big Cove rustlers, "History of Grazing," *Bunchgrass Historian* 17, no. 3 (Fall 1989); "trading took place," Sarah Mason, quoted in ibid.; vigilantes, Frederick Allen, *A Decent Orderly Lynching: The Montana Vigilantes* (Norman: University of Oklahoma Press, 2004).

71 "Scarcely a day," *Union*, November 12, 1870; "large bands of cattle," *Statesman*, August 17, 1872; cattle drives to Dakota, J. Orin Oliphant, "The Cattle Trade from the Far Northwest to Montana," *Agricultural History* 6, no. 2 (April 1932), pp. 659–83; "purchasing companies," *Lewiston* [ID] *Teller*, April 2, 1882.

72 Sheep herds, E. N. Wentworth, *America's Sheep Trails: Histories, Personalities* (Ames, IA: State College Press, 1948), p. 285, and Alexander C. McGregor, *Counting Sheep: From Open Range to Agribusiness on the Columbia Plateau* (Seattle: University of Washington Press, 1982), pp. 28–29; Arthur Cox quotation from "Interviews Obtained from Washington Pioneers Relative to the History of Grazing in the State of Washington," manuscript, Works Progress Administration, Historical Records Survey, 1941, in Manuscripts and Special Collections, Washington State University Libraries, Pullman.

73 A good source of information about Frenchtown is Clem Bergevin, "Canadians and French Aided in This Section," in Nesmith Ankeny, *The West as I Knew It,* pp. 65–71; "on pioneer farms," ibid., pp. 74–76. I am indebted to two personal friends, Claro Bergevin and Lawrence Dodd, both of whom are deeply interested in the history of Frenchtown, Lowden, and adjacent early settlements west of Walla Walla, for sharing with me their knowledge about the packers and farmers of that region.

74 George Lucas house, McGregor, *Counting Sheep*, p. 11; Drumheller brothers, ibid., p. 71, 76, and 175; Broughton ranch, Hobbs, *The Broughtons of Dayton*, pp. 9-18; "Uncle Jim" Kennedy with cattle, Ankeny, *The West as I Knew It*, p. 74. See also William Granger, *An Illustrated History of Klickitat, Yakima, and Kittitas Counties* (Chicago: Interstate Publishing Co., 1904), pp. 588, 612, 769.

75 Virgil Bennington, interview by Al McVay and Vance Orchard, 1973, audiotape in "Horse Era Tape Series," Whitman College and Northwest Archives, Walla Walla, WA; "these hills reach from hell to heaven," *[Pomeroy] East Washingtonian,* June 6, 1914; A. J. Splawn, "A Cattleman's Reminiscences," *The Ranch,* February 13, 1902; Thomas W. Symons, *The Symons Report on the Upper Columbia River and the Great Plain of the Columbia* (1882; reprint, Fairfield, WA: Ye Galleon Press, 1967), p. 111.

76 Holt letters, July 30, 1872; September 12, 1875; and January 21, 1877, from a collection of letters made available by Betty Jean Peatross of Glendale, California.

77 "Large droves of stock cattle," *Union*, April 22, 1871; "Crab Creek . . . well occupied by sheep," ibid., September 30, 1882; crossings information from Lyons Ferry Record Books. A summary of these crossings is in my undergraduate honors thesis of long ago, "The Economic Impact of the Mullan Road on Walla Walla, 1860–1883," Whitman College, Department of History, 1971, pp. 110–32.

78 Pioneer Market ad, *Statesman*, August 25, 1877. For a good description of Ennis and his career, see Bennett, *Walla Walla*, p. 166.

79 Lyons Ferry Record Books.

80 "Quite an emigration to the Union Flat," *Union*, October 18, 1870; "very large trade with the Palouse country," *Statesman*, November 18, 1871; "one to half a dozen wagons," *Union*, March 2, 1872; "a splendid wagon road," *Statesman*, March 2, 1872; "Information for Immigrants," "road is an old one," *Union,* February 28, 1880. The Stine House was an upscale fifty-room hotel, built in 1873 by pioneer blacksmith Fred Stine, at the intersection of the Mullan Road and a major stagecoach route. The Stine was where Walla Walla welcomed President Hayes to dinner in 1880 with a menu printed on white satin bordered in gold (Bennett, *Walla Walla*, p. 82, 83, 104–5).

81 Jerome Peltier, "Robert Franklin Cummins: 'The Youngest 12-Year-Old Freighter in Washington Territory,'" *The Pacific Northwesterner* 40 (1996), p. 2.

82 Ritz sends barrels of flour, "Phillip Ritz," *Walla Walla Union*, April 10, 1953. For information about grain sent overseas, see Donald W. Meinig, "Wheat Sacks out to Sea," *PNQ*, January 1954, pp. 13–16; John B. Watkins, "Wheat Exporting from the Pacific Northwest," *Washington Agricultural Experiment Station Bulletin* 201 (May 1926); and Donald W. Meinig, *The Great Columbia Plain*, 1968 (reprint, Seattle: University of Washington Press, 1995), p. 354; James M. Tattersall, "The Economic Development of the Pacific Northwest to 1920," Ph.D. diss., University of Washington, 1960, p. 60.

83 Ralph Sherman, "The Followers of the Bunch-Grass Hunter," *Pacific Monthly*, February 1906, p. 607.

84 Quartermaster crossings, Lyons Ferry Record Books; Gen. Sherman inspects road, *Statesman*, September 22, 1877; quote re Nez Perce war, General Sherman to Secretary of War George A. McCrary, August 3, 1879, in *Reports and Inspections Made in the Summer of 1877 by Generals P. H. Sheridan and W. T. Sherman of Country North of the Union Pacific Railroad* (Washington, DC: GPO, 1878), p. 49.

85 Senate allocation, *Union*, March 27, 1880; six weeks of crossings, "Report from Captain William H. Penrose in Regard to the Reopening of Mullan Wagon Road in Montana," S. Ex. Doc. 3, pt. 2, 46th Cong., 2d Sess. (1879).

86 Walla Walla in 1880, Schmid and Schmid, *Growth of Cities and Towns;* Dorsey Baker's "strap iron" railroad, Carlos A. Schwantes, *Railroad Signatures Across the Pacific Northwest* (Seattle: University of Washington Press, 1993), pp. 35–38. My father used to take me to see Kennedy's corral, near our ranch in Hooper, Washington, and told me of the colorful cattleman who built it, as well as of Dorsey Baker and his "rawhide" railroad.

87 Spokane in 1880 and 1890, Schmid and Schmid, *Growth of Cities and Towns.*

88 First scientifically engineered road, Harvey Erickson, "Mullan's 1862 Interstate Road," *Pacific Northwesterner* (Fall 1974), and Howard, *Northwest Trail Blazers,* p. 145.

89 Father Cataldo quoted in George W. Fuller, *History of the Pacific Northwest, with Special Emphasis on the Inland Empire*, 2d ed. (New York: Knopf, [illegible]), p. 317.

90 Mullan quote from 1888 interview, quoted in Oscar O. Winther, *The Old Oregon Country: A History of Frontier Trade, Transportation, and Travel* (Lincoln: University of Nebraska, 1969) p. 213.

91 There are always undiscovered traces of the old roads to be found. In 2012, with an intrepid trail scout, Mahlon Kriebel, and my McGregor Ranch colleague Dave Hannas, we found tracks down the steep hills of the remote Stewart's Canyon, where wagons had skidded down a steep slope, and the Indian trail turned pack trail that traversed the gentler slopes in modern-day Whitman County for several miles to the northeast. The later route of the road, farther west across the Palouse River, has visible ruts we'd known of for many years, but these were the first signs we'd seen of the original route Mullan used in 1859.

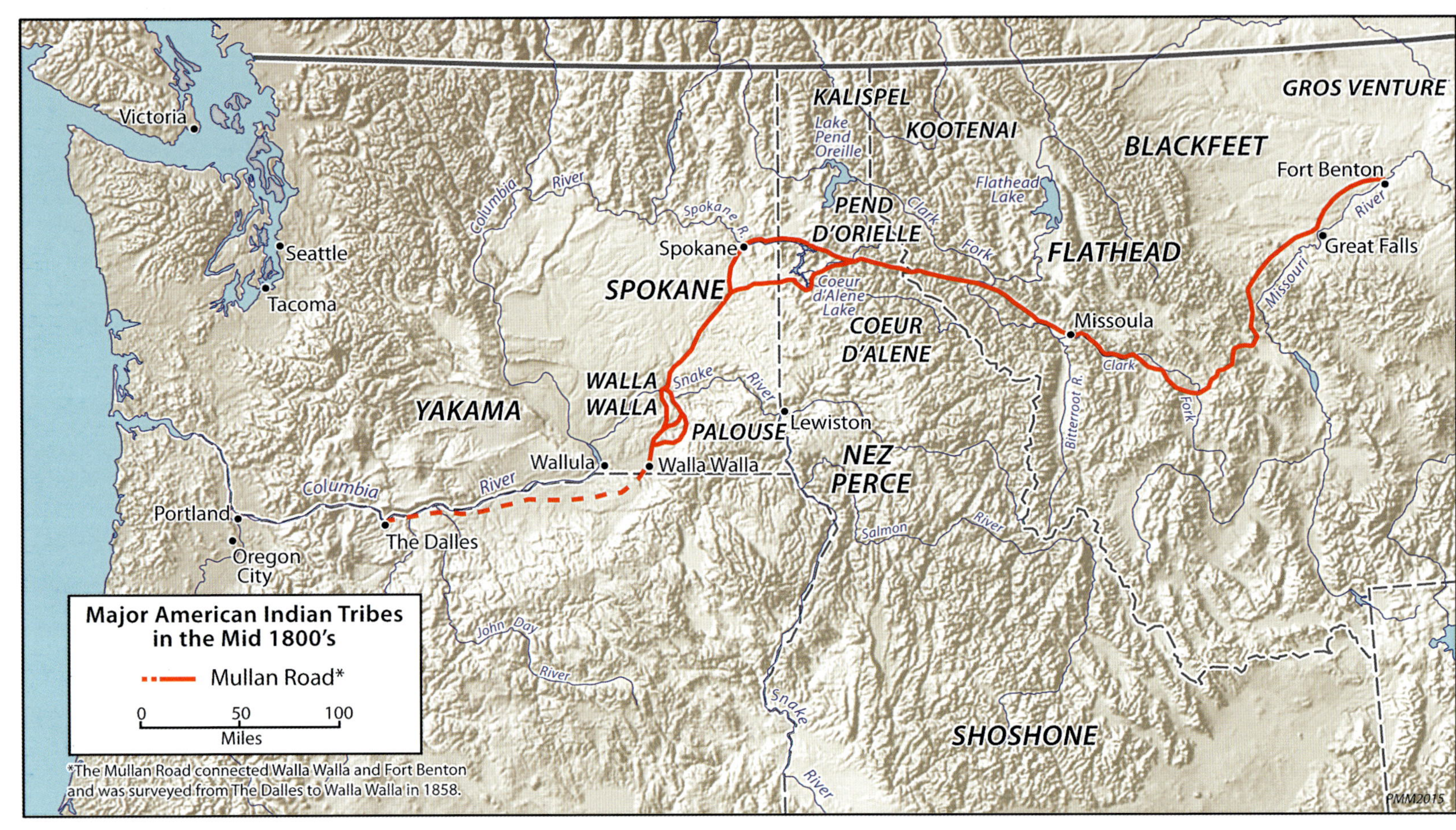

Figure 13-1. *Indian territories in the mid-1860s, showing the main tribes that Mullan and his men encountered during the building of the road* –Map by Philip Mobley

"Through the Indian Country"

John Mullan and Native American Encounters on the Northern Overland Trail

BY RICHARD D. SCHEUERMAN

For Captain Karl A. Scheuerman, and in memory of Missouri River–Fort Benton woodhawk Andrew Sunwold

Editor's note: This essay is based on a transcript of a talk given at the Mullan Road Sesquicentennial Conference, Fort Benton, Montana, May 21, 2010.

"Had the white man been to [the Indians] more just, fate would have proved less harsh."

—John Mullan[1]

"The earth was made to be filled, and made fruitful even to the maximum degree, and if in its subjugation, settlement, and civilization, the fish should disappear from its creeks and rivers, and game from its forests, they were incidents to civilization for which no tribes could claim compensation."

—"Observer," Portland *Oregonian*[2]

On a windswept blacktop road twenty miles west of my hometown of Endicott, Washington, is a rutted trail crossing that for years has transported me back in time. I often pause at this place, visible from the outskirts of tiny Benge, where the main highway to Winona meets the road to Ritzville. Even on some modern maps, the diagonal path seen there is marked "Old Mullan Trail," and while its namesake might be grateful for the twenty-first-century recognition, he would not have appreciated its being called a trail. John Mullan wrote in 1862, the year of the project's completion, that he has been sent to build and construct a *road.* While Mullan might have preferred us to designate his historic route a road, in that remote district of eastern Adams County, it is definitely a trail. I know because in my youth I hauled many a truck-load of grain along that rutted way as I drove south from the Kleweno ranch.

Heritage-minded neighbors like Roy and Karin Cline-smith told me that old-timer Dick Parrish, who was nearing the century mark in my youth, remembered when mule-team freighters still traversed the route in his boyhood there. So I visited with Dick in Benge about his Mullan Road/Trail recollections, and he showed me the evidence of its use in a massive castaway iron mule shoe and other relics he had collected in the vicinity. In 1989, when I returned to the area as principal of Endicott–St. John Middle School, I couldn't help but think that taking our students on annual wagon rides near Benge would do more to educate them about frontier experience than anything we could read in a book. So thanks to the remarkable organizing efforts of a small army of ranchers such as Ray Reich, Stan Reibold, Louis Gaiser, with their wagons and horse teams, as well as hospitable neighbors along the way, including the Clinesmiths and Dick and Paula Coon, we managed to raise more dust and make more noise for several miles each year there than likely had been seen since the days of Captain Mullan.

I was so taken with the experience that when our son, Karl, was in high school, I asked Pat Kleweno if he'd be willing to hire Karl to work the Benge grain harvest, as I had done for his father many years before. In this way, Karl came to know not only the trials and tribulations of "Mullan Road freighting," with its periodic summer encounters with rattlesnakes and thirst, but also the chance to know a majestic land as God fashioned it, a place unkindly named the "channeled scablands" by latter-day regional geographers. To John Strachan, who helped build this section of the road with Mullan in 1859, the land was inhabited by "lynxes, bears of the gray, brown, black, and yellow colors" and "wolves [perhaps coyotes], which prowl around the camping ground and make a fearful noise at night." But here also Strachan noted an abundance of "some of the most beautiful flowers that could possibly be imagined, every tint of color and beauty of form–the *Convolrusus, Frasira, Habenaria, . . .* and *Tropaeolium*."[3]

After several long days in the truck, Karl returned home and told us not about wild animals or wildflowers–rather, he said he had read every available *Sports Illustrated* from cover to cover and was looking for something to occupy his time in between visits by the combine. So I found the same tattered paperback I had read when I was working down there–James Michener's *Caravans*, a novel about Afghanistan. It occurred to me that there were some geographic similarities between our region and the setting in the book, but mostly I wanted to expand Karl's literary horizons and teach him about rural self-reliance. Today he makes us proud serving as a military intelligence officer in the Washington National Guard.

Our family also has an association–an older one–with Fort Benton and that region's Native Americans. My great-grandfather, Andrew (Anderson) Sunwold immigrated to the United States from Norway about 1881 and settled in North Dakota. He worked for a time as a sawyer for the steamboats that plied the Upper Missouri from Bismark to the western-most terminus of river traffic, Fort Benton, Montana, where the Mullan Road began (although this was after the heyday of both fort and road). Before the advent of coal steamers, a boat could consume twenty-five cords of hardwood a day, preferably red cedar or diamond willow, or up to thirty cords of cottonwood. Because the steamers couldn't carry that much wood in addition to their freight to the upriver army outposts and mining camps, the operators carried sawyers like my great-grandfather to cut wood and store it at regular, well-timbered stops along the river.

On one October woodcutting trip upriver, a worker had a dispute with the captain of the boat, perhaps because he couldn't keep up with other men who were stacking wood for the boilers. Tempers flared, and the offender was shoved overboard along with my great-grandfather, who had tried to defend him. The hapless pair found themselves stranded about a hundred miles northwest of Bismarck with little more than the clothes on their back. They built a raft and started downriver, but they hit a snag and lost everything in the water, including their food and money. This happened near present Elk Woods, North Dakota, about fifty miles east of Williston. They had no choice but to go the rest of the distance on foot. At one point, near starvation, the two happened upon an Indian camp, probably Sioux, which they approached with trepidation. Much to their relief, the Indians befriended the desperate wanderers and fed them pemmican. Andrew returned home clad only in his long woolen underwear and was so sunburned that that his wife didn't recognize

Figure 13-2. *The steamboat* Rosebud *chugging its way up the Missouri River to Fort Benton, c. 1880* –Courtesy Overholser Historical Research Center, Fort Benton, MT

him. For the rest of his life, Andrew always spoke respectfully of Indians, as he credited them with saving his life.

Mullan, Stevens, and the "Northern Overland Road"

Just as Native Americans helped my great-grandfather, so were they instrumental in guiding Lieutenant John Mullan during his extensive series of surveys from 1853 to 1860, although some bands still attempted to obstruct his movement in the wake of the Northwest Interior Indian wars of that decade.

Mullan had conceived of his road project as early as the spring of 1854, after serving as a topographical engineer with territorial governor Isaac Stevens's epic northern transcontinental railway survey project. Stevens had ambitiously directed the work of some 240 men in an effort to find the most practical railroad route over the northern Great Plains to the Rocky Mountains and across the Columbia Basin to Puget Sound. Stevens's main exploration party encountered little difficulty until they reached the Rockies in October 1853. Here, Stevens felt compelled to leave a task force of thirteen men in the mountains to provide a more detailed and thorough exploration of this strategic section.

Young Lieutenant Mullan was placed in charge of this detachment, and he thoroughly met his superior's ambitious objective to reconnoiter the rugged Upper Clark Fork region. Mullan's men established as their headquarters Cantonment Stevens, a primitive camp of four log huts on the Bitterroot River about ten miles south of Fort Owen, near present-day Victor, Montana.

Mullan's contingent included twenty-eight-year-old Private Gustavus Sohon, fluent in English, German, and French. Sohon would soon distinguish himself as an expedition artist as well as an interpreter, and as Mullan's right-hand man, who had a way of making friends with the Indians. Through these interactions with local Indians, Sohon acquired fluency in the Interior Salish languages of the Flathead, Pend d'Oreille, and Spokane tribes.

That winter, Sohon began work on what became an outstanding portfolio of landscape drawings, many of which were reproduced in Stevens's and Mullan's published reports as color lithographs. His masterful series of chieftain portraits were drawn during his time with Mullan and at Stevens's 1855 Indian treaty councils. Sohon's images of leaders of the Yakamas, Walla Wallas, Palouses, Nez Perces, Flatheads, Pend d'Oreilles, Blackfeet, Gros Ventres, and other tribes comprise what is considered "the most extensive and authoritative pictorial series on the Indians of the Northwest Plateau in pre-reservation days."[4]

During that winter of 1853–54, Mullan traveled over one thousand miles and crossed the Continental Divide six times. Early in the process he concluded that "a wagon road could be easily and economically constructed from Hell's Gate Ronde to the east of the [Coeur d'Alene] lake."[5] Mullan passed his suggestion on to Stevens and the War Department; Stevens

Figure 13-3. *Gustavus Sohon lithograph of Cantonment Stevens, near present-day Corvallis, Montana, looking westward* –From John Mullan, *Report on the Construction of a Military Road from Fort Walla-Walla to Fort Benton* (1863)

Figure 13-4. *Stevens in council with the Nez Perce Indians, watercolor by John Mix Stanley, c. 1854* –Courtesy Yale University Art Gallery

restated Mullan's conviction that "with a moderate amount of labor, a first-class stage-road could be here constructed and gave the *experimentum crucis* by taking a wagon train through on my return across the mountains in March, 1854."[6] Mullan's work impressed Stevens and significantly contributed to the latter's recommendation to Congress that Mullan receive an appropriation of $30,000 for the proposed road; the funding was approved in 1854. Five years later, Congress authorized an additional $100,000 to complete the road.[7]

A cache of Mullan's letters, long sequestered at Yale's Beinecke Library and recently brought to light by scholar Dan McDermott, indicates the scope and pace of Mullan's ambitious Rocky Mountain endeavors during the winter of 1853–54. In addition to his extensive surveys in every direction, the young lieutenant is recruited by local Flathead leaders to intervene in disputes with neighboring tribes, and he sought a treaty council with the troublesome Blackfeet.

The litany of springtime comings and goings of the mountain peoples that Mullan recounts is remarkable as we learn of happenings at the Hell Gate River camp of Flatheads, Spokanes, and Pend d'Oreilles, and we meet chiefs Victor and Moise, and traders John Owen and Peter Skene Ogden. In Mullan's correspondence he reveals himself as a man willing to risk the disfavor of whites and Indians alike by condemning the widespread practice of gambling, which contributed to much strife. He fully assumed his responsibilities as Stevens's area representative for Indian affairs.

Stevens's and Mullan's valued guides for their work were Native Americans and mixed-bloods, including former Hudson's Bay Company trapper Antoine Plante, of Gros Ventre and French-Canadian ancestry; Charlot (possibly Chief Victor's son) of the Flatheads; Aeneas (Ignace), a highly regarded Iroquois who wandered among the Flatheads; Coeur d'Alene leader Basil; and Gabriel Prudhomme, who had assisted Father DeSmet. Chief Slowiarchy of the Palouses later guided Gustavus Sohon in the summer of 1859, when Mullan's party surveyed through Palouse country eastward to the Bitterroots, along the plateau north of the Snake River.

In spite of the explorers' debt to the Native Americans, Washington territorial representative and former territorial governor Isaac Stevens was zealous about dominating the indigenous population. In 1857 he addressed Congress on several occasions to vindicate his controversial Indian policy of "absolute and unconditional submission" of the tribes, and to castigate the army for not supporting it. For his "edicts" preventing white settlement east of the Cascade Mountains, Stevens branded General John Wool a "dictator of the country." He also maintained that Colonel George Wright's 1856 foray into Yakama country and maladroit offering of gifts to area chiefs was "greatly to be deplored." The army's "long delays" and its insistence on "talking and not fighting," he said, gave "safe conduct to the murderers" of whites in recent hostilities.[8] Even worse, according to Stevens, Wright's failure to apprehend the famed Yakama chief Kamiakin gave the tribal leader control of the whole field of the interior.

Stevens's ire was in part a response to the congressional rejection of his proposed northern transcontinental railroad route in favor of the central route. An irrepressible promoter, Stevens then lobbied strongly for the construction of a military road from Fort Benton to Fort Walla Walla. In this bid he was successful, and the initial appropriation for the "Northern Overland Road" came through in late 1857. Twenty-seven-year-old John Mullan was named chief engineer for the work.

At the same time he was maneuvering in Congress, Stevens also published circulars that were widely distributed in the East, informing prospective immigrants of settlement opportunities in Washington Territory. Effusive newspaper reports celebrated the anticipated "large population . . . soon to be attracted to that region" of the Columbia Plateau.[9] Coastal residents aware of favorable conditions for agriculture and ranching east of the Cascades warmed to the prospects of settlement within the broad swath of prime land along Mullan's intended route, which the lieutenant described in numerous press accounts. The Bitterroot Valley was "capable of grazing immense bands of stock of all kinds," he asserted, while throughout the Palouse Valley, "we possess a rich, fertile, and productive area that needs but the proper means and measures . . . to be turned into public and private benefit."[10] In spite of this enthusiasm, the uncertain conditions in the

Northwest Interior stalled the work, and Mullan and Stevens waited impatiently for the situation to change.

Writing from the Colville mining district in November 1857, Indian agent B. F. Yantis added to the chorus of praise for development, assuring the readers of local newspapers that travel to the gold strikes was now "perfectly safe for Americans in any number," and that "with the exception of Kamiakin, all the principal chiefs" of the Spokanes, Colvilles, Yakamas, Palouses, and other area tribes had expressed "in strongest terms their friendship."[11] But the Indians' abiding desire for peace did not necessarily mean they would accept an unrestrained surge of whites across their lands north of the Snake River. Thus many Indian leaders besides Kamiakin did not share Yantis's cheery sentiments.

Figure 13-5. *The Yamaka-Palouse warrior Kamiakin, a Yakama chief, c. 1860* –Courtesy Manuscripts, Archives, and Special Collections, Washington State University

In March 1858 new gold strikes in the Colville Valley and on Canada's Fraser River inflamed the qualms of the region's Natives, as streams of miners ventured across the Cascades and onto the contested lands. Relations remained tense throughout a troubled spring. The arrival at Fort Colville of tough-minded George Blenkinsop to assist veteran trader Angus McDonald made matters worse. While Hudson's Bay Company personnel and Jesuit missionaries continued to walk safely between both worlds, the occasional American officials who passed through the valley were wary of Indian intentions. Later in the spring of 1858, two French-Canadians headed to the gold strikes were killed in Palouse country, and in another incident, Palouse raiders stole some livestock near Fort Walla Walla.

Stevens's insistence in 1858 that road building commence as soon as possible aroused the resistance of both military officials and Indian leaders in the Northwest. To make preparations for the project, Mullan had been dispatched to Fort Dalles, but on May 15, 1858, Colonel Wright flatly informed him that "no probability" existed for work on the road that year. "Lt. Mullan and his party will remain here [at Fort Dalles] until he hears from Col. Steptoe," Wright wrote in the *Pioneer and Democrat* in May. Wright condemned the road project as a dangerous threat to the region's fragile peace: "In fact it is said that the proposed opening of the road through the Indian country was a primary cause" of the uneasiness, he noted.[12]

Palouse chief Tilcoax, whose legendary horse herds ranged across thousands of square miles from the Walla Walla Valley to north of the Snake River, had harbored ill will against the whites since the days of the Cayuse War, and many whites blamed him for the recent trouble. Father Joseph Joset, the Swiss-born Jesuit ministering to the Coeur d'Alenes, was among those who believed that Tilcoax, not Kamiakin, was most responsible for fomenting hostilities among the region's Indians, and that he "had bribed the Spokanes, and some Kalispels to continue hostilities."[13]

For his part, Tilcoax felt the Americans were responsible for the epidemics that had struck his people, and the only way to rid the recurrent plagues was to drive the whites from the region. According to historian Robert I. Burns, Tilcoax's influence among the disaffected Snake River bands "rivaled that of the great Kamiakin," and he had been advocating war for months as he watched Colonel Edward Steptoe's garrison–sent to subdue the Indians–grow in strength.[14]

The Steptoe Disaster

In an effort to stop the thieving of army livestock with a show of strength, and to reassure the Colville miners who had petitioned for troops, Colonel Steptoe, with five companies of 152 enlisted men, five senior officers, and a contingent of Nez Perce scouts, departed Fort Walla Walla for the Palouse region on May 6, 1858. Interestingly, Steptoe's soldiers were poorly armed, as most of the dragoons had left their sabers behind and carried only short-range musketoons. The actions of Steptoe's ally, Chief Timothy of the Nez Perces, further played into the hands of the enemy. Timothy sent his envoys to tell the opposing bands to fear for their horses and their lives. Timothy had feuded with Tilcoax and spoiled for a fight with him, although he likely directed the Nez Perces to say nothing about this to the soldiers.

Instead of heading north along the Colville Trail through the central Palouse, Steptoe rode northeast on another well-worn trail toward Alpowa on the Snake River, then, upon hearing that hostile Palouses were in the Pine Creek area, headed toward that stream. Timothy may well have been advised of Tilcoax's presence along the *Smakodl* (South Palouse Fork), where many bands gathered each spring to dig roots in Paradise Valley. Although Steptoe's Nez Perce proxies were familiar with the route, the area was essentially *terra incognita* to the colonel and his men.[15] Father Joset, who was familiar with Steptoe's march, later commented that the colonel's decision to confront Tilcoax "would explain the whole puzzle" of why the soldiers used the more remote eastern route northward rather than taking the more direct Colville Trail from Fort Walla Walla.[16] The choice of this peculiar itinerary would prove fateful and led to Steptoe's defeat at the May 17–18 Battle of Tohotonimme (near present Rosalia, Washington), also known as the Steptoe Disaster.[17]

The Wright Campaign

The Steptoe Disaster was a stinging national embarrassment for the military. Throughout the summer of 1858, Army of the Pacific general Newman Clarke directed a torrent of men and supplies to the Department of the Columbia, thoroughly planning and provisioning troops for a strike into the very heart of the disputed region. On July 4 he issued orders directing Colonel Wright to effect "complete submission" of the warring tribes. By August, about 2,200 men, or roughly one-sixth of the nation's entire active forces, had been deployed in the region, with one thousand of the men assigned to Wright, who was orchestrating an extensive pincers movement across the Columbia Plateau. The fifty-four-year-old officer, known to his men as a stern but evenhanded taskmaster, had commanded the Ninth Infantry at Fort Dalles since 1856. Now the experienced officer was the principal architect of a major invasion of the Northwest Interior, sending at least a thousand men to strike from two directions.

Wright called on Major Robert Garnett, a Virginia native and former West Point instructor and commander at Fort Simcoe, to take three hundred men north to the confluence of the Columbia and Okanogan Rivers. Garnett aimed not only to punish Indians who had harassed miners, but also to flush out hostiles and force them eastward to face Wright's larger force.

Meanwhile, the ambitious John Mullan languished at Fort Dalles, the battle plans having halted his road-building expedition. With characteristic single-mindedness, Mullan wrote, "I had no disposition to remain idle during the summer, but, on the contrary, was anxious to become personally cognizant of such topographical facts as would give me a correct idea of the western section of the country through which our road would pass."[18] Rather than let Wright's campaign hinder his own plans, Mullan took it as an opportunity to further his explorations, and he solicited General Clarke to reassign him to Wright's command. In this way and at this pivotal moment in Northwest history, Mullan came to serve as the colonel's topographical engineer, along with valued assistance from topographer-artists Gustavus Sohon and Theodore Kolecki.

Wright's advance party departed Fort Walla Walla on August 7, aiming to secure the Snake River crossing point near the mouth of the Tucannon. Here the men gathered basalt shards and fashioned alder bastion posts to build Fort Taylor. The depot was named for Captain Oliver H. P. Taylor, the slain officer whose remains at the Steptoe battlefield the soldiers hoped to retrieve along with those of their other fallen comrades. The forward guard was soon followed by some seven hundred well-armed troops and support personnel with a contingent of forty friendly Nez Perce scouts under Mullan's command, all having "a very wholesome respect for the Indians" who had so thoroughly defeated Steptoe.[19] Ever attentive to topographical interests, Mullan noted in careful detail, mile by mile, the lay of the land, distances traveled, and campsite locations.

Wright's soldiers marched through the violent summer heat, moving across the charred grasslands that had recently been a "lake of fire," torched by Indians seeking to deprive the soldiers of provender for their animals. The troops crossed the Snake River near the mouth of the Palouse on August 25, but the Indians who usually resided there had moved out of harm's way.

The size and pace of Wright's grand cavalcade gave Kamiakin and other Indian leaders ample time to plot a strategy of response, but time would not necessarily benefit the Indian cause in light of Wright's much larger force and superior weaponry, which included several hundred new long-range rifled muskets, a good number of .54-caliber Harpers Ferry rifles, perhaps several dozen of the recently patented Sharps carbines, and six mountain howitzers. While the factors at Fort Colville had turned a deaf ear to Kamiakin's desperate pleas for ammunition, Hudson's Bay personnel at Fort Vancouver had been doing a booming business supplying Wright's forces with arms and ammo, earning the traders a commendation from the American Secretary of War for their valued support.[20]

Wright's command pressed beyond the Snake River, trailing great plumes of dust. On September 1, the troops emerged onto open prairieland, where the opposing forces finally met at the Battle of Four Lakes. The Indians assembled their forces amid the scattered pines and bunchgrass between Granite and Willow Lakes, while the soldiers thrusted directly into the Indians' contingent, taking positions between a "bald butte" (later named Wright's Hill) and Riddle Hill to the west, where, as Lieutenant Lawrence Kip observed, the hostiles met "the long range rifles now first used by our troops."[21] The Indians were torn down by the soldiers' superior weaponry, and they slowly began to fall back in spite of Qualchan and Kamiakin's appeals to stand their ground. The Indians' resistance held for some three hours before the warriors were forced to scatter in turmoil.[22] Throughout that most fateful week of Kamiakin's life, Donati's Comet flared portentously in the night sky.

Four days later, the fighting bands regrouped ten miles northwest of Four Lakes to challenge the soldiers again on September 5. The Battle of Spokane Plains proved to be the decisive confrontation of Wright's campaign and a defining moment in the region's primal clash of civilizations. The Indians set the dry summer prairie on fire to confuse the advance of Wright's mile-long column, which slithered forth as if possessed by the spirit of Rattlesnake. The Indians' smoke billowed up, seeming like the final exhalation of Native freedom, as the soldiers held ranks and marched into the mayhem. The troops watched the enemy "collecting in large bodies" to the northeast, near present Airway Heights, in numbers Wright estimated at five to seven hundred. The Indians rode parallel to the soldiers' line of march, amassing in larger clusters as the troops rode north.[23]

Chief Kamiakin and his wife Colestah, who was clad in battle dress with her scarlet headscarf, fought together in the battle until a cannon shell shattered the tree under which they had positioned themselves. According to family accounts, a falling branch struck Kamiakin, knocking him from his horse and likely breaking his shoulder. Lieutenant George Dandy, whose company attacked Indian positions in the pines and rocks below Deep Creek, recalled that a howitzer from Lieutenant James White's Company D of Wright's Third Artillery fired the shot that almost claimed the chief's life. Dandy said that "one of his shells tore a limb from a large tree under which some Indians were grouped; among whom . . . was the great chief Kamiakin of the Yakama tribe, who was seriously hurt."[24] The injured chief was taken to the family's camp at the mouth of the Spokane River for safe haven.

Seven hours of unremitting onslaught from the army had pushed the Indians seven miles to the Spokane River, where Wright found safe haven beside a steep bluff on the western bank of a bend of the river located a mile above its confluence with Latah Creek. The soldiers were "entirely exhausted," but Wright commended the "zeal, energy, and perseverance" they displayed throughout "this protracted battle," which proved to be the turning point in the Inland Northwest Indian wars.[25]

After his victory at Spokane Plains, Wright's forces continued east on a punitive campaign against his defeated opponents. Finally, on September 23, during the last of several treaty councils, the colonel dictated terms of submission to the Coeur d'Alene leaders. Over the course of the next several days, however, at his camp on Latah Creek, Wright had the Yamaka Qualchan and at least six Palouse leaders executed, giving rise to the name Hangman Creek. Thus ended the active resistance of the Inland Northwest tribes.

Wright's campaign also served Mullan's related purposes. In Mullan's words, prior to the summer of 1858, the entire region "lying between the Spokane and Snake rivers was only known to me through the reports and maps of others; . . . and to say where the line should or should not be located was no easy matter." Through his wartime experiences that summer, Mullan began contemplating Wright's more northerly route for the road's western approach across the Bitterroots. From the Palouse River, the road would follow an ancient trail through the rocky shrub steppe that skirted the Palouse Hills toward Spokane and Coeur d'Alene country, a route that Mullan, in a later memoir, credited "Sohon and his Indians" for recommending.[26]

Impressions of Battle

John Mullan, Theodore Kolecki, and Gustavus Sohon, all trained in topography and drawing, had joined Wright's command to render graphic images of the campaign. In Mullan and Kolecki's detailed map of the Battles of Four Lakes and Spokane Plains (Figure 13-6), the main battle is shown in step-by-step annotated detail, beginning with "Point of Attack; Flames and Smoke Enveloping the troops" and following through to the evening's "Camp on Spokane River," near the present, aptly named Fort George Wright.[27] The map is a classic of frontier military delineation, and it foreshadows the highly detailed Civil War cartography that would soon be undertaken by Union and Confederate topographers.

Figure 13-6. *Battle map for the 1858 Wright campaign, prepared by John Mullan and Theodore Kolecki* —From John Mullan, *Topographical Memoir and Map of Colonel Wright's Late Campaign Against the Indians of Oregon and Washington Territories* (1859)

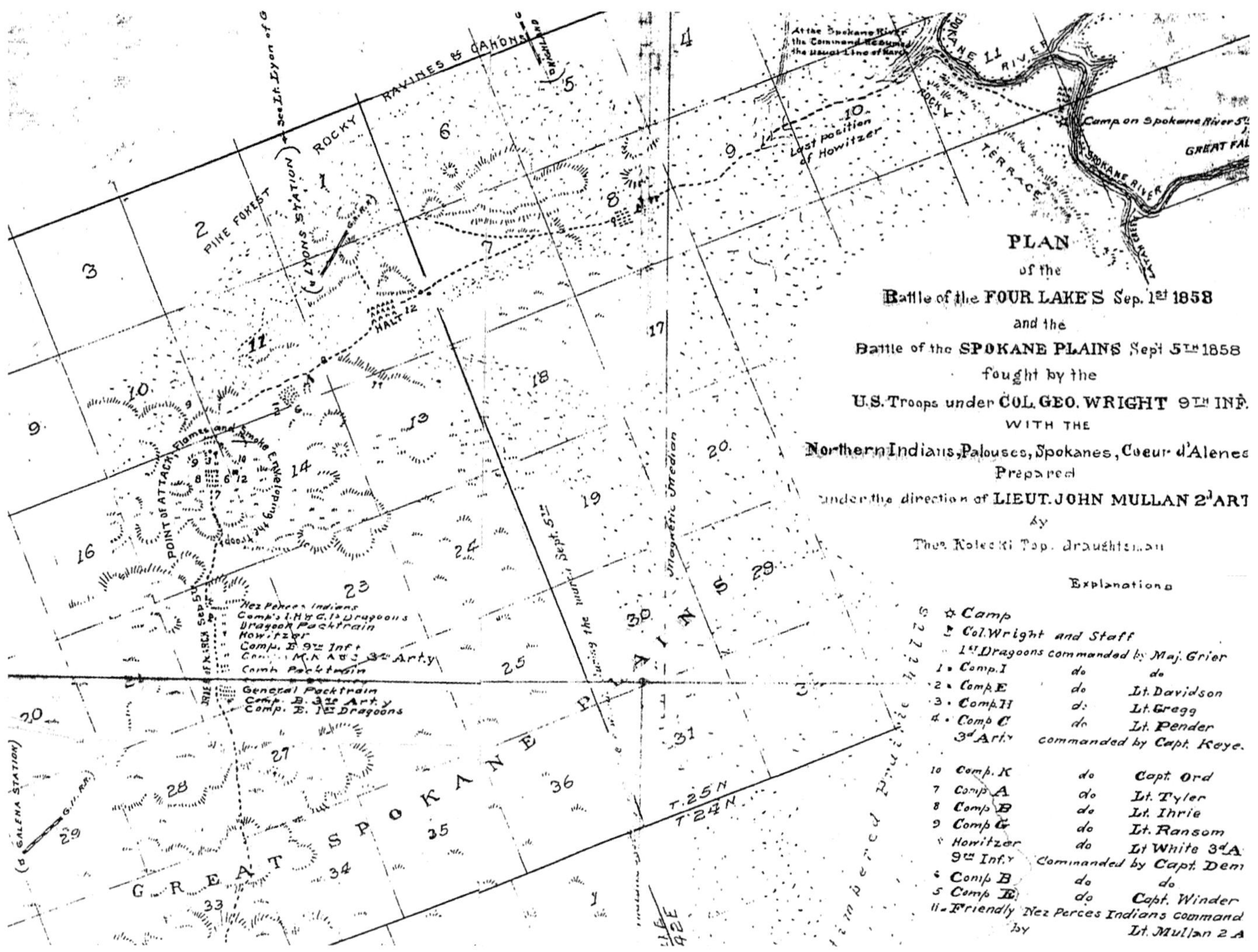

Figure 13-7. *The 1915 plat tracing by W. H. Watson and Garrett B. Hunt* -Courtesy Spokane Public Library

Mullan and Kolecki's map was later helpfully superimposed over an early plat map of Spokane County townships and sections (Figure 13-7), a 1915 project led by Spokane Historical Society chair Garret Hunt with several veterans of the Wright campaign. Mullan's perspective of the battlefields combined with the modern section plat offers significant insight into the sequence of events of that watershed event in Pacific Northwest history. From this, we can accurately locate the initial position of Wright's forces at Four Lakes as well as the Indians' positions to the east, and follow the conflict north to present Airway Heights, where the pivotal Battle of Spokane Plains took place.[28]

In addition to the maps is Gustavus Sohon's striking artistic depiction of the Spokane Plains Battle (Figure 13-8), created from the relative safety of the main pack train, near the present intersection of Rambo and Deno Roads in Spokane. In this image, half a dozen men, likely Wright and his senior aides, appear confidently on horseback atop a slight rise in the foreground, flanked by a line of infantry on each side. Another line of skirmishers, with a company of dragoons to their left, advance in the face of billowing smoke against a cluster of Indians in the distance.[29] As in the previous battles of the Wright campaign, the Indians faced the army's deadly fusillade of rifle and artillery fire. Wright estimated that some

Figure 13-8. *Gustavus Sohon's depiction of the Spokane Plains Battle* –Courtesy Washington State University, Manuscripts, Archives, and Special Collections

eight howitzer shells exploded in the midst of the Indians' positions as they bitterly yielded ground along their northeastern retreat toward the Spokane River.

"Entire Hopelessness"

In the fall of 1858, while most Yakamas and Palouses on the Columbia Plateau stayed in the area and sought a return to normalcy, those with Kamiakin and Tilcoax fled into the Bitterroot Mountains. Tilcoax continued east to the Great Plains, while Kamiakin, Skloom, Lokout, and other members of their band moved north toward Canada. When Kamiakin's band arrived in Pend d'Oreille country, the local Indians refused to help them, fearing they would be punished for harboring hostiles. Despite the cold reception, the fugitives remained until some local Indians stole one of their horses. Disconsolate, the band then moved on to the land of the Kootenai Indians, near the Canadian border.[30]

The Kootenais were equally cool to the exiles, refusing to allow them to settle in their country. Lacking horses, shelter, and food, the destitute band was "in a very hungry situation."[31] From Kootenai country, the refugees traveled nearly two hundred miles southeast to the St. Ignatius Mission in Montana. At the mission, the exiles met Father Adrian Hoecken and Chief Alexander of the Kallspels, both of whom also turned their backs on the indigent band. Father Hoecken knew of the troubles on the Columbia Plateau and feared

Figure 13-9. *Wilson Wewah (descendant of Chief Tilcoax) and the author near the Lyons Ferry Snake River crossing on the Mullan Road, 2007* –From author's collection

that the anti-Catholic sentiment west of the Bitterroots might extend to his mission, especially if he gave quarter to fugitive Natives. Neither he nor Chief Alexander wanted anything to do with Kamiakin, and they ordered him to leave in spite of the desperate circumstances facing the chief's women and children.

Finally, Kamiakin led his group of Yamaka and Palouse refugees into the lands of Flathead chief Victor, a friend, who gave them sanctuary for the winter near present Darby, Montana. However, none of the mountain tribes wanted Kamiakin to remain in the region, fearing reprisals by whites. In short, Kamiakin was an outcast and his people had no home. The prospect of relief came, unexpectedly, in the blustery days of late winter. In late February of 1859, Jesuit pathfinder Father Pierre-Jean DeSmet entered Kamiakin's camp on horseback. A well-known peacemaker, Father DeSmet surely lifted the chief's spirits with his hearty greeting and characteristic grin.

Father DeSmet had been a devoted friend to the Indians for decades. Like others of his time, John Mullan praised DeSmet's "zealous labor" in working with the western tribes; Mullan also credited the padre's book *Oregon Missions* with supplying him with "many geographical and statistical facts."[32] Soon after the war, Father DeSmet accepted an appeal from Columbia Department commander General William S. Harney, who had replaced Newman Clarke in 1858, to find Kamiakin and other fugitive chiefs and sincerely offer an olive branch. DeSmet ventured into the Bitterroot Mountain wilderness in February looking for the chief. He finally learned from other runaway Indians along the trail where Kamiakin was sequestered. Throughout his journey, the peripatetic priest was hospitably received by the local Indians, but he lamented their "state of entire hopelessness."[33] Yet nothing prepared him for the sorry sight of Kamiakin and his band.

DeSmet found the chief and his people in appalling condition, desperately lacking in food and shelter in the face of still-severe weather. "The sight of Kamiakin's children, poverty, and misery in which I found them, drew abundant tears from my eyes," the missionary wrote in his letter to military authorities. "Kamiakin, the once powerful chieftain, who possessed thousands of cattle, has lost all, and is now reduced to abject poverty."[34] The exiles warmly welcomed the weary blackrobe, and Kamiakin readily agreed to counsel with him.

Between February and April 1859, DeSmet frequently visited Kamiakin's austere camp in the Bitterroot Valley and his later one in Thompson's Prairie along the Clark Fork River. Kamiakin and Skloom listened to the padre and spoke "with greatest apparent earnestness," but the two chiefs were undecided about whether or not to risk returning to their homeland. Kamiakin declared that he "never was a murderer, and, whenever he could, he restrained his people against all violent attacks on Whites passing through the country." He conveyed to DeSmet that the Indians wanted "the path of peace," and the priest concluded that if Kamiakin were "allowed to return . . . to his country, it will have the happiest and most salutary effect among the Indian tribes."[35]

Kamiakin and his huddled band remained in the mountains until 1860, then they returned to the vicinity of Lake Coeur d'Alene, where the local Indians had once fought by Kamiakin's side and where the chief had established trust with the Catholic missionaries. Throughout their period of mountain exile, Kamiakin's ménage had not been bothered by soldiers, and no expeditions had pursued them. Reassured by this, they gradually drifted westward, back to their

Figure 13-10. *Gustavus Sohon's watercolor of the St. Ignatius (a.k.a. Pend d'Oreille) Mission, 1862*
–Courtesy John Mullan Papers, Georgetown University Library, Special Collections Research Center, Washington, DC

people's ancestral homeland in the sheltered and remote valley of the Palouse River, settling at "Kamiak's Crossing," north of present Endicott, Washington.[36]

Kamiakin and his clan established their summer camp along the south shore of the Palouse, on a broad flat close to the ford. The chief, who loved gardening, took great pleasure in raising corn, potatoes, and other vegetables at his new home. He and his people once again fished, hunted, collected wood, and tended livestock. Later, Kamiakin's sons fashioned a race course where they could ride the finest horses from the herd the family endeavored to restore. As the seasons changed, the clan followed planting, fishing, and hunting cycles in accordance with the ancient patterns. After years of war, turmoil, and exile, Kamiakin's band recovered and eventually thrived beside the free-flowing Palouse.[37]

Surveys and Road Building

Mullan's road-building crew "turned up the first dirt" on the Tucannon River, in present Washington state, on June 25, 1859, commencing the arduous contruction of their 624-mile wagon road to Fort Benton, in present Montana. A month later, while camped at the mouth of Union Flat Creek, a tributary of the Palouse River, Mullan dispatched topographer Theodore Kolecki and two others to explore several possible transportation routes across the central Palouse Valley, while he and the main crew continued northeast, reaching the Sil-sil-cep-pow-vetsin (Rock Creek), at present Hole-in-the-Ground in Washington (south of present Cheney and west of the Steptoe battlesite), on July 13. (The explorers' term for Rock Creek likely derives from Coeur d'Alene *Slslpsp'ultsn*, or "Foam Swirling on the Water.")[38] Kolecki, meanwhile, passed by Kamiak's Crossing, where he encountered isolated Indian camps. He noted a dramatic rise in the elevation of the Palouse River valley as they moved eastward, with rocky bluffs four to five hundred feet high shielding the crossing and adjacent bottomlands, which were "extremely fertile, covered with tall grass, cottonwood groves, and wild currant bushes."[39]

Later that summer, topographical engineer P. M. Engle surveyed the Lower Snake River Valley while Gustavus Sohon explored possible routes through the Coeur d'Alene and St. Regis River valleys and the Bitterroot Mountains to the east. Traveling upstream on the south side of the river, Engle moved from the recently abandoned Fort Taylor, located opposite the village of Palus, to the lands of the Upper Palouses. He was surprised to find that the Palouses and Nez Perces in this region were cultivating seven large farms "amounting to from 300 to 400 acres." The Indians also farmed along the river banks and on some of the islands in the river. "Besides wheat and corn," Engle noted, "they raise vegetables of different kinds, and gain sufficient crops to encourage them in their labors." Although Engle judged the soil in the valley to be gravelly and sandy, he astutely recorded that "the plateaus on both banks produce fine grass, offering magnificent pasture grounds."[40]

Meanwhile Sohon, on his reconnoiter to the Bitterroots, found densely timbered terrain, scanty grass, and multiple streams that would require crossing. Throughout the summer of 1859, as Mullan's main road-building crew and its armed escort pressed steadily to the northeast, they met no Indians until they crossed the Spokane Trail, in the northern part of Palouse country. Although the Palouse Indians knew that soldiers had crossed their lands and that whites planned to open a road, the surveyors encountered no difficulty from them.

In their reports, both Sohon and Engle remarked on the fertility of the soil on Palouse lands. The whites also viewed the region's rolling bunchgrass hills, once grazed by immense horse herds, as having vast potential for farming. Mullan's report noted that the "undulating prairie of plentiful grass" covered "rich soil" conducive to grain production.[41]

By December 1859, Mullan had reached the St. Regis (Borgia) River valley, where he established his "curious little town" of Cantonment Jordan. In desperate need of winter provisions for his men, Mullan asked Flathead chief Ambrose for assistance. The chief kindly provided 117 horses for a pack train, which was sent to Fort Benton for loading. After a harsh winter, Mullan's road builders resumed their pry-bar and shovel work in March 1860, finally completing the road to Fort Benton at the end of July. On his return to Fort Vancouver later that summer, Mullan himself led the first contingent of men across the entire route.

Mullan and the Native Peoples

In the early 1860s, John Mullan predicted that it was "not at all improbable" that "the grazier and agriculturalist will find at no distant day tracts of land that will amply repay their reclamation." News of the lush earth in the western Indian country attracted white settlers, who would gradually push most Native peoples onto reservations. Mullan, the masterful topographer and strategist, was seldom given to philosophic musings in his writings, but in an 1888 memoir he imparted his hope to see "the country thickly populated" and "fruitful to the maximum degree," even if that meant that "the fish should disappear from its creeks and rivers, and game from its forests."[42]

But at the same time, Mullan lamented the impact his work would undoubtedly have on the Native population, as indicated in the comments on the Coeur d'Alene Mission he included in his official report of 1863:

> I fear that the location of our road, and the swarms of miners and emigrants that must pass here year after year, will so militate against the best interests of the mission that its present site will have to be changed. . . . This . . . is to be regretted; but I can only regard it as the inevitable result of opening and settling the country. . . . [The Indians] can never exist in contact with the whites; and their only salvation is to be removed far, far from their presence. But they have been removed so often that there seems now no place left for their further migration; the waves of civilization have invaded their homes from both oceans, . . . and now that we propose to invade these mountain solitudes, to wrest from them their hidden wealth, where

> under heavens can the Indians go? . . . It is a matter that but too strongly commends itself to the early and considerate attention of the general government.[43]

We can be grateful that Mullan provided substantial ethnographic information in his field notes and reports about the Native peoples of the Columbia Plateau and Rocky Mountains. He and his associates provided the locations of major villages and missions and identified headmen of the Spokanes, Coeur d'Alenes, Flatheads, Pend d'Oreilles, and other tribes. Mullan praised the labors of Catholic missionaries like DeSmet and Joset at a time when their influence was criticized by many in Washington Territory as foreign and counterproductive to American interests. He complained about the corrupt Office of Indian Affairs and suggested that its responsibilities be transferred to the War Department. For both personal and professional reasons, he also twice lobbied, unsuccessfully, for Gustavus Sohon's appointment as a Northwest Indian agent, but although his old friend was a thoroughly qualified advocate for the Native peoples, the entrenched system of executive patronage in the government prevented his receiving the appointment.[44]

In the end, Mullan was fatalistic about the Indians' future in the wake of Euro-American expansion. He often referred to the Indians disappearing as practically a fait accompli. "[T]he only project . . . likely to save a portion" of Natives, Mullan said, was to "take the children and educate them under a proper system."[45] The suggestion foreshadows the Indian Office's nationwide system of Indian boarding schools, established in the late nineteenth and early twentieth centuries at places like Fort Spokane, Puyallup, and Salem. These schools' detrimental impact on the students' cultural identity and emotional well-being, for both the children and their families, became the subject of intense critical study in later years.[46]

Historian Alexander C. McGregor's careful study of the record books at Lyons Ferry and early issues of the *Walla Walla Statesman* dating back to 1869 reveals the considerable immigrant traffic that passed over the Mullan Road at this strategic ferry crossing during the first two decades of the road's use. The lists represent a pioneer "Who's Who," including the enterprising Schwabacher and Oppenheimer brothers, William "Virginia Bill" Covington, S. W. "Wild Goose Bill" Condit, "Uncle Dan" Drumheller, and Ben Burgunder as well as various Rocky Mountain packers whose destinations included the Kootenai, Blackfoot, and Birch Creek mining districts.[47]

In spite of such activity, there were isolated examples of places along the Mullan Road where Indian and white learned to coexist peacefully in close proximity to each other, some lasting well into the twentieth century. The Native residents of Palus came to live peaceably with the passersby. In the 1880s, several of Chief Kamiakin's sons, along with other members of their Palouse band, entered claims adjacent to the Mullan Road under the Indian Homestead Act. Relationships built through mutual trust with pioneer families like the Pettyjohns, Hunters, and McGregors enabled the tribe to acquire an immense tract of over 1,500 acres, from Palouse Falls to the Palouse River's confluence with the Snake; some seventy villagers were still living there as late as 1900. Although the Palouses' holdings had dwindled by 1930, isolated parcels along the Snake River remain under Indian title to this day.

Mullan's views may be understood in the context of his time, an era of American exceptionalism that widely portrayed progress in terms advancing the dominant Euro-American culture. While not everyone shared this attitude, even in mid-nineteenth-century America, most newcomers to

Figure 13-11. *The Snake River crossing at Lyon's Ferry, c. 1940*
–Photo courtesy Robert Eddy

the Northwest came with typical white prejudices and disregarded the rights and property of Native peoples. Most immigrants had scant interest in Indian culture or the work of the Jesuits and other Indian advocates. "For one that comes with a pencil," Henry David Thoreau was then lamenting back in New England, ". . . a thousand come with an axe or rifle."[48]

Euro-Americans generally conceived of themselves as agents of regeneration, making the wilderness useful for private ambition and for beneficent civilization. However, as the Concord sage postulated–in language that men like Pandosy and Kamiakin would have comprehended–true human greatness depends on a balance between primordial vitality and agricultural and commercial progress. Whatever ideas about western life that frontier missionaries and artists contributed through books and sketches, the Native people already knew from experience. Kamiakin likely would have agreed with Thoreau's celebrated observation that places "where the pine flourishes and the jay still screams" were necessary for the human spirit.[49]

Figure 13-12. *Chief Cleveland Kamiakin, c. 1920* –Courtesy Manuscripts, Archives, and Special Collections, Washington State University

Holding Sacredness

In 1956, a century after the defeat of the tribes in the Northwest Interior, Chief Kamiakin's last surviving son, Cleveland Kamiakin, then residing on the Colville Reservation, was invited to address a Grant County Historical Society meeting in Ephrata, Washington, as the guest of state senator Nat Washington. Washington also personally interviewed Cleveland and recorded their sessions in a series of tape recordings that were not publically available until 2011. Although Cleveland was eighty-six years old, his voice emerged strong and deep as he responded, in his native Northern Sahaptin language, to his hosts' questions about local places and events through agency interpreter Harry Nanamkin. Cleveland and another elder, Chief Billy Curlew of the Columbia-Sinkiuse, expressed delight at the opportunity to visit old traditional campsites and to reminisce about times when their ancestors traded buffalo hides to Rocky Mountain tribes for spring roots and salmon. They responded patiently to inquiries about tribal leadership, local place names, and the lives of Chief Moses and Chief Kamiakin. The two men's words also reflected a profound sense of loss as they described the marvels of creation by Coyote (*Spílya*) across the landscape.

In the interview, translated by native Sahaptin speaker Carrie Jim Schuster, Cleveland spoke of a recent walk among the teepee rings that were still visible at Rocky Ford Crossing. "*Íchna áw míimi pawyáninxana* [I am not strong like I used to be]," he said wistfully. "I am an orphan from everywhere. I am separated from all my homelands."[50] Cleveland later told his historical society audience that some things need not change:

> We are gathering here at a special place and it still holds its sacredness [*ahtow´* in the original Sahaptin]. . . . We must live together on the same land; one people to another, face to face. We have families, communities, in friendship on this land. I will not tell [you] how to manage the land, but my food is also here and I hope we can continue to gather it. You use this land for your needs as we have ours. May we live together.[51]

Translator Carrie Jim Schuster, who as a young girl knew Cleveland, observed that the word *ahtow´* "suggests a sacred trust," with environmental consequences if selfishly broken.[52] In his parting words to the audience, Cleveland did not advise his listeners against cultivating the land, but he implied that farming was not his way. He asked them to respect the Native peoples' conviction that the land's natural resources are to be protected as a sacred obligation in order to sustain humanity:

> The desire to get something more than we need and are provided leads persons, families, and even nations to do things that are harmful to the land and to life. This is what Cleveland and our elders meant when they spoke about the "law" to [Governor] Stevens and others who wanted to make the treaties. For the sake of our children and in accordance with these sacred ways we must respect the land and water and not pollute it as is happening now.[53]

Like Schuster and Cleveland Kamiakin, I personally hope that the twenty-first-century heirs to the remarkable accomplishments of individuals like Mullan and Sohon, Kamiakin and Victor, summon similar resolve to live responsibly in these lands they so cherished, to utilize the region's resources in ways that sustain not only our own livelihoods but those of future generations. Challenges to our private and public well-being abound, but no more so than when Lieutenant John Mullan passed through, later reminiscing about how "Night after night I . . . laid out in the unbeaten forest or on the pathless prairies with no bed but a few pine needles, and no pillow but my saddle." Perhaps he was thinking about those—like you and me, like my great-grandfather and my son—who would someday "make their home in this far western land."[54]

---

1 John Mullan, *Report on the Construction of a Military Road from Fort Walla-Walla to Fort Benton* (Washington, DC: Government Printing Office, 1863), p. 14. This source is hereafter cited as Mullan, *Report*, 1863.

2 Quoted in John Mullan, "Memoir," 1888, Richard Scheuerman Papers (Collection MS2010-17), Archives and Special Collections, Washington State University Libraries, Pullman, WA. This source is hereafter cited as Mullan, "Memoir," 1888.

3 John Strachan, *Blazing the Mullan Trail: Connecting the Headwaters of the Missouri and Columbia Rivers and Locating the Great Overland Highway to the Pacific Northwest* (printed serially in the *Rockford* [IL] *Register,* April 1860–April 1861; reprint, New York: Edward Eberstadt & Sons, 1952).

4 John Ewers, "Gustavus Sohon's Portraits of Flathead and Pend Oreille Indians, 1854," *Smithsonian Miscellaneous Collections*, vol. 110, no. 7 (Washington, DC: Smithsonian Institution, 1948), p. 2.

5 Mullan, *Report*, 1863, p. 2.

6 Mullan, "Memoir," 1888.

7 U.S. War Department, *Annual Message from the President and Report of the War Department, 1858* (35th Cong., 2d sess., H. Ex. Doc. 2, serial 998; Washington, DC: James B. Steedman, Printer, 1858), p. 182.

8 Quotations from Stevens's speeches came from *New York Tribune*, January 5, 1857, and *Pioneer and Democrat* (Olympia, WA), Feburary 4 and 27, 1857.

9 *Pioneer and Democrat*, November 27, 1857.

10 Ibid.

11 Ibid.

12 Ibid, May 21, 1858.

13 Remarks re Tilcoax ill will in E. Steptoe to W. Mackall, April 17, 1858, Letters Received, Records of the Commands of the U.S. Army (RG 393), National Archives, Washington, DC; Joseph Joset, "Account of the Indian War of 1858," unpublished manuscript, 1873, Oregon Province Archives, Gonzaga University, Spokane, WA.

14 Robert Ignatius Burns, *The Jesuits and the Indian Wars of the Northwest* (New Haven, CT: Yale University Press, 1966), pp. 205–06.

15 Nesmith Ankeny, "The Story of the Steptoe Battle as Told by Lewis McMorris, Michael Kenney, Thos. Beall, and John McBean," unpublished manuscript, n.d., Washington State Historical Society Research Center, Tacoma, WA.

16 Father Joset quoted in H. M. Chittenden and A. T. Richardson, eds., *Life, Letters, and Travels of Father Pierre Jean DeSmet, S.J., 1811–1873* (New York: Francis P. Harper, 1905), pp. 750–51.

17 For more on the Steptoe Disaster, see Richard D. Scheuerman and Michael Finley, *Finding Chief Kamiakin: The Life and Legacy of a Northwest Patriot* (Pullman. Washington State University Press, 2008) pp. 74–82.

18 Mullan, *Report*, 1863, p. 9.

19 Lawrence Kip, *Army Life on the Pacific: A Journal of the Expedition against the Northern Indians* (New York: Redfield, 1859), pp. 38–39.

20 Facts re Wright's weaponry per War Department, *Annual Message*, pp. 344–45; information re traders per Steven Plucker, "The U.S. Army's Fort Walla Walla: Its Development and Early History," unpublished manuscript, 2007, Richard Scheuerman Papers (Collection MS2010-17), Archives and Special Collections, Washington State University Libraries, Pullman, WA.

21 Kip, *Army Life on the Pacific,* pp. 55–57.

22 War Department, *Annual Message, pp.* 386–90.

23 Quotation from War Department, *Annual Message*, pp. 392–93; see also Dandy Jim, oral history, n.d., L. V. McWhorter Papers (Collection 55), Manuscripts, Archives, and Special Collections, Washington State University Libraries, Pullman, WA; and Kip, *Army Life on the Pacific*, pp. 63–66; see also Burns, *The Jesuits,* pp. 294–97.

24 George B. Dandy, "Reminiscences," 1907, Wm. Compton Brown Papers, 1830–1963 (Collection 196), Manuscripts, Archives, and Special Collections, Washington State University Libraries, Pullman, WA.

25 George Wright, "Report of the Indian War of 1858," in *Message from the President of the United States*, House Ex. Doc. 2, 35th Cong., 2d sess. (Washington, DC: Government Printing Office, 1858), pp. 392–93; and George B. Dandy, quoted in Garrett B. Hunt, *Indian Wars of the Inland Empire* (Spokane, WA: Spokane Community College, n.d.).

26 Mullan, *Report*, 1863, p. 9; Mullan, "Memoir," 1888.

27 John Mullan, "Plan of the Battle of the Four Lakes, Sept. 1, 1858, and the Battle of the Spokane Plains, Sept. 5, 1858" (map), Washington State Library, Olympia.

28 Mullan and Kolecki also mapped in detail Steptoe's battle on Pine Creek, based on their visit to the area with a company of Wright's soldiers to gather the remains of their fallen comrades. Thanks to the careful study of Kolecki's meticulous notes of the Steptoe campaign by historians Paul D. McDermott and Mahlon Kriebel, we have also come to know only recently important details about that route and events surrounding his defeat.

29 Gustavus Sohon, "Battle on the Spokane Plain–Col. G. Wright in Command and Against Combined Forces of the Indians, 1858" (drawing), National Anthropological Archives, Smithsonian Museum Support Center, Suitland, MD.

30 Andrew J. Splawn, *Ka-Mi-Akin: Last Hero of the Yakimas* (Portland, OR: Kilham, 1917), p. 121; and Emily Peone, oral history interview by the author, October 31, 1982.

31 Pierre DeSmet, in H. M. Chittenden and A. T. Richardson, eds., *The Life, Letters and Travels of Father Pierre Jean DeSmet, S.J., 1811–1873*, vol. 3. (New York: Francis P. Harper, 1905), p. 968.

32 Mullan, *Report*, 1863, p. 53.

33 Pierre DeSmet to Alfred Pleasanton, May 25, 1859, Records of the Commands of the United States Army, National Archives and Records Administration, Seattle, WA.

34 P. DeSmet to A. Pleasanton, May 25, 1859, Letters Received, Records of the Commands of the U.S. Army (RG 393), National Archives, Washington, DC.

35 Ibid.

36 Arthur Kamiakin, oral history interview by the author, June 16, 1972; Peone, oral history.

37 A. Kamiakin, oral history.

38 Gary B. Palmer, *Hntmikhw'lumkw: This Is My Land* (Plummer, Idaho: Coeur d'Alene Tribe, 1987), p. 94.

39 Kolecki quotation in Mullan, *Report*, 1863, pp. 104–05.

40 Engle quotations in Mullan, *Report*, 1863, p. 95.

41 Mullan, *Report*, 1863, pp. 94–105. The Palouse exploration reports were authored by Gustavus Sohon, February 15, 1860 (Union Flat to the Coeur d'Alene Mission), Theodore Kolecki, February 8, 1860 (Rock Creek to Steptoe Butte), and P. M. Engle, March 7, 1860 (Red Wolf's Crossing to Union Flat).

42 Mullan, "Memoir," 1888.

43 Mullan, *Report*, 1863, p. 53.

44 Information re Mullan's lobbying for Sohon appointment from Paul D. McDermott, interview by the author, February 5, 2008.

45 Mullan, *Report*, 1863, p. 50.

46 See, e.g., Clifford E. Trafzer et al, eds., *Boarding School Blues: Revisiting American Indian Educational Experiences* (Lincoln, NE: Bison Books, 2006).

47 Alexander C. McGregor, "The Economic Impact of the Mullan Road on Walla Walla, 1860–1883" (master's thesis, Whitman College, 1971), pp. 110–30.

48 Henry David Thoreau, *The Writings of Henry David Thoreau*, ed. Bradford Torrey and Francis H. Allen (Boston: Houghton, Mifflin, 1906), vol. 3, p. 133.

49 Ibid., vol. 1, p. 179.

50 Cleveland Kamiakin, interview by Nat Washington, October 1956 (sound recording), Oral History Collection, Ellensburg Public Library, Ellensburg, WA; and Cleveland Kamiakin, Grant County Historical Society Speech, Ephrata, WA, October 8, 1956 (sound recording), Office of History and Archaeology, Colville Confederated Tribes, Nespelem, WA. Both recordings translated by Carrie Jim Schuster, November 30, 2008.

51 Ibid.

52 Ibid.

53 Ibid.

54 Mullan, "Memoir," 1888.

Precision and Poetry

The Utility and Elegance of the Mullan Report

BY J. WILLIAM T. YOUNGS

In his life and in his writing, John Mullan displayed two traits not often combined in one person–serious-minded practicality and thoughtful eloquence. As a West Point–educated engineer, he mastered the many details of building a military road, including determining routes, assigning tasks, arranging supplies, and seeking funding. Added to these abilities was his hard-driving commitment to building a highway through seemingly impassable terrain. Put together, you have the soldier-engineer who would complete the first wagon road across the northern Rockies. But if you look at a photograph of John Mullan in his later years, with his long flowing beard and a faraway look in his eyes, you see another person–a philosopher or a mystic, someone who realizes that there is more to life than carving side cuts through the mountains.

Mullan's 1863 report on the building of the military wagon road from Fort Benton to Walla Walla–soon to be known as the Mullan Road–is enriched by both qualities.[1] It is the work of an exacting reporter and a literary craftsman. He begins his account with a cogent explanation of the national policies that led to the work in 1853 on several military surveys in the West. In one of his characteristically long but precise sentences, he explains:

> The necessity felt by the government for a more thorough and satisfactory knowledge in detail of the geographical and topographical character of the country that lay between the Mississippi river and the Pacific ocean, looking especially towards the location and construction of a Pacific railroad, called into the field, in the spring of 1853, under authority from Congress, several corps of engineers and explorers, whose mission it was to supply this desired information within certain limits of time and means.[2]

During the summer of 1853, several survey parties set out for the West. One of these, an army corps directed by Washington territorial governor Isaac Stevens and including John Mullan, moved rapidly from St. Paul to the Rocky Mountains. This section of terrain was "accurately and rapidly explored" with ease. But then, like Lewis and Clark almost fifty years before, the Stevens party ran into a challenge: the Bitterroot Mountains. In Mullan's expressive phrasing, "the late autumn of 1853 found our labors of exploration truly only begun."[3]

At that point, Governor Stevens decided to return to "the States"–as the East was then known–leaving John Mullan in charge of a small party of men assigned to explore this rugged region. Mullan's description of his situation displays his ability to write prose that is evocative as well as lucid:

> We now entered the more difficult section of the Bitter Root range of the Rocky Mountains, where the lateness of the season, the difficulty of the country, the importance of our mission, the scarcity of our supplies, the meagerness of the information we then possessed, and the necessity felt for a more detailed and thorough exploration of the Rocky Mountain section, since proved to be the key of the work, together with other equally cogent reasons existing at the time, all conspired to influence Governor Stevens to leave in the mountains a small party for the winter of 1853, for further explorations, and thus supply that information which a lack of time did not allow him to collect at an earlier period.[4]

As we can see in this passage, Mullan strengthens his depiction of the challenge he and his men faced with a strong narrative line. The obstacles and imperatives, he tells us, included lateness, difficulty, importance, scarcity, meagerness, and necessity. Reading along, one is drawn to ask, "What will happen next?" And as both report and as literature, the account does not disappoint.

Mullan goes on to tell us that he and his men erected for their 1853-54 winter camp "comfortable, though rude, log huts" and, later, that they had been driven by necessity to abandon their wagon in favor of pack trains.[5] He also provides a pithy overview of the task at hand: "My mission in the winter of 1853 was to solve the problem of a proper connection, through a practicable mountain pass, of the plains of the Missouri with the plains of the Columbia, between the 45th and 48th degrees of north latitude, for either wagon road or railroad lines, and to make my reports thereon."[6] Beyond describing the situation, Mullan *evokes the scene.* They were working in a region "whence flow the sources of the Columbia and Missouri rivers in a network of babbling brooks."[7] Mullan might simply have written that they were traveling at the headwaters of those great rivers, which would have been accurate but abstract, but instead he paints for his readers a more vivid picture. "Headwaters" is a geographer's term; "babbling brooks" is a poet's.

Mullan reminds his readers that the railroad route would be determined in part by the points at which steamboats on the Missouri and Columbia Rivers could no longer navigate. The railroad would take up the cross-country journey from those points. Mullan clearly summarizes the challenge:

> The question of the navigation by stream of the upper Columbia and Missouri rivers was, at that date, in the infancy of its discussion, but was deemed of such importance that its practical solution was looked upon by the friends and advocates of the railroad line as the sine qua non to the full and speedy settlement of the country between them, as well as the all-important aids in the construction of a Pacific railroad via a northern route.[8]

In the next paragraph, Mullan describes the actual craft then at work on those waters, underscoring the fact that steam navigation on both rivers was in its infancy. On the Columbia, several smaller "steam-craft" carried cargo and passengers as far as The Dalles, he notes. As for the Missouri, Mullan reports in a lyrical phrase that on this river was only "a solitary steamer, engaged in the fur trade, that made its annual trips from St. Louis till it crept along the waters of the Missouri to a region where the red man alone walked."[9] Speaking as an engineer, Mullan emphasizes the challenge of "testing by steam the further practicability of each of these rivers . . . in order to attain the minimum land transit." His task was soon expanded to include reconnaissance to determine not only a railroad line but also "a continuous practicable wagon-road line" between the heads of steam navigation on the Missouri and the Columbia, "be those heads where they may."[10]

The winter survey party needed, of course, to operate as geographers, exploring possible routes across the mountains to join the heads of navigation on the two great rivers. In that regard, information was scarce. Their only map was "left us by Lewis and Clark" in 1805; and beyond that, they had only a few "addenda" provided by members of the Hudson's Bay Company and "scraps of information which the chance traveler or sojourner in the country felt disposed to offer."[11] Later in his report, Mullan mentions that the party gathered information from "hunters and trappers and those familiar with the country from a long residence in the mountains."[12] The presence of only a few non-Indian wanderers highlights the fact that they were on a frontier far from government agencies. These early "sojourners" carried their goods on pack animals, and so they were not accustomed to contemplating locations for wagon roads "in their thoughts or discussions."[13] Thus, they could provide only hints about the terrain, leaving the army to figure out where a road might be built.

Still, sketchy information was better than none. "During the winter of 1853," Mullan writes, "many were the conversations held with whomsoever could give us information of the geography of the country."[14] One senses that any stranger passing the army's little log encampment during the winter of 1853-54–or any person found in the mountains–was greeted warmly and questioned on possible routes. Some of Mullan's best information came from "an old Iroquois Indian called Aeneas, now resident in the Bitter Root valley, whose wanderings amid the mountains had often thrown him with parties traveling with wagons at the southward . . . ." The old man's valuable experience "render[ed] him capable of judging of the requisites for a wagon road." Aeneas told the soldiers that "a line could be had through a gorge-like pass in the Coeur d'Aléne mountains."[15] Much work lay ahead in testing this route, but the Indian's information helped focus the army's investigations on a specific area that was potentially passable by wagons.

Mullan's report continues with an overview of the various routes tested when warmer weather returned in the spring and summer of 1854. Soon after finishing the survey, he submitted to Congress his proposal to build a wagon road. But further work on the road was delayed by the outbreak of Indian conflicts in the Northwest, which began in 1855. The hostilities, Mullan writes, "converted each man's home in Oregon and Washington into a block-house, caused the portals of the upper Columbia to be opened awide, and brought especially to the attention of the people a hitherto neglected and little known region."[16] At first the eruptions occurred east of the Cascades, but then came warfare in the interior, including the 1858 defeat of Colonel Steptoe in what came to be known variously as the Battle of Steptoe Butte, the Battle of Pine Creek, or the Battle of Tohotonimme. Mullan describes, in marvelously sensory language, the change brought about by the war at the western terminus of the incipient road: "A military occupation of the Walla-Walla valley became necessary," he writes, "and the rude mud walls of old Walla-Walla, that year after year had listened to nought but the jargon of Indians and traders, caught a new sound in the tramp of the march of civilization."[17]

Thus for a time, the business of road-building gave way to the work of defeating the Indians, accomplished in 1858 under the leadership of Colonel George Wright. Mullan then began the actual construction of a road across eastern Washington. Along the way he got to know the pioneer homesteads of the region: "Already have each and all these valleys become the comfortable homes of the pioneer farmer and grazier, where the hand of industry, adding daily to the wealth and prosperity of the country, gives a new beauty, by the erection of school-houses and churches, those barometers of the intelligence and morality of a people."[18]

Mullan also notes historical landmarks along the route. At one point he camped at "Hangman's Creek," the site of one of the most notorious episodes of the Indian wars in the Northwest. For Mullan, this was not simply a spot to rest his men, but a place that evoked the complex relationship between Indians and whites on the frontier. "We encamped this day on the banks of the Nedlwhuald," he writes, "and at the same point where General Wright hung Qualtian [sic], the noted Yakima chief, and several other Indians; from which fact the stream is known to many as Hangman's creek. Poor creatures! their doom, although in this instance a just one, is, nevertheless, pitiable; had the white man been to them more just, fate had proved less harsh."[19] Here, as in many other places in his report, Mullan goes beyond the necessity of providing a routine account to reflect on the broader course of history.

The road-building across eastern Washington was comparatively easy, with little grading required and few rivers to cross. But negotiating with the Indians provided a continuing challenge, as indicated in this engaging passage:

> It was on this day that we met, for the first time, a band of the Coeur d'Aléne Indians, under Pierre, a few miles from our camp; they were well mounted, anxious to learn our mission, our line of direction, and plied us with many questions as to our ultimate ends and objects. A few of these Indians were also met by Mr. Sohon in his exploration toward the head of the Palouse. Discretion and prudence directed that our course towards them should be both frank and honest. I invited them to accompany me to my camp, gave them to eat and to smoke, and afterwards explained to them in detail our mission and object; they left, apparently satisfied, and with a promise to preserve friendly relations in future.[20]

As with other descriptions by John Mullan, this one provides vibrant details and dramatic elements that go beyond the necessities of a mere report. We learn that the Indians were well mounted, how Mullan entertained them, and how he practiced diplomacy. Additionally, although he refers to the possibility of "friendly relations in the future," he concludes this entry with the notation: "They are wily fellows, and great caution is necessary in all intercourse with them."[21]

Animated passages such as these are recurrent in the early pages of the Mullan report. The first thirty-six pages comprise his narrative of building the road. The next forty-eight pages, the remainder of his personal account (which is followed by one hundred pages of various correspondence and accountings), contain less descriptive entries organized under thirty subheadings. About half of these consist of his observations, looking forward, about the special challenges of building a transcontinental railway and various branch lines through the Northwest. The other entries constitute a well-organized compendium on a variety of topics regarding the character of the land along the Walla Walla to Fort Benton road. The topics range from geography to natural history to settlement along the route.

Three of the first four subdivisions provide information for travelers using the Mullan Road. The first section, "General Division of the Route," is a brief overview of the terrain crossed by the completed road and the efforts the builders had made to traverse it. "Itinerary of the Route" is a day-by-day guidebook for emigrants and freighters; it contains such data as road conditions and places to find forage for livestock. "Recommendations for Travelers" offers suggestions on what equipment to take along. And "Advice to Emigrants by this Route" includes a brief review of other trails that join the Mullan Road. Together these sections provide useful knowledge for anyone considering a journey along the road. Most of the material in these sections is presented straightforwardly, without the stylistic flourishes that appear elsewhere in the report. Nevertheless, Mullan still goes beyond simple reporting to help readers envision the route and its environment.

The first section, "General Division of the Route," is very brief, but it gives us a bird's-eye view of the topography through which the Mullan Road was built. The first 180 miles eastward from Walla Walla are mainly "open, level, or rolling prairie," and the final 324 miles consist of "open timbered plateaus, with long stretches of prairie." The road through much of this country was constructed at almost the rate of march of the party. But the middle of the route was another matter. This section, Mullan writes, "involved one hundred and twenty miles of difficult timber-cutting, twenty-five feet broad, and thirty measured miles of excavation, fifteen to twenty feet wide."[22] While laying the road across the prairie posed only "slight obstructions" for the builders, in the mountains, "sterner engineering problems" were encountered.[23] Other parts of Mullan's report reveal just how "stern" those problems could be and how high the cost was; several crew members suffered frostbite, one was injured by an explosion, and another drowned at a river crossing. Even on days without drama or tragedy, there were for Mullan's men the burdens of felling timber, building bridges, and moving themselves and their animals from camp to camp.

The next section of Mullan's report, his "Itinerary of the Route," carefully sets forth goals for each day's travel and suggests camping sites along the way. In this segment he is all business, giving little description of the various campsites, but he does provide useful information about what to expect on each day of the journey. The pace was determined partly by the difficulty of the terrain between camps and partly by where wood, water, and forage were available. On the first day out from Walla Walla, Mullan suggests traveling only 7.5 miles, perhaps to allow time to break in the train's men, livestock, and equipment at the beginning of the journey, but also to bring the travelers to the crossing of Dry Creek, where they would find "wood, water, and grass at camp."[24]

Modern guides for RV campers give information on the availability of hookups (notably, water and electricity) at various sites, along with such amenities as showers and wi-fi. Mullan was equally solicitous of the needs of the campers of his time. In addition to pointing out specific locations with adequate wood, water, and grass, he also cautions readers to plan for stretches where these necessities were scarce. On the second day out from Walla Walla, for example, at Touchet Bridge, Mullan advises travelers to stock up on wood, since none would be available for the next two days.[25] Later, when on the Snake River, he says, they should gather driftwood on the riverbanks to prepare for another treeless section.

In such ways, Mullan's itinerary, while designed as a guide for emigrants, also enables today's readers to picture life on the trail in that era. One site, he reports, is favored with a spring three hundred yards to the north of the road. Another has good grazing "on hills on the left bank" of the river.[26] Keeping in mind that traveling parties were only as strong as their animals, Mullan often suggests places to rest mounts and livestock.

On the fourteenth day, Mullan's itinerary has travelers arriving at the Coeur 'd'Alene Mission, which after two weeks on the road must indeed have seemed like "a St. Bernard in the Coeur d'Aléne mountains"–his phrase. Up to this point, travelers would have seen a handful of compatriots at ferry crossings on the Snake and Spokane Rivers, but otherwise no signs of settlers. The mission, Mullan says, is a "good place to rest animals for a day or two, and which is by all means advisable, as you now enter the timber, where camp-grounds have to be specially selected, and the animals should be well rested. Vegetables may be had at the mission."[27]

Mullan himself stopped often at the Coeur d'Alene Mission while exploring and building the road, and he came to admire the Jesuit missionaries. He praises the priests by name for their efforts among the Indians, complimenting Father DeSmet especially, "for he truly is the great father of all Rocky Mountain missionaries." With a few well-chosen words, Mullan invokes the long years of work the Jesuit fathers had devoted to the tribes of the region, reminding the reader of "their zeal and toil amid the fastnesses and solitudes of the Rocky Mountains."[28] He contrasts the Jesuits' dedication with the indifference of the government bureaucracies that squandered millions of dollars intended for the Indians. The missionaries, he points out, live and travel among the Natives and "actually take care of the Indians when sick and educate them when well."[29]

One interesting tidbit the modern reader will find in Mullan's "Itinerary" is the cost of crossing the Snake, Spokane, and Bitterroot Rivers by ferry. Ferry operators–who were among the earliest entrepreneurs on the frontier–generally charged four dollars per wagon and team, and fifty cents per passenger and per each "loose stock"–cattle, horses, mules, and oxen loaded aboard without a wagon in tow.[30] In today's dollars these prices convert to about $160 per wagon and $20 per passenger or animal. Many travelers must have swum or forded their loose stock, as Mullan often advised, to avoid the charges. But sometimes, as in the case of the Bitter Root River (Mullan's name for what is now considered part of the Clark Fork River), Mullan adds a caveat: "The stream is fordable in very low water, but I would advise all strangers to cross in the ferry-boats, as the ford is a dangerous one, except to those who know it well."[31]

Finally, after four weeks on the road, travelers would come to a welcome site in western Montana, at today's Missoula. Mullan details the plethora of opportunities at this oasis: "Move to Higgin's & Worden's store, . . . road excellent; wood, water, and grass here; good place to rest animals for a day or two; blacksmith's shop at Van Dorn's, and supplies of all kinds can be obtained, dry goods, groceries, beef, vegetables, and fresh animals, if needed." Six days later, after more camping in the wilderness, wayfarers will encounter posts of civilization much more frequently. On the thirty-fourth day, a 13.5-mile journey would bring them to "Gold creek or American Fork of Hell's Gate river; road excellent; wood, water, and grass at camp; supplies of all kinds to be had here, dry goods, groceries, fresh beef, animals, and possibly vegetables." Only one day later, at Deer Lodge, as Mullan reports, "fresh beef to be had here from settlers."[32] The next few days, however, offered little to cheer voyagers except a sufficiency of the essentials provided by the land–wood, grass, and water. Finally, on day forty-seven, if they had followed the itinerary without delays, travelers would arrive at Fort Benton after braving 624 miles on the road.

Though devoid of picturesque language, the Itinerary nonetheless suggests to the imagination the long journey of men and livestock through a region that was still principally a wilderness. Mullan adds depth to this picture in his next section, "Recommendations for Travelers." He advises beginning in St. Louis, purchasing a wagon and harnesses and shipping these up the Missouri by steamboat to Fort Benton. Horses, mules, or oxen for the journey could be purchased at the fort, he points out. While in St. Louis, travelers should also acquire a complete outfit for camping. For "mess furniture," Mullan recommends buying several tin or iron "kettles" [pots] to fit "one inside the other"–an arrangement familiar to any modern camper–as well as "tin plates and cups and strong knives and forks" and a "covered oven," i.e., a Dutch oven. For provisions he suggests "brown sugar, coffee, or tea, bacon, flour, salt, beans, sardines, and a few jars of pickles and preserved fruits," particularly dried apples. These items "will constitute a perfect outfit in this department."[33] He later added vinegar and yeast powder to the list. The core of the traveler's diet, following Mullan's advice, would be one pound of flour and one pound of bacon per day.

Mullan gives further guidance on choosing saddles, saddle blankets, and bridles for the pack animals. In addition, each horse or mule should be provided with a picket rope "from thirty-five to forty feet long" and a hefty "one inch in diameter." For packing loads, Mullan recommended using a Mexican stuffed packsaddle, or "apperajo" (aparejo). Each load, he cautions, should not exceed two hundred pounds. Mullan continues with his recommended routine and procedures: "Starting at dawn and camping not later than 2 p.m. I have always found the best plan in marching. Animals should not go out of a walk or a slow trot, and after being unloaded in camp they should always be allowed to stand with their saddles on and girths loose, for at least fifteen minutes, as the sudden exposure of their warm backs to the air tends to scald them."[34] In his one flight of fancy in this otherwise cut-and-dried set of recommendations, Mullan makes a comparison that was more acceptable, perhaps, in the nineteenth century than it would be in the twenty-first. On getting the most out of one's beasts of burden, he writes, "They should be regularly watered, morning, noon, and night. Never maltreat them, but govern them as you would a woman, with kindness, affection, and caresses, and you will be repaid by their docility and easy management."[35]

Mullan then turns his attention to the wagons: "If you travel with a wagon, provide yourself with a jackscrew, extra tongue, and coupling pole; also, axle grease, a hatchet and nails, auger, rope, twine, and one or two chains for wheel locking, and one or two extra whippletrees, as well as such other articles as in your own judgment may be deemed necessary."[36] This passage suggests the hard work required to keep a wagon fit for travel, as does this one: "If your wagon tires become loose on the road, caulk them with old gunny sacks, or in lieu thereof, with any other sacking; also, soak the wheels well in water whenever an opportunity occurs."[37]

Mullan completes his tract with advice on camping equipment, suggesting a "light canvas tent, with poles that fold in the middle by a hinge," something he had "always found most convenient."[38] Travelers do not really need tables and chairs on the trail, he writes, "but if deemed absolutely necessary, the old army camp stool, and a table with a lid that removes and legs that fold under, I have found to best subserve all camp requisites." Mullan concludes his "Recommendations for Travelers" with this sage and perennial counsel: "Never take anything not absolutely necessary. This is a rule of all experienced voyageurs."[39]

Moving from suggestions for traveling on the road, Mullan turned to descriptions of the terrain itself, with sections on "Agricultural Lands," "Grazing Capabilities," "Mineral Wealth," "Building Materials," "Rivers and Water-Courses," and the "Mountain System." Many of the passages in these sections show not only Mullan's aptitude for faithful depictions, but also his ability to capture the mood of the times, as displayed in his evocative account of the restless miners who would soon be using the road to reach camps in Idaho and Montana. His portrayal of these men deserves to be quoted at length as a prime example of Mullan's blending of precision and poetry:

> It has been only during the last three years that sanguine expectations have existed of mineral wealth to any great extent being found east of the Cascade mountains or even in the northern portion of the Rocky Mountains proper. As soon as the Frazier river mines in British Columbia were discovered and the question of [the] route by which to reach them in the shortest time and cheapest means was discussed, we found the eastern portion of Washington Territory ramified by hundreds of gold seekers in quest of this new Eldorado. This was in 1858. Gold was soon after discovered in the Wenatchee, Nachees, Okinagen, Simalkameen, and Clark's Fork, and worked till the Indians drove the miners from the country. In the succeeding year Captain Pierce, with a boldness and a judgment worthy every commendation, explored the Bitterroot mountains, and the discovery of the Nez Perces' gold mines was the result. The wanderings of the miners in this region southward led to the Salmon river discoveries. But the gold miner, who is the most restless of mortals, did not rest content until he had crossed Snake river and discovered the Powder river mines; and not even then content, he opened up the Grand Ronde, Boisé, and Burnt river mines, in East Oregon, and elated at his success returns to his friends in

West Oregon, taking the head of John Day's river in his route only to discover the richest quartz [lodes] to be found outside of California.

His companions on Frazier's river heard of this new gold field and they too must visit it, taking on their route the mouth of Clark's Fork and Spokane river, only there to discover gold which has since been taken out by the pound. This news spreads and the Powder river miner tracks it to the latter point, and thence to the Kootenay, where he is amply repaid for his toil and travel. While this is being done the Salmon river miner is not content with making twenty dollars per day, but he too must follow the Salmon river till it becomes a silvery thread in the mountains and crosses the range to discover the wealth of the Deer Lodge mines. As the weary emigrant crossing the plains hears of this Eldorado so near himself, he too must journey thither only to discover the rich gold mines in Big Hole and Beaver Head valleys. Nor does it stop here, but the restless adventurer in St. Louis contracts the gold fever and threads the Missouri to Fort Benton in quest of the Deer Lodge mines, and while en route discovers the Prickly Pear gold mines. Thus, working like beavers, have the miners and emigrants crossed and recrossed the mountains during the last three years, ramifying in every direction until they have opened a gold region which, to-day, is sending to our mints a wealth equal to that of all California in her palmiest days. . . .

Wonderful have been the effects of this great alchemist in that quarter. It has transmuted sluggishness into activity, has brightened the dullest vision, elevated the industrious and frugal laborer, silenced the sceptic and caviller, and struck a new blow in behalf of a northern Pacific railroad route. The trade and travel along the upper Columbia, where several steamers now ply between busy marts, of themselves attest what magical effects three years have wrought.[40]

Of course, along with his happy ruminations on the march of civilization and his bursts of colorful prose, John Mullan wrote many passages that simply outline the day-to-day tasks of building a wagon road. He writes, for example, of one river crossing:

We immediately set the whip sawyers in the timber to get out the necessary lumber and some men to burning tar, and, being provided with the necessary oakum, we built two flat-boats, forty-two feet long, twelve feet broad, and two feet deep–one for the St. Joseph's and the other for the Coeur d'Aléne. The lat[t]er, when completed, was rowed down into the lake, and thence up the Coeur d'Alene river to the point selected for its crossing. While this was being accomplished, the divide between the two streams was examined, the road marked out, and several parties placed at work upon it.[41]

In this passage, we see Mullan recording nothing but the humdrum facts. But elsewhere, his writing once again takes on a lyrical, even whimsical tone. In a letter included with the report, he wrote:

In winter it is snowing frequently by quarter inches as a whole day's labor, to be destroyed the following day by a warm sun. If any snow should remain, the temperature of the air near zero, or even sometimes 40° below zero, the freezing point of mercury, the result of an excessive clear and calm night will freeze the snow to an ice crust for the deer to walk on without snow-shoes, and, on the morrow, to enjoy the bright sun with a temperature of 70° and more.[42]

One can imagine the smile on his face as he wrote the part about the deer going without snowshoes!

John Mullan's talents as engineer and leader are manifest in the road that came to bear his name, and his penchant for precision is apparent in much of his report. But what is the source of his capacity for evoking beauty, reflecting on broader history, and indulging now and then in playful images to describe ordinary phenomena? Ryan Shaw, an expert on the railroad surveys of Mullan's time, argues that numerous explorers, geographers, and intellectuals of the time were influenced by the philosophy of the great Prussian scholar Alexander von Humboldt, whose followers included John Frémont and Isaac Stevens as well as Henry David Thoreau and Ralph Waldo Emerson. Humboldt's fusion of science and romanticism colored the views of many thinkers in the nineteenth century. "The philosophical study of nature rises above the requirements of mere delineation," wrote Humboldt, "and does not consist in the sterile accumulation of isolated facts." Mullan's writing, which moves so gracefully between precision and poetry, exemplifies this scientific yet *romantic* mind-set, an outlook that shaped the perception of so many early explorers and builders of the West.[43]

1 In this article I quote extensively from John Mullan, *Report on the Construction of a Military Road from Fort Walla-Walla to Fort Benton* (Washington, DC: Government Printing Office, 1863). Unless otherwise indicated, all page numbers refer to this document. Most of the report, including all of the pages referenced in this article, can be viewed online at http://www.narhist.ewu.ed/mullan report/mullan report home.html.

2 Mullan, *Report*, p. 2.

3 Ibid.

4 Ibid.

5 P. 3.

6 Ibid.

7 P. 2.

8 P. 3.

9 P. 4.

10 Ibid.

11 P. 3.

12 P. 4.

13 Ibid.

14 Ibid.

15 P. 5.

16 P. 8.

17 Ibid.

18 P. 12.

19 P. 14.

20 P. 15.

21 Ibid.

22 P. 37.

23 Ibid.

24 Ibid.

25 Ibid.

26 Ibid.

27 P. 38.

28 P. 53.

29 P. 52.

30 P. 37.

31 P. 38.

32 P. 39.

33 P. 40.

34 Ibid.

35 P. 41.

36 Ibid.

37 P. 40.

38 Ibid.

39 P. 41.

40 Pp. 44–45.

41 P. 16.

42 P. 182.

43 Alexander von Humboldt, quoted by Ryan Shaw, "John Mullan and 'Humboltian Exploration,'" in paper delivered at Mullan Road Conference, Walla Walla, WA, April 14, 2012.

BIBLIOGRAPHY

Note: Archival, manuscript, newspaper, oral interview, and cartographic resources are not included in this bibliography. The interested reader is referred to the endnote citations accompanying each individual essay.

Allen, Frederick. *A Decent Orderly Lynching: The Montana Vigilantes*. Norman: University of Oklahoma Press, 2004.

Ambrose, Stephen E. *Duty, Honor, Country: A History of West Point.* Baltimore: Johns Hopkins University Press, 1966.

Ankeny, Nesmith. *The West as I Knew It.* Lewiston, ID: R. G. Bailey Printing Co., 1953.

Axline, Jon. *Conveniences Sorely Needed: Montana's Historic Highway Bridges, 1860-1956.* Helena: Montana Historical Society Press, 2005.

Bell, William Gardner. *Secretaries of War and Secretaries of the Army: Portraits & Biographical Sketches.* Washington, DC: Center of Military History, United States Army, 1992.

Bemis, Samuel F. "Captain John Mullan and the Engineers' Frontier." *Washington Historical Quarterly* 14 (July 1923).

Bryant, Jerry. "Daniel Lyons and His Ferry." *Bunchgrass Historian* 23, no. 2 (1996).

Bryant, John Henry. "Letter to family in New York, 1870." *Seattle Genealogical Society Bulletin* 39, no. 1 (Autumn 1989). Reprinted in *Mullan Chronicles* 7, no. 3 (Summer-Fall 1998), newsletter of the Mineral County Museum.

Burlingame, Merrill. *The Montana Frontier.* Helena, MT: State Publishing Co., 1942.

Burns, Robert Ignatius. *The Jesuits and the Indian Wars of the Northwest.* New Haven, CT: Yale University Press, 1966.

Carhart, Tom. *Sacred Ties: From West Point Brothers to Battlefield Rivals, a True Story of the Civil War.* New York: Berkley Books, 2010.

Chittenden, H. M., and A. T. Richardson, eds. *Life, Letters, and Travels of Father Pierre Jean DeSmet, S.J., 1811-1873.* 4 vols. New York: Francis P. Harper, 1905.

Cohen, Paul E. *Mapping the West: America's Westward Movement, 1524–1890.* New York: Rizzoli, 2002.

Coleman, Louis C., and Leo Rieman. *Captain John Mullan: His Life, Building the Mullan Road, As It Is Today and Interesting Tales of Occurrences along the Road.* Montreal, Canada: B. C. Payette, 1968.

Crackel, Theodore J. *West Point: A Bicentennial History.* Lawrence: University Press of Kansas, 2003.

Cullum, George W. *Biographical Register of the Officers and Graduates of the U. S. Military Academy, from 1802 to 1867. Rev. Ed., With a Supplement Continuing the Register of Graduates to January 1, 1879.* New York: J. Miller, 1879.

Delo, David Michael. *Peddlers and Post Traders: The Army Sutler on the Frontier.* Salt Lake City: University of Utah Press, 1992.

Drumheller, Daniel. *'Uncle Dan' Drumheller Tells Thrills of Western Trails in 1854.* Spokane: Inland American Printing Co., 1925.

Dupuy, R. Ernest. *Sylvanus Thayer, Father of Technology in the United States.* West Point, NY: Association of Graduates, U.S. Military Academy, 1958.

Egan, Timothy. *The Big Burn: Theodore Roosevelt and the Fire that Saved America.* Boston: Houghton Mifflin Harcourt, 2009.

Ehrenberg, Ralph. *Geographical Exploration and Mapping in the 19th Century: A Survey of the Records in the National Archives.* Reference Information Paper no. 66. Washington, DC: National Archives and Records Administration, 1973.

Erickson, Harvey. "Mullan's 1862 Interstate Road." *Pacific Northwesterner* 18, no. 4 (Fall 1974).

Ewers, John C. *Artists of the Old West.* New York: Doubleday, 1973.

Ewers, John C. "Gustavus Sohon's Portraits of Flathead and Pend d'Oreille Indians, 1854." *Smithsonian Miscellaneous Collections* vol. 110, no. 7. Washington, DC: Smithsonian Institution, 1948.

Fleming, Thomas J. *West Point: The Men and Times of the United States Military Academy.* New York: Wm. Morrow, 1969.

Fletcher, W. F. *The Era of Chief Old Bones.* N.p.: Privately printed, 1994.

Forman, Sidney. *Cadet Life Before the Mexican War.* West Point, NY: United States Military Academy Print Office, 1945.

______. *West Point: A History of the United States Military Academy.* New York: Columbia University Press, 1950.

Freeman, Otis W. "Early Wagon Roads in the Inland Empire." *Pacific Northwest Quarterly* 45, no. 4 (October 1954).

Friis, Herman R. "The Documents and Reports of the United States Congress: A Primary Source of Information on Travel in the West, 1783–1861." In *Travelers on the Western Frontier.* Edited by John F. McDermott. Urbana: University of Illinois Press, 1970.

Fuller, George W. *History of the Pacific Northwest, with Special Emphasis on the Inland Empire.* 2d ed. New York: Knopf, 1946.

Gates, Paul W. "Public Lands Disposal in California." *Agricultural History* 49:1 (January 1975).

Goetzmann, William M. *Exploration and Empire: The Explorer and Scientist in the Winning of the American West.* New York: Knopf, 1966.

Greever, William S. *Bonanza West: The Story of the Western Mining Rushes, 1848–1900.* Moscow: University of Idaho Press, 1963.

Hanchett, Leland J., Jr. *Montana's Benton Road.* Cave Creek, AZ: Pine Rim Publishing, 2008.

Hardin, Martin. "Across the New Northwest in 1860, Part II." *The United Service* 7, no. 2 (August 1882).

Hardin, Martin D. "Up the Missouri and Over the Mullan Road." *The Westerners, New York Posse Brand Book* 5, nos. 1 and 2 (1958).

Hewitt, Randall Henry. *Across the Plains and Over the Divide: A Mule Train Journey from East to West and Incidents Connected Therewith.* New York: Broadway Publishing Co., 1908.

Hobbs, Richard S. *The Broughtons of Dayton: Family and Business in the Northwest Heartland.* Cambridge, MA: Winthrop Group, 2010.

Hoss, Bishop Elijah Embree. *David Morton: A Biography.* Nashville, TN: Publishing House of the Methodist Episcopal Church, South, 1916.

Howard, Helen Addison. "Captain John Mullan." *Washington Historical Quarterly* 25:3 (July 1943).

______. *Northwest Trail Blazers.* Caldwell, ID: Caxton Printers, 1963.

Interstate Publishing Co. *An Illustrated History of Klickitat, Yakima, and Kittitas Counties.* Chicago: Interstate Publishing Co., 1904.

Irvine, Caleb E. *Medicine Tree Hill: A Legend.* Helena: State Historical Society of Montana, 1907.

Jackson, W. Turrentine. "Across the Northern Plains: The Mullan Road, 1853–1869." In *Wagon Roads West: A Study of Federal Road Surveys and Construction in the Trans-Mississippi West, 1846–1869.* Berkeley: University of California Press, 1952.

Jakle, John A. "Time and Space, and the Geographic Past: A Prospectus for Historical Geography." *American Historical Review* 76 (October 1971).

Jefferson, Thomas. *The Writings of Thomas Jefferson*, Definitive Edition, vol. 27. Edited by Albert Bergh. Washington, DC: Thomas Jefferson Memorial Association, 1905.

Johansen, Dorothy, and Frank Gill. "History of the Oregon Steam Navigation Company." *Oregon Historical Quarterly* 38, no. 2 (March 1937).

Johnson, Randall A. "The Mullan Road." *Pacific Northwesterner* 39, no. 2 (1995).

Kautz, Augustine V. "From Missouri to Oregon in 1860: The Diary of August V. Kautz." *Pacific Northwest Quarterly* 37, no. 3 (July 1946).

______. *Missouri to Oregon in 1860.* Fairfield, WA: Ye Galleon Press, 1995.

Kip, Lawrence. *Army Life on the Pacific: A Journal of the Expedition against the Northern Indians.* New York: Redfield, 1859.

Larsen, Lawrence H., and Barbara J. Cottrell. *Steamboats West: The 1859 American Fur Company Missouri River Expedition.* Norman, OK: Arthur H. Clark, 2010.

Lass, William E. *A History of Steamboating on the Upper Missouri River.* Lincoln: University of Nebraska Press, 1962.

______. *Navigating the Missouri: Steamboating on Nature's Highway, 1819–1935.* Norman, OK: Arthur H. Clark, 2008.

Leeson, Michael. *History of Montana, 1739–1885.* Chicago: Warner, Beers & Co., 1885.

Lepley, John G. *Blackfoot Fur Trade on the Upper Missouri.* Missoula, MT: Pictorial Histories Publishing, 2004.

Mayer, David. "'Necessary and Proper': West Point and Jefferson's Constitutionalism," in *Thomas Jefferson's Military Academy: Founding West Point.* Edited by Robert M. S. McDonald. Charlottesville: University of Virginia Press, 2004.

McBride, Sr. Genevieve. *The Bird Tail.* New York: Vantage Press, 1974.

McDermott, Paul D., and Ronald E. Grim. "Early Images of the Northwest: The Works of Gustavus Sohon." *Columbia: The Magazine of Western History* 21, no. 4 (Winter 2007–8).

______. *Gustavus Sohon's Cartographic and Artistic Works: An Annotated Bibliography.* Philip Lee Phillips Society Occasional Paper Series, no. 4. Washington, DC: Geography and Map Division, Library of Congress, 2002.

______, and Philip Mobley. *Eye of the Explorer: Views of the Northern Pacific Railroad Survey, 1853–54.* Missoula, MT: Mountain Press Publishing, 2010.

McDonald, Christine, and Robert McDonald. "West from West Point: Jefferson's Military Academy and the 'Empire of Liberty'." In *Light and Liberty: Thomas Jefferson and the Power of Knowledge.* Charlottesville: University of Virginia Press, 2012.

McGregor, Alexander C. *Counting Sheep: From Open Range to Agribusiness on the Columbia Plateau.* Seattle: University of Washington Press, 1982.

_________. "The Economic Impact of the Mullan Road on Walla Walla, 1860–1883." *Pacific Northwest Quarterly* 65, no. 3 (July 1974).

McIntosh, Clarence F. "The Chico and Red Bluff Route." *Idaho Yesterdays* 6, no. 3 (Fall 1962).

McLaughlin, Patrick D. *Transportation in Nineteenth Century America: A Survey of the Cartographic Records in the National Archives of the United States.* Reference Information Paper no. 65. Washington, DC: National Archives and Records Administration, 1973.

Meinig, Donald W. *The Great Columbia Plain: A Historical Geography, 1805-1910.* Seattle: University of Washington Press, 1968.

______. "Isaac Stevens: Practical Geographer of the Early Northwest." *Geographical Review* 45 (October 1955).

______. *Transcontinental America, 1850-1915.* Vol. 3 of *The Shaping of America: A Geographical Perspective on 500 Years of History.* New Haven, CT: Yale University Press, 1998.

______. "The Walla Walla Country: A Century of Man and the Land, 1805–1910." Ph.D. dissertation, University of Washington, 1953.

______. "Wheat Sacks Out to Sea: The Early Export Trade from the Walla Walla Country." *Pacific Northwest Quarterly* 45 (January 1954).

Morrison, James L. *"The Best School in the World": West Point, the Pre–Civil War Years, 1833–1866.* Kent, Ohio: Kent State University Press, 1986.

Mullan, John. "Journal from Fort Dalles O.T. to Fort Walla Walla W.T. July 1858, Lieut. John Mullan U.S. Army." Edited by Pal Clark. *Sources of Northwest History*, no. 18 (May 1932), State University of Montana, Missoula.

______. *Miners' and Travelers' Guide to Oregon, Washington, Idaho, Montana, Wyoming and Colorado via the Missouri and Columbia Rivers.* New York: Wm. M. Franklin, 1865.

______. "Remarks of John Mullan, 7 May 1863, on the Geography, Topography and Resources of the Northwestern Territories." *Proceedings of the American Geographical & Statistical Society* 2, no. 1 (1863–64).

______. *Report on the Construction of a Military Road from Fort Walla-Walla to Fort Benton.* S. Ex. Doc. 43, 37th Cong., 3d sess., serial 1149. Washington, DC: Government Printing Office, 1863.

______. *Reports to Hon. George Stoneman, Governor of California, on Certain Claims of the State of California Against the United States.* Sacramento, CA: State Printing Office, 1886.

______. *Topographical Memoir and Map of Colonel Wright's Late Campaign Against the Indians of Oregon and Washington Territories.* S. Ex. Doc. 984, 35th Congress, 2d sess., serial 32. Washington, DC: Government Printing Office, 1859.

______, P. M. Engle, et al. *Military Road from Fort Benton to Fort Walla Walla: Letter from the Secretary of War, Transmitting the Report of Lieutenant Mullan, In Charge of the Construction* [illegible]. Washington, DC: Government Printing Office, 186[illegible].

Nash, Gerald. "The California Land Office, 1858–1898." *Huntington Library Quarterly* 27 (August 1964).

Nicandri, David. *Northwest Chiefs: Gustave Sohon's Views of the 1855 Stevens Treaty Councils.* Tacoma: Washington State Historical Society, 1986.

Oliphant, J. Orin. "The Cattle Trade from the Far Northwest to Montana." *Agricultural History* 6, no. 2 (April 1932).

______, ed. "The Recollections of Ben Burgunder." *Washington Historical Quarterly* 17, no. 3 (July 1926).

Overholser, Joel. *Fort Benton: World's Innermost Port.* Helena, MT: Falcon Press, 1987.

Palladino, Lawrence Benedict. *Indian and White in the Northwest, or a History of Catholicity in Montana.* Baltimore: John Murphy & Co., 1894.

Pappas, George S. *To the Point: The United States Military Academy, 1802–1902.* 1st ed. Santa Barbara, CA: Praeger Publishers, 1993.

Partoll, Albert J. "Frank L. Worden, Pioneer Merchant, 1830–1887." *Pacific Northwest Quarterly* 40, no. 3 (July 1949).

Patterson, Ida Smith. *Montana Memories: The Life of Emma Magee in the Rocky Mountain West, 1866–1950.* 1981. Reprint, Pablo, MT: Salish Kootenai College Press, 2011.

Peltier, Jerome. "Robert Franklin Cummins: The Youngest 12-Year-Old Freighter in Washington Territory." *The Pacific Northwesterner* 40, no. 2 (1996).

Petersen, Keith C. *John Mullan: The Tumultuous Life of a Western Road Builder.* Pullman: Washington State University Press, 2014.

Phillips, Paul C., ed. *Forty Years on the Frontier as Seen in the Journals and Reminiscences of Granville Stuart, Gold-Miner, Trader, Merchant, Rancher and Politician.* Cleveland: Arthur H. Clark Co., 1925.

Price, Edward T. *Dividing the Land: Early American Beginnings of Our Private Property Mosaic.* Chicago Geography Research Paper no. 238, Department of Geography, University of Chicago, 1955.

Raynolds, W. F. *Report of Brevet Colonel W. F. Raynolds, U.S.A., Corps of Engineers, on the Exploration of the Yellowstone and Missouri Rivers in 1859-60.* S. Ex. Doc. 77, 40th Cong., 2d sess., serial 1317. Washington, DC: Government Printing Office, 1868.

Regulations Established for the Organization and Government of the Military Academy at West Point, New York. New York: Wiley & Putnam, 1839.

Richards, Kent D. *Isaac I. Stevens: Young Man in a Hurry.* Pullman: Washington State University Press, 1993.

Robison, Ken. *Fort Benton.* Charleston, SC: Arcadia Publishing, 2009.

Schafft, Charles. "General Garfield and the Flatheads." *Council Fire and Arbitrator* 5, no. 1 (January 1882).

______. "Sketch of a Life–Charles Schafft." Edited by Vivian Paladin. *Montana: The Magazine of Western History* 26, no. 1 (Winter 1976).

Scheuerman, Richard D. "Through the Indian Country: John Mullan and the Northern Overland Road." *Columbia Magazine* 24, no. 2 (Fall 2010).

Schlicke, Carl P. *General George Wright: Guardian of the Pacific Coast.* Norman: University of Oklahoma Press, 1988.

Schmid, Calvin F., and Stanton E. Schmid. *Growth of Cities and Towns, State of Washington.* Olympia: Washington State Planning and Community Affairs Agency, 1969.

Schultz, James Willard. *William Jackson, Indian Scout.* 1926. Reprint, Springfield, IL: William K. Cavanagh, 1976.

Schwantes, Carlos. *Railroad Signatures Across the Pacific Northwest.* Seattle: University of Washington Press, 1993.

Scott, Leslie M. "The Pioneer Stimulus of Gold." *Oregon Historical Quarterly* 18, no. 3 (September 1917).

Sherman, Ralph. "The Followers of the Bunch-Grass Hunter." *Outing Magazine* 47 (February 1906).

Skelton, William. "West Point and Officer Professionalism, 1817–1877." In *West Point: Two Centuries and Beyond.* Edited by Lance Betros. Abilene, Tex.: McWhiney Foundation Press, 2004.

Sobel, Dava. *Longitude: The True Story of a Lone Genius Who Solved the Greatest Scientific Problem of His Time.* New York: Walker Books, 2007.

Splawn, Andrew J. "A Cattleman's Reminiscences." *The Ranch*, February 13, 1902.

______. *Ka-Mi-Akin: Last Hero of the Yakimas.* Portland, OR: Kilham Stationery and Printing Co., 1917.

Stevens, Hazzard. *The Life of Isaac Ingalls Stevens.* Boston: Houghton Mifflin, 1900.

Stevens, Isaac I. "Narrative and Final Report of Explorations for a Route for a Pacific Railroad." In U.S War Department, *Reports of Explorations and Surveys* . . . . Vol. 12, book 1. Washington, DC: Thomas H. Ford, Printer, 1860.

Stewart, Earl K. "Transporting Livestock by Boat Up the Columbia, 1861–1881." *Oregon Historical Quarterly* 51, no. 1 (December 1949).

Strachan, John. *Blazing the Mullan Trail: Connecting the Headwaters of the Missouri and Columbia Rivers and Locating the Great Overland Highway to the Pacific Northwest*. 1860–1861. Reprint, New York: Edward Eberstadt & Sons, 1952.

Stuart, Granville. *Forty Years on the Frontier*. Edited by Paul C. Phillips. 1925. Reprint, Lincoln: University of Nebraska Press, 2004.

Sunder, John E. *The Fur Trade on the Upper Missouri, 1840–1865*. 1865. Reprint, Norman: University of Oklahoma Press, 1993.

Swan, James G. *The Northwest Coast, or Three Years' Residence in Washington Territory*. Seattle: University of Washington Press, 1972.

Symons, Thomas W. *The Symons Report on the Upper Columbia River and the Great Plain of the Columbia*. 1882. Reprint. Fairfield, WA: Ye Galleon Press, 1967.

Taft, Robert. *Artists and Illustrators of the Old West, 1850–1900*. New York: Charles Scribner, 1953.

Tattersall, James M. "The Economic Development of the Pacific Northwest to 1920." Ph.D. dissertation, University of Washington, 1960.

Thoreau, Henry David. *The Writings of Henry David Thoreau*. Edited by Bradford Torrey and Francis H. Allen. Boston: Houghton, Mifflin, 1906.

Thrower, Norman J. W. *Maps and Man: An Examination of Cartography in Relation to Culture and Civilization*. Englewood Cliffs, NJ: Prentice-Hall, 1972.

United States Military Academy. *The Centennial of the United States Military Academy at West Point, New York, 1802–1902*. Washington, DC: Government Printing Office, 1904.

U.S. War Department. *Annual Message from the President and Report of the War Department, 1858*. H. Ex. Doc. 2, 35th Cong., 2d sess., serial 998. Washington, DC: James B. Steedman, Printer, 1858.

______. *Report from Captain William H. Penrose in Regard to the Reopening of Mullan Wagon Road in Montana*. Sen. Ex. Doc. 3, part 2, 46th Cong. 2d Sess., 1879.

______. *Reports of Explorations and Surveys, to Ascertain the Most Practicable and Economic Route for a Railroad from the Mississippi River to the Pacific Ocean, Made under the Direction of the Secretary of War, in 1853–5*. 12 vols. Washington, DC: Thomas H. Ford, Printer, 1855–60.

______. *Reports and Inspections Made in the Summer of 1877 by Generals P. H. Sheridan and W. T. Sherman of Country North of the Union Pacific Railroad*. (Washington, D.C., 1878)

Wagoner, Jennings L., Jr., and Christine Coalwell McDonald. "Mr. Jefferson's Academy: An Educational Interpretation." In *Thomas Jefferson's Military Academy*. Edited by Robert M. S. McDonald. Charlottesville: University of Virginia Press, 2004.

Watkins, Albert. "The Oregon Recruit Expedition." *Mid-West Quarterly* 1, no. 1 (October 1913).

Watt, James W. *Journal of Mule Train Packing in Eastern Washington in the 1860s*. Fairfield, WA: Ye Galleon Press, 1978.

Wayland, Francis. "Report to the Corporation of Brown University." In Richard Hofstadter and Wilson Smith, eds., *American Higher Education: A Documentary History*. Chicago: University of Chicago Press, 1961.

Weisel, George F. *Men and Trade on the Northwest Frontier*. Missoula: Montana State University Press, 1955.

Wetterman, Robert P. "West Point, the Jacksonians, and the Army's Controversial Role in National Improvements." In *West Point: Two Centuries and Beyond*. Edited by Lance Betros. Abilene, Texas: McWhiney Foundation Press, 2004.

Winther, Oscar O. "Early Commercial Importance of the Mullan Road." *Oregon Historical Quarterly* 46 (March 1945).

______. *The Old Oregon Country: A History of Frontier Trade, Transportation, and Travel*. Stanford, CA: Stanford University Press, 1950.

Willey, Dan Allen. "Building the Mullan Road." *Sunset* 24, no. 6 (June 1910).

Wischmann, Lesley. *Frontier Diplomats: The Life and Times of Alexander Culbertson and Natoyist-Siksina'*. Western Frontiersmen Series. Norman, OK: Arthur H. Clark, 2001.

______, and Andrew Erskine Dawson. *This Far-Off Land: The Upper Missouri Letters of Andrew Dawson*. Norman, OK: Arthur H. Clark, 2013.

Woody, Judge Frank H. "A Sketch of the Early History of Western Montana." *Contributions to the Historical Society of Montana*, vol. 2. Helena, MT: State Publishing Co., 1896.

INDEX

ABOUT THE EDITORS

Paul D. McDermott completed graduate studies in physical geography, cartography, and historical geography at the Universities of Washington and Maryland. After thirty-four years of teaching geography and cartography at Montgomery College in Rockville, Maryland, he retired as a professor emeritus at the school, and he continues to write articles and books on historical geography and related topics. Having grown up in the state of Washington, he has a particular interest in the historical geography of the Pacific Northwest.

Ronald E. Grim studied cultural and historical geography at the University of Maryland. After a long career at the National Archives and the Library of Congress, he currently serves as curator of maps at the Norman B. Leventhal Map Center at the Boston Public Library. He has written extensively on historical geography and historical cartography.

Philip Mobley studied geography and cartography at Kutztown [Pennsylvania] and Akron [Ohio] Universities. Currently a professional cartographer, he also teaches cartography classes for Global View, Inc.

The three authors' previous collaborations include *Eye of the Explorer: Views of the Northern Pacific Railroad Survey, 1853–54*, published in 2010 by Mountain Press Publishing.

ABOUT THE CONTRIBUTORS

Kim Briggeman, a lifelong Montanan, grew up in Bonner, near Missoula. A graduate of the University of Montana, he has written for the *Missoulian* newspaper for more than thirty-five years. He has also penned numerous history articles for various publications and served on the organizing committee for the 2014 Mullan Road Conference in Missoula. Honors include the 2010 Dorothy Ogg Award from the Missoula Historic Preservation Commission and a 2014 Heritage Keeper Award from the Montana Historical Society. He lives in Missoula with his wife, Linda.

Alexander C. McGregor grew up on a ranch near the deeply rutted traces of the Mullan Road, where his family has raised livestock and grain since territorial days. After earning history degrees from Whitman College and the University of Washington, he taught history at both institutions before returning to his family's agricultural business. He currently serves as managing general partner of McGregor Land and Livestock Company, and as president of the McGregor Company, dedicated to helping farm families in the Northwest, which was chosen as National Family Business of the Year in 2008. He is also a board member of the Washington State Historical Society. His first book was *Counting Sheep: From Open Range to Agribusiness on the Columbia Plateau* (University of Washington Press, 1982), and he recently coauthored, with Richard D. Scheuerman and John Clement, *Harvest Heritage: Agricultural Origins and Heirloom Crops of the Pacific Northwest*, (Washington State University Press, 2013). He and his wife, Linda, live in rural Washington state near Pullman.

Keith C. Petersen is Idaho's State Historian and the associate director of the Idaho State Historical Society. The author of several books and numerous articles on the history of Idaho and the Northwest, he has twice received the Outstanding Book on Idaho Award. His most recent work is the acclaimed full-length biography *John Mullan: The Tumultuous Life of a Western Road Builder*, published by Washington State University Press in 2014. He lives in Moscow, Idaho.

Don Popejoy, a native and current resident of Spokane, Washington, graduated from Eastern Washington University in 1973. He currently teaches history and geology at Spokane Falls Community College. He also develops and leads history tours for the Road Scholar educational travel program (formerly called Elderhostel) and other programs. In 2000 he organized the first annual Mullan Road convention. He is also coauthor of *Early Spokane: 1800–1945*, with Penny Hutten, published by Arcadia Press in 2010.

Ken Robison, a retired U.S. Navy captain, currently serves as historian at the Overholser Historical Research Center in Fort Benton, Montana. A native Montanan, he graduated from the University of Montana, then served in naval intelligence for thirty years. He is author of five books and numerous articles on Montana history. In 2010 the Montana Historical Society honored him with a Montana Heritage Keeper Award.

Richard D. Scheuerman was raised on a farm in eastern Washington. After teaching for twenty-five years in Washington public schools, he began teaching and writing at Seattle Pacific University's Graduate School of Education. He holds degrees in history, Russian, and education and has written several books and articles on regional history. His honors include the Governor's Award for Excellence in Education and the Robert Gray Medal for contributions to historical scholarship. He lives in Spanaway, Washington.

Maj. Ryan L. Shaw, United States Army, presently serves as an army strategist at NORAD and U.S. Northern Command. A native of Idaho, he commanded a cavalry troop in Operation Iraqi Freedom and taught American History at West Point. He is currently a Ph.D. candidate at Yale University.

William "Montana Bill" Weikel is the senior engineer at Weikel Consulting and a historical surveying reenactor. He also serves as an officer of the Surveyors Historical Society. He is a resident of Missoula, Montana.

J. William T. Youngs received his degrees from Harvard and University of California at Berkeley. Currently a professor of history at Eastern Washington University in Cheney, he is the author of five books and numerous articles on a variety of history topics.